International Scientific and Technical Conference "Metrology, Standardization, Quality: Theory and Practice" (MSQ-2017)

Journal of Physics: Conference Series Volume 998

Omsk, Russia
14 – 16 November 2017

ISBN: 978-1-5108-6167-1
ISSN: 1742-6588

Printed from e-media with permission by:

Curran Associates, Inc.
57 Morehouse Lane
Red Hook, NY 12571

CURRAN ASSOCIATES INC. proceedings.com

Some format issues inherent in the e-media version may also appear in this print version.

Printed by Curran Associates, Inc. (2018)

Institute of Physics
Dirac House, Temple Back
Bristol BS1 6BE UK

Phone: 44 1 17 929 7481
Fax: 44 1 17 920 0979

techtracking@iop.org

Additional copies of this publication are available from:

Curran Associates, Inc.
57 Morehouse Lane
Red Hook, NY 12571 USA
Phone: 845-758-0400
Fax: 845-758-2633
Email: curran@proceedings.com
Web: www.proceedings.com

Table of contents

Volume 998

"Metrology, Standardization, Quality: Theory And Practice" (MSQ-2017)

14–16 November 2017, Omsk, Russian Federation

Accepted papers received: 21 March 2018
Published online: 13 April 2018

Preface

Papers

IOP Conf. Series: Journal of Physics: Conf. Series **998** (2018) 011001 doi:10.1088/1742-6596/998/1/011001

International scientific and technical conference "METROLOGY, STANDARDIZATION, QUALITY: THEORY AND PRACTICE" (MSQ-2017)

was aimed at the exchange and generalization of Russian and world experience in the following fields:
• improving normative, technical and other documentation;
• developing new high-precision methods and means of measurement;
• developing technological standards in various spheres to correctly assess the accuracy and the impact on the products (processes) quality.

The problems discussed at the conference included the following:
I. Metrological assurance of processes and products quality:
• measurement and testing in scientific research and industry;
• certification of measurement techniques and calibration of measuring instruments;
• measurements in thermo physics, radio engineering and telecommunication systems.
II. Standardization and quality management:
• standardization, certification, quality management in scientific research and industry;
• assessment of integrated protection systems quality and information facilities;
• information technologies in processes, certification and quality management.

Conference organizers:
Omsk state Technical University http://omgtu.ru
Federal State Autonomous Educational Institution on Further Vocational Training, «Academy for Standardization, Metrology and Certification (Educational)» http://en.asms.ru

Organizing committee:
Chairperson:
Anatoly Kosykh, rector of the Omsk State Technical University
Co-Chair:
Viktor Shalay, president of the Omsk State Technical University

Members of the Organizing Committee:
Bekin Zhenatov,
Vice-rector on science Omsk State Technical University
Hristo Radev, professor, Sofia University "St. Kliment Ohridski"
A. Bataev, Professor, Rector of the Novosibirsk State Technical University
V.Voronin, professor, Rector of the Academy of Standardization, Metrology and Certification
V. Vavilov , professor, Institute of NDT, Tomsk Polytechnic University
I. Zhivotkevich, professor, autonomous non-commercial organization "Institute for testing and certification of weapons and military equipment"
V. Teleshevsky, Professor, Moscow State Technical University "STANKIN"
P. Chubikov, Doctor of Technical Sciences, Professor, Rector of Tomsk Polytechnic University
L. Varepo, professor, Omsk State Technical University
V. Glukhov, professor, Omsk State Technical University
V. Nikonenko,
Director general of the scientific-production Enterprise "Etalon"
D. Svetlichny, director of the State Regional Center for Standardization, Metrology and Test in the Omsk Region
O. Babenko, head of Information and Patent Department, Omsk State Technical University

IOP Conf. Series: Journal of Physics: Conf. Series **998** (2018) 011002 doi:10.1088/1742-6596/998/1/011002

Peer review statement

All papers published in this volume of *Journal of Physics: Conference Series* have been peer reviewed through processes administered by the proceedings Editors. Reviews were conducted by expert referees to the professional and scientific standards expected of a proceedings journal published by IOP Publishing.

Study of sandy soil grain-size distribution on its deformation properties

L B Antropova, A V Gruzin, M I Gildebrandt, L D Malaya, and V B Nikulina

Omsk State Technical University, 11, Mira av., Omsk, 644050, Russia

E-mail: lubashka_2010@mail.ru

Abstract. As a rule, new oil and gas fields' development faces the challenges of providing construction objects with material and mineral resources, for example, medium sand soil for buildings and facilities footings of the technological infrastructure under construction. This problem solution seems to lie in a rational usage of the existing environmental resources, soils included. The study was made of a medium sand soil grain-size distribution impact on its deformation properties. Based on the performed investigations, a technique for controlling sandy soil deformation properties was developed.

1. Introduction

According to the existing regulations, the soil base works have to be carried out at the early stages of buildings and facilities construction [1]. For example, when constructing oil and oil products storage tank, one needs to perform an operation cycle on soil base arrangement. Specially prepared medium sandy soil is used for this application. But this requirement cannot always be met due to remoteness of the sandy soil with a necessary grain-size distribution from the object under construction. The necessary materials delivery may require substantial material and time expenses, causing a significant increase in the costs of the constructed object. Apparently, the sustainable use of the existing local resources, soil included, may significantly affect the decrease in time and material costs at the construction stage of oil and gas industry facilities.

2. Problem statement

The dependence of the existing technologies for building and facilities soil bases arrangements is problematic to decrease. This problem solution appears to lie in developing an auxiliary specialized technique aimed at controlling the deformation properties of the present sandy soil. The main objective of the performed investigation is to specify the impact nature of the sandy soil grain-size distribution on its deformation properties. To achieve the formulated task, the following problems are to be solved:

- to study deformation properties of medium sandy soil;
- to study deformation properties of separate sand fractions;
- to develop the technique for altering the soil deformation properties on the basis of the obtained results.

3. Laboratory investigations

At the early investigation stage, the type of the sandy soil was defined according to the existing classification [5]. A sieve analysis is used to determine the grain-size distribution of the sandy soil.

This analysis is performed by screening the sandy soil through a column of laboratory sieves (figure 1a) arranged in the order of mesh size increase from 0.1 mm to 10 mm. The weight percentage of the separated fractions determines the type of sand (figure 1b).

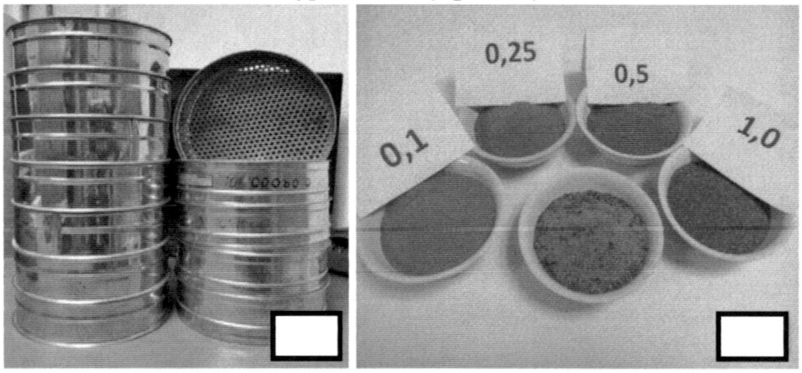

a) b)

Figure 1. Preparing the sandy soil to compression tests:

a) laboratory sieves; b) fractions of the sandy soil.

In previously performed investigations the nature of sandy soil humidity effect on its deformation properties was determined [7]. Laboratory tests were performed on automated test complex ASIS, and they allowed characterizing the humidity effect on the deformation properties of separate sandy soil fractions.

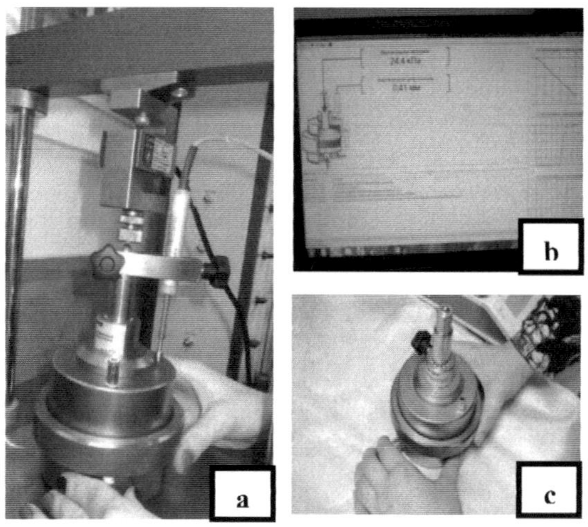

Figure 2. Automated test complex ASIS.

Automated test complex ASIS includes:

- an axial loading device (figure 2a);
- measuring system ASIS (figure 2b);
- an odometer (figure 2c).

A special loading programme was developed to carry out compression tests for a sandy soil sample. The maximum load on the sandy soil sample under the test is equivalent to the pressure under a

50000 m^3 tank bottom during its hydraulic tests. The test procedure includes soil sample step loading with a time exposure at each step, and subsequent step unloading of the investigated sample.

4. Laboratory findings

In the course of the performed investigations a technique was developed, that included the following stages:

- preparatory stage (soil fractionating, determining the soil bulk density and humidity, adding necessary amount of water into soil and its mixing);
- principal stage (soil humidity control, compression tests);
- final stage (laboratory data processing, conclusions and recommendations).

Medium sandy soil from the Irtysh river floodplain was used as initial soil. During compression tests, the dependencies of sandy soil humidity effect on its deformation properties were determined. Figure 2 represents the dependence for vertical deformation of the medium sand sample with the height of 25 mm on its humidity.

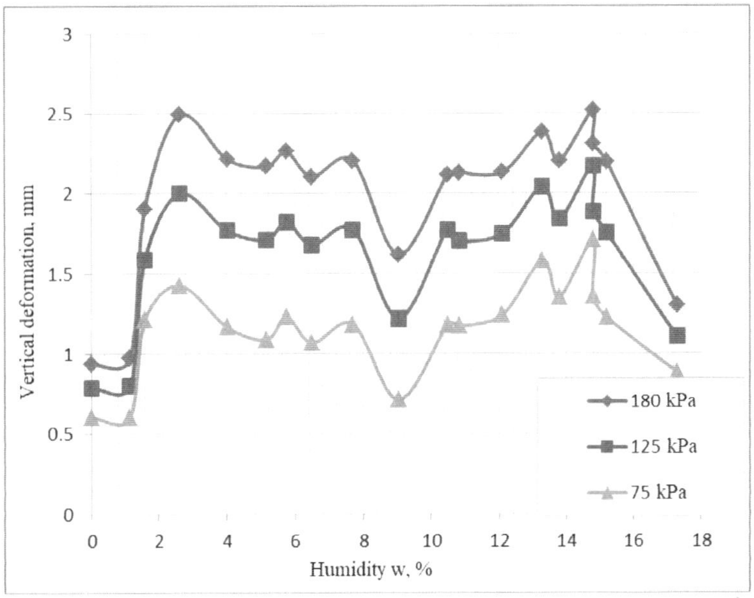

Figure 3. The results of compression tests performed on the medium sandy soil.

As the diagram shows, there is a humidity range with noticeable maximal vertical deformation of the sandy soil sample. The sharp increase is observed at the range from 0 % to 3 %, while at humidity more than 15 % there is a decrease.

These dependencies were received for sandy soil fractions with the grain sizes of 0.5-1.0 mm, 0.25-0.5 mm, 0.1-0.25 mm and for the fractions less than 0.1 mm. The resulting characteristics will enable the synthesis of the soil with predetermined mechanical properties.

Based on the data from [8, 9], the technique was proposed to adjust the deformation properties of loose dispersive soil having a number of advantages in comparison with traditional ones, namely, the decrease in material and time expenses. This technique calls for application of neither expensive components nor specialized equipment.

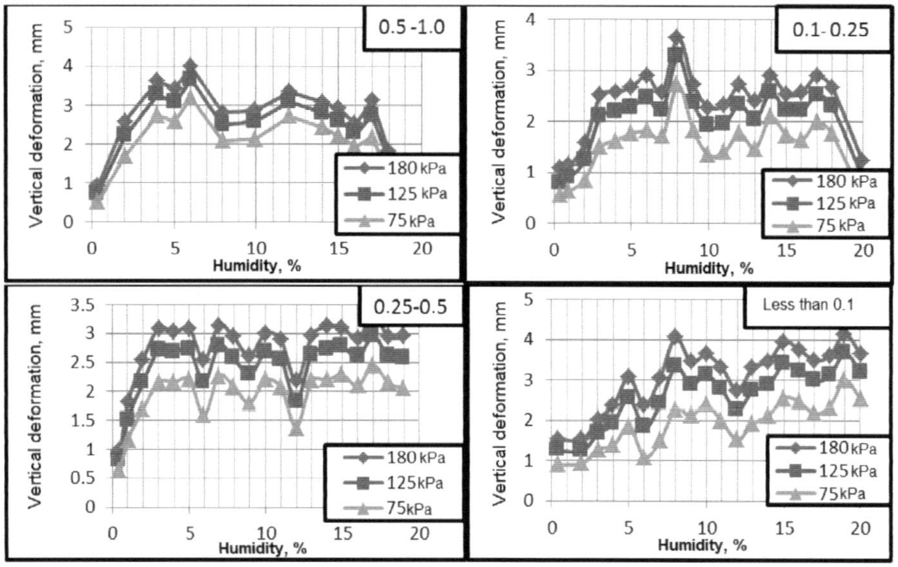

Figure 4. The dependence of sandy soil vertical deformation on humidity for its different fractions.

The proposed technique consists in fractionating sandy soil, determining deformation properties of each one, and then creating a composite mixture with predetermined mechanical properties.

5. Conclusions

Therefore we may draw the following conclusions:

- the technique for performing laboratory tests is developed;
- the deformation properties of medium sand soil as well as its separate fractions were studied;
- the sand soil deformation properties control technique was developed;

The developed technique has been granted a patent [10].

References

[1] SP 22.13330.2011. *Soil Bases of Buildings and Structures.* Revised edition of SNiP 2.02.01-83* Russian

[2] SP 45.13330.2012. *Earthworks, Grounds and Footings.* Revised edition of SNiP 3.02.01-87. Russian

[3] RD 16.01-60.30.00-KTN-026-1-04. 2010. *Design Norms of Vertical Steel Oil Storage Tanks with Capacity of 1000 to 50000 Cubic Meters.* (Moscow, JSK "Transneft") Russian

[4] GOST RF 31385-2008. *Vertical Cylindrical Steel Tanks for Oil and Oil-Products. General Specifications.* Russian

[5] GOST RF 25100- 2011. *Soils. Classification*. Instead of GOST 25100-95. Russian

[6] GOST RF 12536-2014. *Soils. Methods of Laboratory Granulometric (Grain-Size) and Microaggregate Distribution.* Instead of GOST 12536-79

[7] Gruzin A V, Tokarev V V, Shalai V V, Logunova Yu V 2015 *Proc. Eng.* **113** pp158–168

[8] Antropova L B, Gildebrandt M I, Sirotin A D, Gruzin A V 2015 *Future Generation: The Viewpoint of Young Scientists* 129–131

[9] Antropova L B, Gildebrandt M I, Gruzin A V 2016 *Challenges of Geology and Soil Development* p 877–879

[10] Gruzin A V, Antropova L B, Rusanova A D, Katunin A V, Gildebrandt M I, Sirotin A D 2017 *The Technique for Controlling Deformation Properties of Loose Dispersive Soil.* RF Patent 2621799 C1 Russian

IOP Conf. Series: Journal of Physics: Conf. Series **998** (2018) 012002 doi:10.1088/1742-6596/998/1/012002

Bootstrap and Counter-Bootstrap approaches for formation of the cortege of Informative indicators by Results of Measurements

M V Artemenko[1], I E Chernetskaia[2], N M Kalugina[3] and E N Shchekina[1]

[1]Department of biomedical engineering, South-West State University, 94, 50 years old of October, Kursk, 305040, Russia
[2]Department of computer facilities, South-West State University, 94, 50 years old of October, Kursk, 305040, Russia
[3]Department of Space instrument making and communication systems, South-West State University, 94, 50 years old of October, Kursk, 305040, Russia

E-mail: artem1962@mail.ru

Abstract. This article describes the solution of the actual problem of the productive formation of a cortege of informative measured features of the object of observation and / or control using author's algorithms for the use of bootstraps and counter-bootstraps technologies for processing the results of measurements of various states of the object on the basis of different volumes of the training sample. The work that is presented in this paper considers aggregation by specific indicators of informative capacity by linear, majority, logical and "greedy" methods, applied both individually and integrally. The results of the computational experiment are discussed, and in conclusion is drawn that the application of the proposed methods contributes to an increase in the efficiency of classification of the states of the object from the results of measurements.
Keywords: object of research, the information capacity of measured characteristics, bootstrap method, self-organization modeling.

1. Introduction

The development of computing and intellectual resources of modern decision support tools in scientific research and industry allows applying the achievements in this field for the analysis of complex and open systems by conducting adequate simulation. The simulation model implies a certain architecture and mathematical apparatus that describes the interaction of model elements, the functioning of internal and external interfaces. In the general case, the cybernetic simulation model describes the behavior of an object in the form of a certain multipolar network characterizing a set of values of the variously measured characteristics to ensure observability and controllability of the object.

At the initial stage of modeling, there are problems associated with complex technologies with inadequate reliability. The article [1] describes that the behavior of complex objects is characterized by a set of indicators ordered by a certain principle - a cortege, consequently, the result depends not

only on the measurement method, but also on the algorithm that determines the sequence of measurement of the indicators.
Therefore, the problem of the formation of a cortege of informative features is actual.

2. Task

Research of various information sources shows that nowadays methods of forming a set of informative features of a modeling object are based on the application of the following methods and algorithms [2]: *Full* – a complete search of various combinations of characteristics to achieve an acceptable classification effect, *Add* and *Del* – consecutive addition and exclusion of features; *AddDel* – combining of algorithms *Add* and *Del*; *Prob* – each feature of the object is determined by weights and the procedures of the indicated algorithms are applied; *fractal analysis* - used for tensor data; *Grad* - is similar to the algorithm *AddDel*, but the inclusion and exclusion of characteristics in the final informative set is not "one by one", but "jointly".

Since the cortege implies a certain order, it is suggested that the criterion of informative sign of the characteristic be taken as the sorting criterion. The complexity of its definition is due to the fact that the mentioned algorithms function adequately in conditions of statistically representative volumes of training and examination samples. If the above condition is violated, the efficiency of the algorithms decreases drastically.

In this connection, there arises the problem of optimal formation of a tuple in conditions of different volumes of training material based on the application of measures characterizing the informative capacity of the feature from the point of view of certain classification possibilities for determining the state of the object in the process of performing the measurements. The application of the cortege will allow rationalizing the technology of the measuring process.

3. Theory

Let's allocate three main situations, characteristic for the decision of problems of classification, approximation and extrapolation:

1. The volume of data satisfies the statistical criteria of representativeness. In this case, standard methods for calculating informative capacity are used (Kullback's method, Student's t-test, classification errors, etc.), adapted to the subject domain in question.

2. The volume of information far exceeds the statistical threshold, which significantly increases the likelihood of obtaining false conclusions on typical statistical criteria. This situation arises quite often when measuring the characteristics of an object recorded with a high sampling frequency. In this case, it is recommended to reduce the sample size by applying convolution algorithms or reducing the dimensionality of the feature space. The drawback of this approach is the possible loss of information. In this regard, it is recommended that the sample be selected as a minimum for the information assessment method used in the future and gradually increase its volume to an acceptable value for the adequacy of the estimate. Let's call this approach a "counter-bootstrap".

3. The sample size is much smaller than the statistical threshold. In this case, it is recommended to use the bootstrap methodology for bringing the volume to the required level (considered, for example, in [4]) and apply the "law of small numbers" [5].

The theory of self-organization and the system approach allows us to combine the last two approaches to assess the informative capacity of the measured characteristics of an object, using the integration of properties of methods that require a large amount of information and allow obtaining acceptable results on the basis of a small number.

First approach is used in Rasch measurement [6], and second approach - in Group Method of Data Handling (GMDH) [7]. We denote as $InfRasch(x_i)$ the quantitative evaluation of the informative capacity of the analyzed measurable or latent attribute.

In contrast to G. Rasch's models, GMDH self-organization modeling algorithms, proposed by Ivakhnenko AG, work more adequately in conditions of small-volume samples, - especially at the stage of exploration analysis. Thus, we denote as $InfMGDH(x_i)$ the quantitative evaluation of the informative capacity for GMDH.

As any artificial modification of the data sample size generally reduces the measure of confidence in them, it is recommended to use correction coefficient RRC (the reduction ratio of confidence):

$$RRC = por + (1 - por) \cdot \cos\left(\frac{\pi \cdot d}{2\sqrt{2}}\right) \quad (1)$$

where por is the threshold value RRC, determined by the researcher (recommended values are $0 \leq por \leq 1$); d - the distance between the characteristic points of the original and modified sample in the plane of mathematical expectation and variance of the reduced values ($0 \leq d \leq \sqrt{2}$) (the reduced values of the elements of the sets of recorded characteristics are determined by normalizing the values of sets by linear transformations in the range$[e, 1 - e], e = \frac{1}{N}$, where N – initial sample size).

We denote these coefficients, respectively, as: RRC_b и RRC_{cb}.

Guided by the hypothetical-deductive method of scientific research and the concepts of methodological invariant, we propose the following technology of calculations $InfMGDH(x_i)$ [8]. At the first stage, the structural-parametric identification of the approximating polynomial $App\{X\}$ [6] for each characteristic of the initial set {X} according to the equation:

$$Y(Z) = A_0 + \sum_{k=1}^{K} (A_k \cdot \prod_{i=1}^{N} z_i^{p_{i,k}}), \quad (2)$$

where: $Z=\{z_1, z_2,\ldots,z_N\}$ is the set of arguments, $Y(Z)$ is the response function, K is the number of terms in a polynomial, A_k, $p_{i.k}$ – model's parameters, N – number of arguments.
The identification procedure is repeated N times for each object state class w_l , $l = \overline{1, L}$ (L is the number of classes) with the consequent formation of a sets $\{Z\} = \{X\} - x_j$ and responses $Y(Z) = x_j$.

As a result, sets of mathematical models for alternative classes w_l:

$\{App\{X\}\}_{M_l}^{w_l}$, $M_l \leq N$, $M_l \neq 0, R^2[App\{X\}] \geq R^2_{lim}$, where R^2_{lim} is the threshold value of determination coefficient $R^2[App\{X\}]$. If, as a result of selection, an empty set is formed ($M_l \neq 0$), then models are successively included in it as the values of the determination coefficients decrease (the threshold value increases R^2_{lim}). The minimum volume of many mathematical models is recommended not less than 3.

After that, matrices are formed for each alternative class $\{Rn\}^{w_l}$, the number of rows in which are equal, respectively, *Ml*, columns - the number of indicators of the set {X}, the value of matrix elements is calculated by successive application of equations:
At first, the fraction of influence of each term k in classes is determined by the equation:

$$V_k^{w_l} = \frac{\overline{A_k^{w_l} \cdot \prod_{i=1}^{N} x_i^{p_{i,k,w_l}}}}{\sum_{j=1}^{L} (\overline{A_j^{w_l} \cdot \prod_{i=1}^{N} x_i^{p_{i,j,w_l}}})}, \quad (3)$$

where $\overline{(ZZ)}$ determines the median of ZZ values as the most resistant to artifacts.

At second, for each argument included in the k-th term, the weight of the multiplier is calculated by the equation:

$$M_{i,k}^{w_l} = \frac{|p_{i,k,w_l}| \cdot |\ln(\bar{x}_i)|}{\sum_j^N |p_{i,k,w_l}| \cdot |\ln(\bar{x}_i)|}. \quad (4)$$

Next, the values of the additive-multiplicative influence of the exponent x_i are calculated on the response function $Y(Z)$ for each alternative class:

$$AM_{x_i}^{w_l} = 1 - \prod_{k=1}^{K} \left(1 - V_k^{w_l} \cdot M_{i,k}^{w_l}\right). \quad (5)$$

«Relative error of differences» (RED) ε is introduced (recommended values are $0{,}01 \leq \varepsilon < 0{,}1$) and the values of the magnitude of the multiplier effect are recalculated at $[AM_{x_i}^{w_l}, \varepsilon]$ by the equation:

$$[AM_{x_i}^{w_l}, \varepsilon] = \begin{cases} AM_{x_j}^{w_l}, \text{если } (1-\varepsilon)\cdot AM_{x_j}^{w_l} < AM_{x_i}^{w_l} \leq (1+\varepsilon)\cdot AM_{x_j}^{w_l} \\ AM_{x_i}^{w_l}, \ otherwise \end{cases}, j = \overline{1,N}, j \neq i \quad (6)$$

For each class, the characteristics of x_i are ordered as the values decrease$[AM_{x_i}^{w_l}, \varepsilon]$ and l sets of characteristics for classes $\{X\}^{w_l}$ and the corresponding sets of ranks $\{Rn\}^{w_l}$ are formed. $j = \overline{1,N}$ – number of elements of sets of ranks, and elements are calculated by the equation:

$$Rn_j^{w_l} = \begin{cases} N, \ \text{для } j = 1 \\ Rn_j^{w_l} - 1, \quad \text{для } (i \neq 1) \& ([AM_{x_i}^{w_l}, \varepsilon]_j \neq [AM_{x_i}^{w_l}, \varepsilon]_{j-1}) \\ Rn_j^{w_l}, \quad \text{для } (i \neq 1) \& ([AM_{x_i}^{w_l}, \varepsilon]_j = [AM_{x_i}^{w_l}, \varepsilon]_{j-1}) \end{cases} \quad (7)$$

The set $\{Rn\}^{w_l}$ forms the final set of informative characteristics for a certain volume $NI \leq N$, which includes elements $XI_i = (x_i / InfMGDH(x_i)) \ / \ i = \overline{1,NI}$, imported from the source set {X} according to $\{Rn\}^{w_l}$ sequentially, as the rank values decrease. The final informative capacity $InfMGDH(x_i)$ is calculated as:

$$InfMGDH(x_i) = \frac{\max_{l=\overline{1,L}}(Rn_{x_i}^{w_l})}{\max_{j=\overline{1,NI}}(\max_{l=\overline{1,L}}(Rn_{x_i}^{w_l}))} \ . \quad (8)$$

Using values of $InfRasch(x_i)$ and $InfMGDH(x_i)$ for integrating of results of bootstrap and counter-bootstrap, we obtain the value of the integral information $Inf(x_i)$ of the characteristicx_i: $Inf(x_i) = F(InfRasch(x_i), InfMGDH(x_i), RRC_b, RRC_{cb})$. Based on the research objectives, the data structure and the semantic load of the analyzed features of the modeling object, it is proposed to apply the following calculation equations $Inf(x_i)$, where: $a = RRC_b \cdot InfRasch(x_i)$, $b = RRC_{cb} \cdot InfMGDH(x_i)$:

- additive-weight aggregation is recommended in the case of noncollinear features, which were selected by models of G. Rasch and GMDH):

$$Inf(x_i) = \frac{a+b}{RRC_b + RRC_{cb}}, \quad (9)$$

- majoritarian aggregation is recommended when the characteristics are collinear, namely:

a) "careful" aggregation:

$$Inf(x_i) = \min(a, b); \quad (10)$$

b) "tense" aggregation:

$$Inf(x_i) = \max(a, b); \quad (11)$$

c) "greedy" aggregation, with the help of an operation «softmax» (according to R. Sutton [9]):

$$Inf(x_i) = \max\left(\frac{\exp\left(\frac{a}{RRC_b + RRC_{cb}}\right)}{\exp\left(\frac{a+b}{RRC_b + RRC_{cb}}\right)}, \frac{\exp\left(\frac{b}{RRC_b + RRC_{cb}}\right)}{\exp\left(\frac{a+b}{RRC_b + RRC_{cb}}\right)}\right), \quad (12)$$

d) Boolean aggregation:

$$Inf(x_i) = \frac{1}{1+\alpha} \cdot (a + b + \sqrt{a^2 + b^2 - 2 \cdot \alpha \cdot a \cdot b}), \quad (13)$$

where: α is the parameter that takes a value from the interval (-1,1].

Since the formation of a set of informative features according to the proposed methods is carried out according to the ordered values of information indicators, the problem of the formation of the sought-for cortege is solved.

IOP Conf. Series: Journal of Physics: Conf. Series **998** (2018) 012002 doi:10.1088/1742-6596/998/1/012002

4. Results

In order to study the possibilities of the control methods considered, a computer experiment was carried out: the object was in two alternative classes, n=10; 30 и 100 signs were synthesized and signs were correlated and uncorrelated among themselves (determined by the value of the pair correlation R). Number of observations m=500. To simulate the characteristics of the object was used randomization according to the uniform distribution law, the artificial error was carried out at the level of 20%, and all attributes had a classification efficiency at the level of 0,65-0,7. Three options for the formation of training and examination samples were considered: $m1<m$ ($m1$=0.2m); $m1>m$ ($m1$=3m); $m1 \approx m$ ($m1 = 0.8m$). The results of the computational experiment are given in Table 1.

Table 1.
The results of the computational experiment of formation of cortege of informative signs of an object

sample size	Collinearity R	aggregation type	number of informative signs $n1$	Classification efficiency
$m1$=0.2*m	0.8 – 0.9	additive-weight	6; 21; 34	0.66; 0.66; 0.68
	0.8 -0.9	"careful"	6; 21; 35	0.66; 0.66; 0.66
	0.8 – 0.9	"tense"	5; 20; 32	0.68; 0.64; 0.72
	0.8 – 0.9	"greedy"	5; 20; 32	0.66; 0.66; 0.66
	0.8 – 0.9	boolean	7; 22; 34	0.61; 0.72; 0.79
	0 – 0.2	additive-weight	3; 8; 16	0.69; 0.62; 0.71
	0 – 0.2	"careful"	3; 8; 14	0.61; 0.72; 0.72
	0 – 0.2	"tense"	3; 8; 16	0.71; 0.72; 0.71
	0 – 0.2	"greedy"	3;10; 26	0.71; 0.68; 0.7
	0 – 0.2	boolean	5; 12; 22	0.61; 0.62; 0.59
$m1$=3*m	0.8 – 0.9	additive-weight	10; 30; 100	0.96; 0.96; 0.97
	0.8 -0.9	"careful"	9; 28; 98	0.95; 0.98; 0.98
	0.8 – 0.9	"tense"	8; 26; 86	0.98;0.95; 0.97
	0.8 – 0.9	"greedy"	10; 30; 100	0.99; 0.99; 0.99
	0.8 – 0.9	boolean	9; 30; 98	0.99; 0.99; 0.99
	0 – 0.2	additive-weight	10; 28; 96	0.98; 0.95; 0.97
	0 – 0.2	"careful"	8; 26; 92	0.96; 0.95; 0.98
	0 – 0.2	"tense"	7; 18; 72	0.96; 0.95; 0.98
	0 – 0.2	"greedy"	10; 28; 100	0.99; 0.99; 0.99
	0 – 0.2	boolean	10; 30; 100	1; 1; 1
$m1$=0.9*m	0,8 - 0,9	additive-weight	9; 26; 95	0.95; 0.95; 0.95
	0.8 -0.9	"careful"	8; 24; 92	0.96; 0.95; 0.98
	0.8 – 0.9	"tense"	8; 24; 92	0.92; 0.95; 0.96
	0.8 – 0.9	"greedy"	10; 30; 100	1; 1; 1
	0.8 – 0.9	boolean	10; 30; 100	1; 1; 1
	0 – 0.2	additive-weight	9; 25; 94	0.92; 0.94; 0.91
	0 – 0.2	"careful"	9; 22; 92	0.9; 0.9; 0.86
	0 – 0.2	"tense"	8; 25; 94	0.87; 0.9; 0.9
	0 – 0.2	"greedy"	9; 29; 98	0.93; 0.98; 0.99
	0 – 0.2	boolean	9; 28; 97	0.99; 0.99; 0.99

In Table 1, the values for different dimensions of the characteristic space are listed through the ";" sign.

5. Discussion

The analysis of results of a computing experiment allows to draw the following conclusions:

- When using a statistically representative volume as a training and examination sample, the results of applying different approaches to the formation of a tuple of informative features do not significantly improve the quality of classification and, therefore, are unproductive.
- A similar effect is observed when the sample size is much higher than the representative threshold, but the requirements for computational resources (memory, speed) are increasing, therefore it is recommended to use the counter-buster method.
- The use of a set of proposed procedures for assessing the informative capacity for the formation of an informative tuple of characteristics makes it possible to obtain an acceptable classification result in the case of small sample use (statistically unrepresentative); the best way is the additive-weighted, "greedy" and boolean aggregations.

6. Conclusion

The offered technology of numerical nonparametric assessment of informational content of the measured signs of an object

The proposed technology of numerical non-parametric estimation of informational content of measured features of an object is recommended for practical use in conditions of weakly structured and / or statistically insufficient amount of data specific for the study of complex dynamic objects belonging to the class of open, complex systems.

The attribute cortege formed on the basis of estimates of information gives possibility to synthesize adequate mathematical models for solving the classification problems of objects based on the values of the measured features for monitoring and controlling the object in conditions of both a lack and an overabundance of measurement results characterizing the object.

References

[1] Taleb Nassim Nikolas 2017 Chernyj lebed'. Pod znakom nepredskazuemosti (Moscow: KoLibri, Azbuka-Attiku) Russian

[2] Kolesnikova S N 2009 *Tomsk State university Bulletin Upravlenie, vychislitel'naja tehnika i informatika* **1(6)** p 48-54 Russian

[3] Visa Koivunen and Emad Mozafari 2016 *50th Asilomar Conference on Signals, Systems and Computers Year (IEEE Conference Publications)* pp 343–348

[4] Schmitt R, Fritz P and Lose J 2011 *Int. J. Metrol. Qual. Eng.* **2 (1)** p 19–24

[5] Amos Tverski, Daniel Kaneman 2005 Vera v zakon malyh chisel (Kaneman D, Slovik P, Tverski A. Prinjatie reshenij v neopredeljonnosti: pravila i predubezhdenija) (Har'kov: Gumanitarnyj centr Russian) p 40-41

[6] Juan Ramon Oreja-Rodrigues, Vanessa Yanes-Estévez, 2010 *Proc. Environmental scanning: Dynamism with rack and stack from Rasch model (Management Decision)* **48 (2)** p 260-276

[7] GMDH – Examples of Applications *Group Method of Data Handling (GMDH)* 2010 available at: http://www.gmdh.net/GMDH_ exa.htm

[8] Artemenko M V, Podval'nyj E S, Starcev E A 2016 Biiomedicine radioelectronics (*Biomedicinskaja radiojelektronika)* **9** p 38-44

[9] Richard S. Sutton, Andrew G 2016 Barto Reinforcement Learning: An Introduction (MIT Press)

The Papers Printing Quality Complex Assessment Algorithm Development Taking into Account the Composition and Production Technological Features

Kh A Babakhanova[1], **L G Varepo**[2], **I V Nagornova**[3], **E B Babluyk**[3] **and A P Kondratov**[3]

[1] Tashkent Institute of Textile and Light Industry, 5, ul. Shohzhakhon, Tashkent, 100100, Uzbekistan
[2] Omsk State Technical University, 11, Mira av., Omsk, 644050, Russia
[3] Moscow Polytechnic University, 38, B. Semenovskaya str., Moscow, 107023, Russia

E-mail: halima300@inbox.ru

Abstract. Paper is one of the printing system key components causing the high-quality printed products output. Providing the printing companies with the specified printing properties paper, while simultaneously increasing the paper products range and volume by means of the forecasting methods application and evaluation during the production process, is certainly a relevant problem. The paper presents the printing quality control algorithm taking into consideration the paper printing properties quality assessment depending on the manufacture technological features and composition variation. The information system including raw material and paper properties data and making possible pulp and paper enterprises to select paper composition optimal formulation is proposed taking into account the printing process procedure peculiarities of the paper manufacturing with specified printing properties.

1. Introduction

In the modern market environment while producing the printing and packaging products the competitiveness defined by the quality depending on the production capabilities and regulations is strongly influenced by the printing carrier (material) and paint interaction.

The current standard [1] regulates the printed offset products quality indicators requirements. This way provides only limited amount of reference paper types which makes the printouts quality assessment in the small and medium-volume manufacturing environment difficult.

Given a large variety of the printing system components, the lack of new materials printing and technical properties data and taking into consideration the consumers' expectations about the printed image quality, the offset printing quality forecasting problem is supposed to be of great economic importance and relevant. The problem solving requires the evolution and development of the known and new quality indicators assessment methods both printing system components and a printout.

2. Problem statement

The authors of the papers [2, 3] indicate that during the printing ink transfer, the structure characteristics of the print material, the shapes of the pores and the holes being present on its surface play a dominant role. However, in papers [2, 3] the quantitative assessment data of the ink on paper transfer coefficients in the contact surface are not provided.

The fundamental principles on the role of the structure and its influence on the paper properties and printout quality were set forth in papers [4-7]. It should be noted that in the offset printing process, when paper wetting, the surface fibers swell and as a consequence its roughness slightly increases thereby reducing the printout intensity. In addition, large paper surface irregularities (cardboard) can lead to the large offset rubber deformations, and this contributes to the printout printing (elements) sizes increase. When offset printing on the rough paper the reduced printout intensity is observed, this is due to the fact that various surfaces types reflect light differently. Roughness can theoretically cause the local paper inhomogeneity on moisture.

The expert evaluation methodology according to which the expert group members assess 10 printout specimens quality by means of the pairwise comparison was developed by the authors of the paper [8]. The expert evaluations results are processed by using the fuzzy sets theory. However, it should be noted that experts are able to distinguish and identify the printouts quality differences obtained in many ways only with test elements magnification which is various for various apparatuses, and the printed test element (letter) quality perception character without magnification can be entirely changed when evaluating with a high magnification.

The printouts examination method to determine the quality level taking into account the paper quality fluctuations at the micro level was developed by the authors [9]. The method application is also difficult as it requires the special equipment (TAPIO Paper Machine Analyzer and Tapio Printability Sensor) designed for unprinted points quantity rapid measurement. On the basis of the variables of different origin combined influence on the printouts quality conducted studies the authors [11-14] emphasize among other factors the paper sheet density irregularity impact on the paint penetration.

Therefore, the wrong choice of paper type caused by the lack of reliable data on its composition and surface characteristics can be one of the spoilage primary reasons and as a consequence the impossibility to satisfy customer and consumer requirements ones. The proposed solutions represent, in most cases, the individual attention to studying the printing (material specific factor influence on the printing quality.

The produced paper range expansion and making its properties meet the international standards requirements is the topical scientific and practical problem. In recent years in the Republic of Uzbekistan new types of paper have been developed including the paper made from cotton lint (CL). The results of the runability individual parameters interrelation with the printing quality indicators were represented by the authors [15-18].

Modern quality management is based on the fact that quality management activity cannot be effective after the products are produced, this activity should be carried out both at the preproduction stage and during the manufacturing process.

This proves the necessity of developing the paper printing properties assessment models making possible pulp and paper enterprises select paper composition optimal formulation taking into account the printing process procedure peculiarities of the paper manufacturing with specified printing properties. The high priority objective is printing quality on papers complex assessment algorithm development taking into account the composition and production technological features.

3. Results and discussion

To forecast the printing quality before the printing process on papers adding SFM (secondary fibrous materials), the multilevel system being the development of the conception «raw material - paper - printout», which is presented in Figure 1, is proposed.

According to the scheme analysis the regularities of the interrelation between the printing quality and (paper properties) define:

- fibrous semi-finished products obtaining and processing methods;
- paper printing and technical characteristics depending on the composition;
- printouts printing quality indicators.

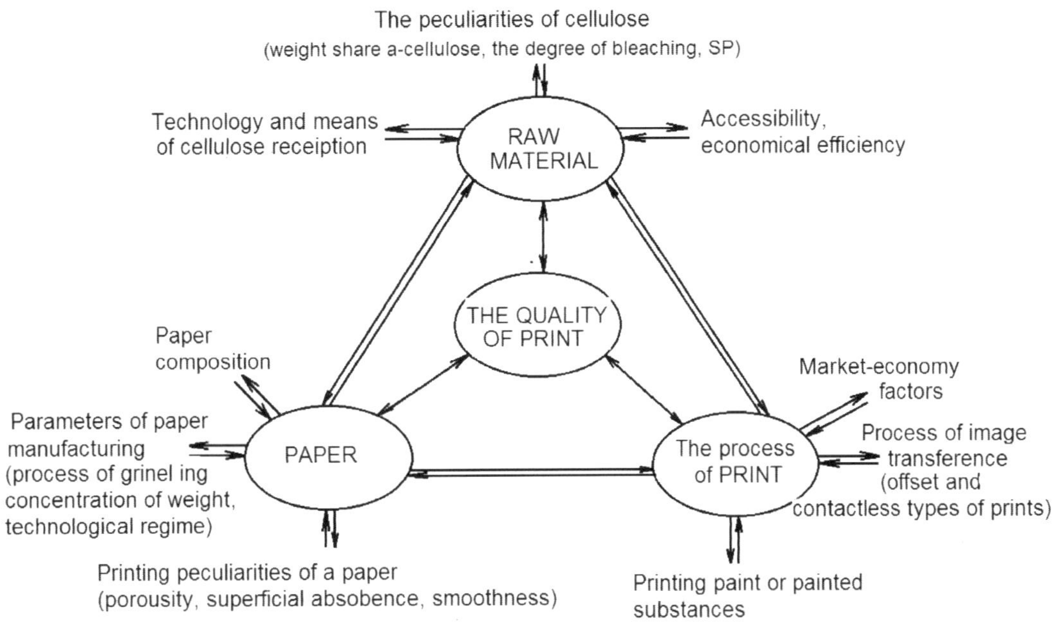

Figure 1. The «raw material - paper - print» system elements interrelation scheme.

Considering the fact that (paper printing properties) varies as a result of the introducing other semi-finished products into the composition, in this paper to improve the paper (printing properties unability based on cellulose from CL () and to give it specified (printing properties), fibrous semi-finished products from wastes (SFM) were selected taking into consideration the fibres structure.

For the comprehensive printing quality assessment the assessment model was developed and the algorithm was recommended, the scheme of which is shown in Figure 2. The model reliability was defined by the multiple rank correlation method.

The developed algorithm practical implementation is shown in Figure 3 which represents the comprehensive (printing properties assessment results adding SFM in the form of the radar chart with markers. This makes possible to illustrate the indicators variability simultaneously in several directions.

The paper properties comprehensive assessment diagram analysis revealed that paper whiteness is within the standard limits (from 72 to 91%). The influence investigation results of adding SFM on nontransparency values defining the paper printing properties confirmed the paper suitability for the multicolor products one-side printing, while for the double-sided printing without image show-through and strike-through the additional sizing or composition or filler quality changing is necessary.

Porousness identical values provided by kanaf cellulose (KC), licorice cellulose (LC) and natural silk waste (NCW) except wheaten cellulose (WC) adding into the paper pulp characterize the pores volume reduction and the reaction surface accessibility. The application of WC as an additive requires the composition careful selection or fibrous material processing optimal performance choosing.

When forming the paper layer, NCW short fibers fill the gaps between the CC long and flexible fibers (paper No. 2). In this case, the printout optical density equal to 1.5 is achieved at a paint layer thickness of 1.0 μm, whereas on paper No. 1 (CC 100%) this result is achieved at a paint layer thickness of 1.5 μm (Figure 4).

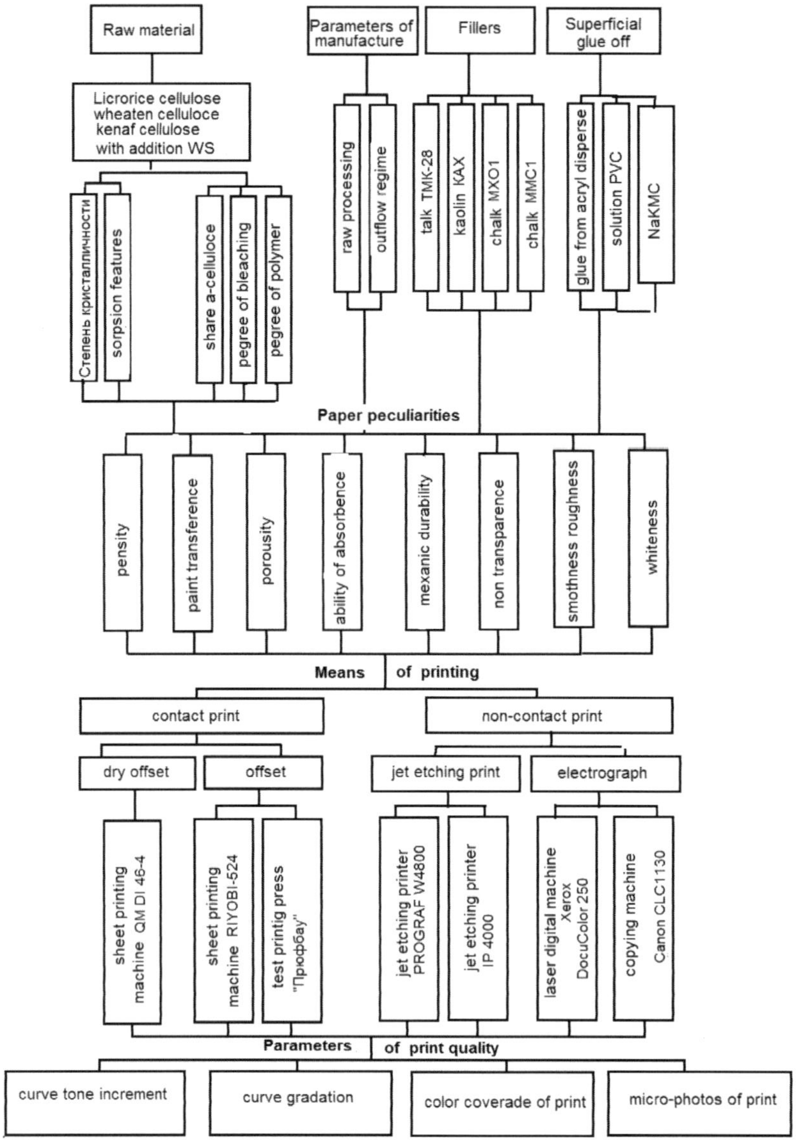

Figure 2. The comprehensive printing quality assessment algorithm.

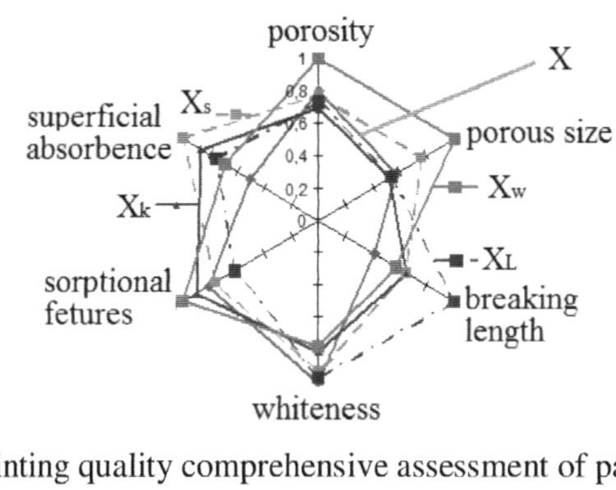

Figure 3. The printing quality comprehensive assessment of paper adding BBM.

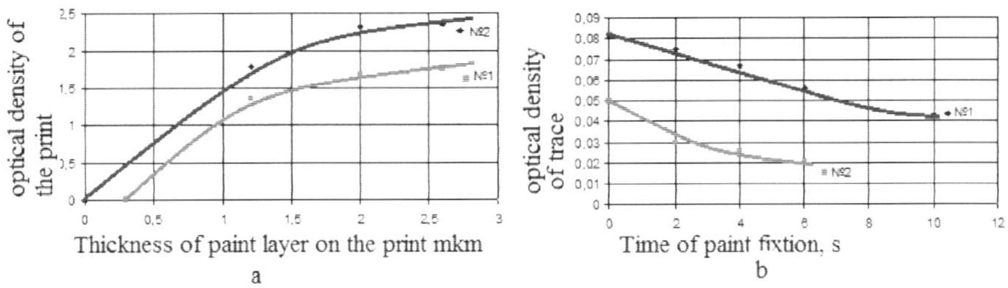

No 1 - a paper on the basis of CC (100 %), No 2 - a paper with addition NSW

Figure 4. The dependence of optical density of the print from thickness of a paint layer on the paper (a) and kinetics of primary fastening of a paint layer on the paper (b).

The study results analysis showed that printed on papers with the addition of SFM printouts optical density meets the international standard requirements [1].

4. Conclusion

- The printing quality complex assessment algorithm based on the «raw material - paper - printout» system approach application, providing the specified paper printing properties (runability) with the addition of recycled fibrous materials (obtained in the Territory of the Republic of Uzbekistan) was developed.
- The recycled materials addition into the paper pulp was found out to reduce the primary ink fixation rate providing the ink fixation condition on the paper surface.
- The optical densities values of the printouts printed on papers with the addition of SFM are relatively equal to the machine-finished paper optical densities values (for example, offset paper No 1) according to the densitometric standards ISO 12647-2.

5. References

[1] ISO 12647-2:2004 Graphic technology – Process control for the production of half-tone colour separations, proof and production prints – Part 2: Offset lithographic processes, 2012
[2] Nordstrom J-E and Gron J 1998 Thin-film transfer in printing and coating PFT Report **39** p 16
[3] Kalela E 2005 New busimess opportunities for printed media: Coated papers for coldest printing / Dissertation for the degree of Doctor of Science in Technology Helsinki: Helsinki University of Technology p 55
[4] Kulak M I 2002 Fractal Mechanics of Materials p 304
[5] Varepo L G, Novoselceva V P and Bobrov V I 2006 The paper composition influence on its printing properties *J. Vestnik MGUP* **7** p 103-107 Russian
[6] Trapeznikov E V, Varepo L G and Kolozova O A The paper surface profile analytical representation algorithm implementation *J. Izvestija vysshih uchebnyh zavedenij. Problemy poligrafii i izdatelskogo dela* 2013 No **3** p 61-65 Russian
[7] Varepo L G and Golunov A V 2011 Influence of surface geometry of paper on colour rendering of impression *J. Advanced Materials Research* **174** 366-369
[8] Lim P, Steinlage Ralph C and Gantner Thomas E 1990 Application of fuzzy set theory for evaluating non-impact printing image qualities *Int. Print. and Graph. Arts Conf., Vancouver* p 79-83
[9] Erho T, Anttila M, Makkonen T, Lammila R, Sarela S and Erho T 1997 A new method of estimating print quality *83 rd Annual Meeting of Technical Section of CPPA* p 267-27
[10] Danby Roger 1994 Print quallty advancement through co-operation (partnerships) *Int. Print. and Graph. Arts Conf., Halifax* p 153-158

[11] Varepo L G, Panichkin A V and Bobrov V I 2013 The finite-difference approximation of the ink transfer equation system in the offset sheet printing process *J. Izvestija vuzov Severo-kavkazskij region. Tehnicheskie nauki* **2** p 13-17 Russian

[12] Varepo L G, Brazhnikov A Yu, Volinsky A A, Kondratov A P and Nagornova I V 2017 Control of the offset printing image quality indices *IOP Conf. Series: Journal of Physics: Conf. Series* **858** p 1-5

[13] Varepo L G, Trapeznikova O V, Trapeznikov E V and Nagornova I V 2016 *Proc. Eng.* **152** p 478-481

[14] Varepo L G, Panichkin A V, Trapeznikova O V 2016 The numerical calculation of the viscous incompressible *IOP: Materials Science and Engineering* **124** p 1-6

[15] Ahmedova Z A and Babahanova H A 2013 The paper composition influence on the paper quality *Problemy tekstilja* **2** p 98-102 Tashkent

[16] Mirzaeva M B and Babahanova H A 2013 Factors affecting the paper specific properties of a particular purpose *J. Problemy poligrafii i izdatelskogo dela* **2** p 13-17 Russian

[17] Varepo L G and Babahanova H A 2014 The paper composition impact on the technical properties parameters **3** (133) p 251-252 Russian

[18] Babahanova H A 2015 Paper with the addition of the local recycled fibrous materials *Tehnologii XXI veka* **5** p 28-29 Tashkent

Intelligent sensor in control systems for objects with changing thermophysical properties

O A Belousov[1], D Yu Muromtsev[1], M P Belyaev[2]

[1] Tambov State Technical University, 106, Sovetskaya str. Tambov, 392000, Russia
[2] The Zhukovsky - Gagarin Air Force Academy, 54 A, Old Bolsheviks str., Voronezh, 394064, Russia

E-mail: jiour314@gmail.com

Abstract. The control of heat devices in a wide temperature range given thermophysical properties of an object is a topical issue. Optimal control systems of electric furnaces have to meet strict requirements in terms of accuracy of production procedures and efficiency of energy consumption. The fulfillment of these requirements is possible only if the dynamics model describing adequately the processes occurring in the furnaces is used to calculate the optimal control actions. One of the types of electric furnaces is the electric chamber furnace intended for heat treatment of various materials at temperatures from thousands of degrees Celsius and above. To solve the above-mentioned problem and to determine its place in the system of energy-efficient control of dynamic modes in the electric furnace, we propose the concept of an intelligent sensor and a method of synthesizing variables on sets of functioning states. The use of synthesis algorithms for optimal control in real time ensures the required accuracy when operating under different conditions and operating modes of the electric chamber furnace.

1. Introduction

Control of heat devices in a wide temperature range requires consideration of changes in thermophysical properties of the device walls and in the treated products. The examples of such devices are electric furnaces, in which the temperature range of dynamic modes reaches thousands or more degrees.

The systems of optimal control of electric furnaces have to meet strict requirements regarding the accuracy of maintaining operational conditions and energy consumption. These requirements can be satisfied only if the optimal control actions are calculated using the dynamics model, which adequately describes the processes occurring in the furnace.

The dynamics model and the values of its parameters depend largely on thermal diffusivity and heat capacity of the whole set of substances participating in heat transfer processes. The dependence of thermophysical characteristics of various materials on temperature varies considerably. In addition, when operating the furnace, the type and volume of the treated products can vary. This makes it impossible to obtain a single mathematical model of dynamics, ensuring the required accuracy of processes determined by the temperature values of the treated products and other factors.

IOP Conf. Series: Journal of Physics: Conf. Series **998** (2018) 012004 doi:10.1088/1742-6596/998/1/012004

2. Problem statement

For a large number of situations [1,2] encountered in real operation of the furnace, there are three basic types of energy-saving control problems with restrictions on product quality.

2.1. Problem 1. For the known model of object dynamics

$$\dot{z} = A_1 z(t) + B_1 u(t-\tau_1), t \in [t_0, t_k] \quad (1)$$

given constraints

$$\forall t \in [t_0, t_k] : u(t) \in [u_l, u_u] \quad (2)$$

$$z(\bullet) = (z(t), t \in [t_0, t_k]) \in (\bullet) \quad (3)$$

it is required to determine the optimal control that transfers the object from the initial state to the final one, i.e.

$$z(t_0) = z^0 \rightarrow z(t_k) = z^k, \quad (4)$$

with a minimum of the functional

$$J = \int_{t_0}^{t_k} u^2(t)dt, \quad (5)$$

here z is vector of phase coordinates; u, u_l, u_u are control, the lower and upper boundaries of its change; A_1, B_1 are matrix of model parameters; τ_1 is lag time on the control channel; t_0, t_k are start and end of the control time interval; $Z(\bullet)$ is allowable region to change the trajectory of phase coordinates.

Thus, in this problem the array of the original data

$$R = (A_1, B_1, u_l, u_u, z^0, z^k z(\bullet), t_0, t_k) \quad (6)$$

at the initial time t_0 is fully known. The solution of problem (1) - (5) according to (6) is performed using the algorithm contained in the knowledge base of the control system [1, 6].

2.2. Problem 2. In contrast to Problem 1, the parameters of the model (1) need to be clarified. Therefore, for the time interval $[t_0, t_0 + \Delta t]$, the control $u_1(t)$ obtained by solving Problem 1 is used. According to the registered values $u_1(\bullet) = (u_1(t),\ t \in [t_0, t_0+\Delta t])$ and $z(\bullet) = (z(t),\ t \in [t_0, t_0+\Delta t])$ the model parameters A_1, B_1, τ_1 are corrected. Next, for the new values A_2, B_2, τ_2 we find

$u_2^*(t),\ t \in [t_0+\Delta t, t_k]$.

2.3. Problem 3. If changes occur in the furnace, for example, in the case of opening the door and increasing (decreasing) the amount of loading, it is necessary to identify the type and parameters of the model (1). Under these conditions, the problem of combined synthesis of optimal control is solved, i.e. on the time interval $[t_0, t_0 + \Delta t_1]$ c $\Delta t_1 > \Delta t$ the trajectories $u_1(\bullet) = (u_1(t),\ t \in [t_0, t_0+\Delta t_1])$ $z(\bullet) = (z(t),\ t \in [t_0, t_0+\Delta t])$ are registered; they are used to determine the model type is and to calculate the values of its parameters. Then, for the model obtained we find $u_3^*(t),\ [t_0, \Delta t_1, t_k]$.

3. Methods

To solve the problem of determining the current situation and identifying the type of the problem to be solved, the concept of intelligent sensor (IS) [2,4,7] was used in the system of energy-efficient control of dynamic modes in the electric furnace. Changing situations in the furnace are mainly related to the changes in the range and volume of treated products, opening of the furnace door, accompanied by intense heat exchange with the environment. In addition, process regulations for different types of products provide for significant changes in temperature modes.

The main function of the intelligent sensor is to receive information (output signal) about the possibility of using a "working" dynamics model or the need to correct it, depending on the data received from the group of physical sensors (PS) and from the knowledge base (KB) of the control system [3,4].

The group of PSs making up the lower level of the IS includes three temperature sensors - PST1 (temperature in the vicinity of the treated product), PST2 (temperature in the chamber near the wall), and PST3 (temperature of additionally loaded products), two weight sensors - PSW1 (the weight (mass) of products in the furnace), FSW2 (the weight (mass) of the additionally loaded product), PS τ (the operating time of the furnace with the door open), i.e. the duration of the loading process. Figure 1 shows the layout of physical sensors in the CHO electric furnace.

Figure 1. Layout of physical sensors in the furnace.

According to its purpose, the IS detects changes in the course of heat transfer processes in the furnace chamber. To fulfil this function the IS performs computational operations with elements of artificial intelligence. Processing of the information from the PS is carried out using the system of production rules. As a result of the processing, one of the following solutions is taken: to save the model without changes, to correct the parameters of the model without changing its form, to perform computational operations to identify the type and parameters of the model. In accordance with the chosen solution, the control device implements one of the following algorithms: calculation of the optimal control (OC) using the previous model of dynamics (Problem 1); solution of a particular problem of combined synthesis of the OC that provides correction of model parameters (Problem 1); solution of the general problem of combined synthesis of the OC, i.e. simultaneous identification of the model of the dynamics of the object and calculation of the OC (Problem 1).

Figure 2 shows a block diagram of an energy-saving control system with an intelligent sensor.

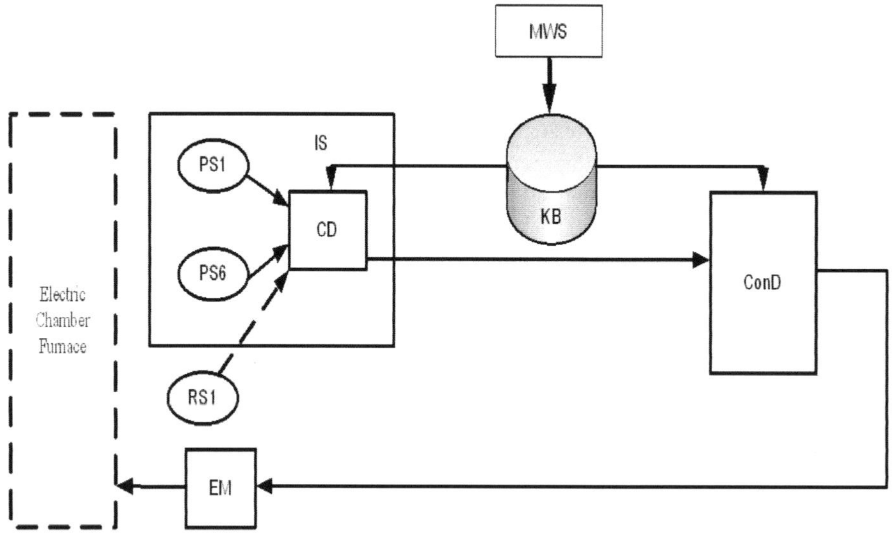

Figure 2. A block diagram of energy-saving control system with intelligent sensor, where CD is a computing device for the IS, ConD is a control device, EM is an executive mechanism, KB is a knowledge base, RS is a reference sensor, PS is a physical sensor, MWS is MiniWebServer.

The use of an intelligent sensor greatly simplifies and simultaneously enhances the ability of the control interface for various situations encountered in the operation of the furnace, and allows developing control systems for a group of devices.

In accordance with the decision made, an algorithm is developed to calculate the control actions implemented by the executive mechanism. The results of this impact are compared with the estimated values.

The algorithmic support of the system was designed and its operability was checked in the FuzzyTECH environment.

Figure 3 shows a variant of technical implementation of the system for a group of electric chamber furnaces. The system provides for recording the temperature inside the chamber and outside the furnace, solving the problems of identifying the model of dynamics and synthesis in real time of energy-saving control actions. The structure of the energy-saving control system includes Miniwebserver (MWS), industrial controller, industrial Ethernet switch, I/O terminal, workstation with expert system.

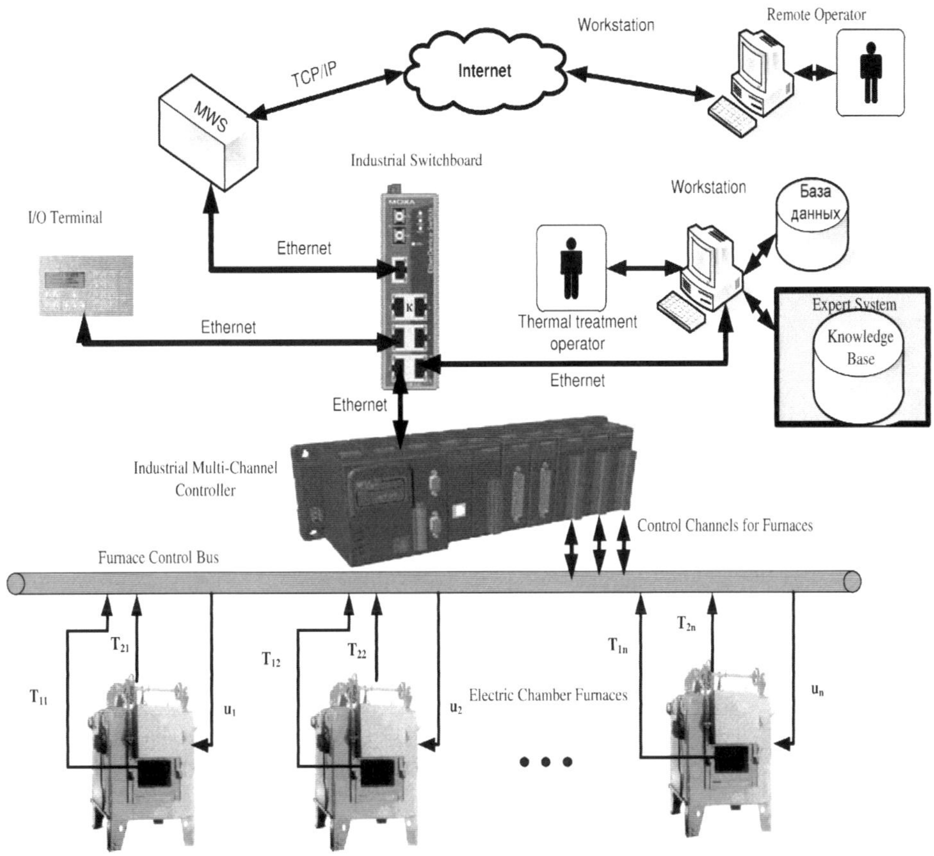

Figure 3. Technical implementation of the plant by a group of furnaces with a multichannel controller: T11 ,..., T1n is temperature inside the furnace chamber; T21,...,T2n is temperature outside the furnace chamber; u1,...,un are control actions.

The expert system contains information received from the experts on the analysis of energy-efficient control for specific quadruples, as well as the information on the functions of fuzzy-set accessories, identification algorithms, etc. The database contains information on the results of previously solved problems of energy-efficient control.

A generalized algorithm combining the methods of synthesizing variables and fuzzy logic is used to control furnaces. There is an adaptation mode with automatic correction of model parameters and fuzzy set membership functions; when the adaptation mode is completed, the corrected parameters are recorded into the controller memory for subsequent use.

4. Results and Conclusions

For centralized control of thermal conditions in furnaces, it is possible to combine the hardware and software platform into an industrial Ethernet network, as well as to have access to the global Internet network. This allows for the transfer of data on the furnace group operation or heat treatment sections to an expert system with continuous monitoring and storing of the data, and changing or correcting the heat treatment process in a remote access mode.

The main results of using the intelligent sensor in conjunction with a generalized control algorithm include:

- optimal control of the furnace;

- optimal distribution of tasks (furnace load) by the criterion of energy consumption, i.e. the choice of such a distribution, at which the total energy consumption will be minimal;
- prompt recalculation of the planned task distribution given the changes of industrial situations, including those in the assortment and volumes of manufactured products, process equipment failures;
- full utilization of the accumulated (residual) heat in the furnace during the fulfillment of the planned task, i.e. the control system takes into account the information about temperatures in furnaces and the availability of products expected for heat treatment.

The field of application of the results obtained is quite extensive and is not limited only to electric furnace furnaces. The developed control system is universal, easy to operate, reliable, maintainable, adaptable to tough manufacturing conditions, and has high noise immunity, which increases its reliability in general. This system is also scalable for various enterprise infrastructures based on specific needs.

The broadest possibilities appear when using remote access modes and controlling spatially distributed objects. Remote access significantly increases the scope for large-scale process control. This is especially true for companies that have a number of offices and can easily monitor and manage all the technological processes of their subsidiaries from the headquarters of the parent company.

Acknowledgments
This work was supported by a grant from the Russian Foundation for Basic Research, project No. 17-08-00457-a.

References

[1] Belousov O A 2005 *Devices and systems. Monitoring, control, diagno*stics **4** 31–3
[2] Belousov O A 2005 *Automation in industry* **5** 32–4
[3] Muromtsev D Yu 2008 *Automatic Control and Computer Sciences.* **42** (3) 163-8
[4] Belousov O A 2005 *Industrial control systems and controllers* **7** 33– 5
[5] Muromtsev Yu L, Muromtsev D Yu and Pogonin VA 2010 *Theoretical bases of energy-saving control: monograph* (Moscow: JaNUS – K)
[6] Muromtsev D Yu and Pogonin V A 2015 *Eng. and Appl. Sci.* **10** 37-9
[7] Muromtsev D Yu 2011 *Control Systems and Information Technology* **4 (46)** 82-7
[8] Muromtsev D Yu 2005 Methods and algorithms for synthesis of energy-efficient control of technological objects Monograph (Tambov: Nobelistika)
[9] Muromtsev Yu L and Tyurin I V 2007 Control problems **5** 69-75
[10] Matveykin V G, Putin S V and Romanov A D 2002 Transactions of TSTU **8 (3)** 410 – 12

Investigation of the Bitumen Modification Process Regime Parameters Influence on Polymer-Bitumen Bonding Qualitative Indicators

P S Belyaev, S V Mishchenko, V P Belyaev, O A Belousov and V A Frolov

Tambov State Technical University, 106, Sovetskaya str, Tambov, 392000, Russia

E-mail: bps800@yandex.ru

Abstract. The objects of this study are petroleum road bitumen and polymeric bituminous binder for road surfaces obtained by polymer materials. The subject of the study is monitoring the polymer-bitumen binder quality changes as a result of varying the bitumen modification process. The purpose of the work is to identify the patterns of the modification process and build a mathematical model that provides the ability to calculate and select technological equipment. It is shown that the polymer-bitumen binder production with specified quality parameters can be ensured in apparatuses with agitators in turbulent mode without the colloidal mills use. Bitumen mix and modifying additives limiting indicators which can be used as restrictions in the form of mathematical model inequalities are defined. A mathematical model for the polymer-bitumen binder preparation has been developed and its adequacy has been confirmed.

1. Introduction

Modern requirements to the quality and durability of highways in conditions of constant increase in traffic intensity and growing loads on the vehicles axis necessitate the use of new higher quality building materials. Road bitumen, which in a relatively recent past satisfied the needs of the road industry, today does not meet modern requirements and is increasingly replaced with polymer bituminous binders (PBB) [1 - 3]. For the PBB production, oil bitumen is subjected to the modification process by polymeric materials [4 -7]. The most widespread among them were thermoplastic styrene-butadiene-styrene [4, 8 - 11], which have a relatively high cost. The bitumen modification process was carried out using expensive colloid mills, the failure of which is often cause temporary and financial losses. In [12], the studies results of bitumen modification by thermoplastic elastomer, low density polyethylene (LDPE) and an adhesive additive with a mathematical model for the dependence of the PBB indexes on the mixture components content were presented. Bitumen modification was carried out on the developed laboratory blade mixers of various volumes. Studies have shown a significant quality dependence of the obtained PBB from the regime variables of the bitumen modification process. To solve the optimal equipment design problem for the PBB production with the proposed composition, it is necessary to develop a mathematical model for the process of combining bitumen with the complex modifier components in the presented type apparatus.

2. Formulation of the problem

The task was to investigate hydrodynamic regimes of combining bitumen with modifiers in a paddle mixer, to determine the influence of regime variables of the process on the qualitative indices of the obtained PBB, to investigate the features of the state of the components distributed in the bitumen, to compile a mathematical model for the process of obtaining the PBB and to verify its adequacy.

3. Theory

To obtain the PBB, it was suggested to use a traditional kind of mixing equipment. This is a vertical, cylindrical mixer, the rotation axis of which coincides with the axis of the apparatus. Due to the fact that the thermoplastic elastomer and adhesion additives dissolve in the bitumen [9] and the polyethylene does not dissolve, the apparatus was calculated from the viewpoint of mixing the mutually insoluble liquids: bitumen and LDPE. It was hypothesized that in order to obtain the specified qualitative parameters of the PBB, it is necessary to provide such a hydrodynamic regime in the apparatus so that during the dispersal of the LDPE in the bitumen, the size of the thermoplastic droplets does not exceed the specified values.

Experimental studies were carried out on geometrically similar laboratory mixers with paddle stirrers [8], capable to create a meridional (axial) circulation flow and characterized by the ratio $G_d \geq 1.5$, where

$$G_d = \frac{D}{d_m}; \tag{1}$$

D is the apparatus diameter; d_m is the stirrer diameter.

The range of hydrodynamic regimes is determined by the Reynolds number Re, calculated from the flow velocity, the apparatus dimensions and the medium viscosity:

$$Re = \frac{n \cdot d_m^2 \cdot \rho}{\mu} \tag{2}$$

where ρ is the mixture density; n is the mixer speed; μ is the mixture dynamic viscosity.

The density ρ and the dynamic viscosity μ of the mixture are calculated by the following relationships [13]:

$$\rho = \rho_d \times \phi + \rho_c \times (1 - \phi), \tag{3}$$

$$\mu = \frac{\mu_c}{1-\phi} \cdot \left(1 + 1.5 \cdot \phi \frac{\mu_d}{\mu_c + \mu_d}\right), \tag{4}$$

where ρ_d is the disperse phase density; ρ_c – is the continuous phase density of; ϕ is the dispersed phase concentration; μ_c is the continuous phase dynamic viscosity; μ_d is the dispersed phase dynamic viscosity.

In apparatus with internal devices, the circumferential flow characteristics must satisfy the moment equilibrium requirement:

$$M_r = M_d + M_i, \tag{5}$$

where M_r is the torque; $M_d = M_w + M_b$ is the moment of body device resistance; M_w and M_b are the moment of resistance on the walls and bottom of the apparatus; M_i is the moment of resistance forces arising on internal devices.

To equalize the circumferential velocity values along the apparatus radius, we used reflective partitions satisfying the condition:

$$\sum_{z_{on}} \zeta_{on} f_{on} > 0.1 \cdot D \cdot H, \tag{6}$$

where z_{on} is the internal partitions number; ζ_{on} is the coefficient of the baffle plate hydraulic resistance; $f_{on} = h_{on} \cdot b_{on}$ is the projection area of the partition on the meridian plane; h_{on} is the immersion height of reflecting partitions in the medium; $\cdot b_{on}$ is the baffles width; H is the fill height of the device.

In this case, under turbulent conditions ($Re > 1000$), the differences between the circumferential velocity local values in the main part of the apparatus volume do not exceed 10 - 15% [14] and we can assume that the velocity is constant and equal v_{cp}. Then, to calculate the resistance forces moment arising on the walls of the apparatus body, we can take the expression [15]:

$$M_{корп} = \frac{\pi}{2.2} \cdot \frac{\lambda}{Re^{0.25}} \cdot \gamma \cdot G_d^{2.75} \cdot v_{cp}^{1.75} \tag{7}$$

where λ is the coefficient of resistance of the device body; γ is parameter of the filling height of the device.

When $G_d \geq 2$ the value λ is constant and equal to 0.095 [15], at lower values G_d it increases with decreasing distance between the blades of the stirrer and the wall of the apparatus and can be calculated from the approximation equation [15]:

$$\lambda = \frac{G_d}{(20.35 \cdot G_d - 19.1)}; \tag{8}$$

$$1.05 \leq G_d \leq 2.$$

The filling height parameter γ for devices with free liquid level is calculated by the formula [15]:

$$\gamma = 4\frac{H}{D} + 1 . \tag{9}$$

The resistance forces moment arising on internal devices can be found by the formula [15]:

$$M_i = G_d \cdot z_{on} \cdot \left[\zeta_{on} \cdot f_{on} \cdot \frac{v_{cp}^2 \cdot r_{on}}{(0.5 \cdot D)^3} \right], \tag{10}$$

where $r_{on} = 0.5 \cdot (D - b_{on})$ is the partitions arrangement radius.

Torque transmitted to the liquid during rotation of the agitator blades [15]:

$$M_r = z_m \cdot \zeta_m \cdot k_N \tag{11}$$

where z_m is the number of stirrers on the shaft; ζ_m is the coefficient of mixer resistance; k_N is the binding power coefficient with the characteristics of the fluid circumferential flow in the apparatus.

The relationship between the power expended and the mixing conditions has the form [15]:

$$N = 3.87 \cdot \zeta_m \cdot z_m \cdot k_N \cdot \rho \cdot n^3 \cdot d_m^5 . \tag{12}$$

With full geometric similarity of industrial apparatus and laboratory mixers, as well as in accordance with the values of the criterion Re, the calculating the power, method using formula (12), is completely reliable and provides the required accuracy [15].

The average droplet diameter of the disperse phase d_k can be calculated by the formula [15]:

$$d_k \approx 0.13 \cdot \left(\frac{\sigma}{\rho_c} \right)^{0.6} \cdot \left(\frac{N}{\rho_c \cdot V} \right)^{-0.4}, \tag{13}$$

where σ is the interfacial tension; $V = 0.25\pi D^2 H$ is the stirred medium volume.

The condition for a disperse phase uniform distribution along the height of the apparatus is formulated as a restriction on the Peclet number [14]:

$$Pe = \frac{\omega_{oc} \cdot H}{D_T} \leq 0.3 , \tag{14}$$

where ω_{oc} is the rate of particles or droplets deposition (emergence); D_T is the macroscale turbulent transport coefficient (turbulent diffusion).

The coefficient of turbulent diffusion D_T with respect to apparatus with baffles is determined by the equation [15]:

$$D_T = 0.435 \cdot n \cdot d_m \cdot D \cdot \sqrt[3]{\frac{z_M \cdot \zeta_m}{G_d \cdot \gamma}}. \tag{15}$$

To calculate the rate of deposition (emergence) of small droplets, we use the Hadamard-Rybczynski equation [14]:

$$\omega_{oc} = \left(\frac{d_k^2 \cdot |\rho_d - \rho_c| \cdot g}{18 \cdot \mu_c}\right) \cdot \left(\frac{3 \cdot (\mu_c + \mu_d)}{2 \cdot \mu_d + 3 \cdot \mu_c}\right), \tag{16}$$

where g is the acceleration of gravity.

4. Experimental results

Investigation the regime variables influence of the oil bitumen BND 90/130 modification process with a complex modifier [12] was carried out in laboratory mixers with a volume of $3.0 \cdot 10^{-3}$ and $5.0 \cdot 10^{-3}$ m^3 [8].The process temperature was maintained at 160° C, mixing time 60 minutes [12]. Investigation the influence of the hydrodynamic mixing regime on the produced PBB properties was carried out by changing the rotational speed of the mixer working elements. Table 1 presents the results of changes in the penetration, softening temperature, ductility and elasticity of the PBB, depending on the rotational mixers speed.

Table 1. Dependence the polymer-bitumen bending physical-mechanical indicators from rotation frequency of working bodies in mixer with volume $3.0 \cdot 10^{-3}$ m^3.

Rotatio frequency (min^{-1})	Penetration at 25 °C, (× 10 mm)	Ductility, (mm)	Softening point, (°C)	Elasticity, (%)
200	39	740	59	57
400	55	500	62	64
800	62	705	73	83
1200	64	730	72	84
Initial bitumen	114	765	46	-

With an increase in the rotation speed of the working bodies, an increase in the penetration index occurs, which stabilizes at a frequency of 800 rpm. This is due to the decrease in the LDPE particles size due to a more intensive dispersion in the bitumen. The softening point reaches its maximum values at a speed of 800 rpm. The bitumen elasticity reaches its maximum value at a frequency of 1200 rpm, but has a slight increase in comparison with the frequency of 800 rpm. The highest ductility values were observed at a rotation speed of 200 rpm. The samples were stretched into a filament, which is typical for the initial road bitumen. However, these results were not stable: often the filament broke at ductility below 200 ... 300 mm, which was caused by the presence of inclusions from the LDPE, which at the given rotation frequency were not dispersed to the required level. Thus, the most preferred was the hydrodynamic regime, at which the rotational speed of the working bodies was 800 rpm. Further increase in the rotational speed of the working bodies was not deemed expedient.

On the basis of the mathematical model equations (1) - (16) presented above, calculations were made of the average LDPE particle size reached at 800 rpm and the conditions for the uniform distribution of the dispersed phase along the apparatus height. The following initial data for calculation was used: bitumen density and dynamic viscosity ρ_c = 1000 kg·m^{-3} [16] and μ_c = $9 \cdot 10^{-2}$

IOP Conf. Series: Journal of Physics: Conf. Series **998** (2018) 012005 doi:10.1088/1742-6596/998/1/012005

Pa·s [17]; LDPE density and dynamic viscosity, respectively: ρ_d = 920 kg·m^{-3}, μ_d = 7.9·10^2 Pa·s [18].

From (3) and (4), the calculated mixture density ρ = 996.8 kg·m^{-3} and the dynamic viscosity μ = 9.9·10^{-2} Pa·s were obtained. Substituting the obtained values into equation (2), taking into account that $n = 13.3$ s^{-1} and d_m = 0.095 m, we get the criterion value $Re = 1177$ corresponding to the turbulent regime in the mixer [13 - 15].

If inequality (6) is fulfilled, and taking into account that the baffle hydraulic resistance coefficient ζ_{on} = 2, we can assume that the value of the circumferential velocity is equalized to the value v_{cp}. Substituting equations (7), (10), (11) in the equation of moments equilibrium (5) and solving it numerically by taking into account the resistance coefficients values of the blade mixer ζ_m = 0.88 and the apparatus body λ = 0.125 , we obtain v_{cp} = 0.29 m s^{-1}.

The power consumption for mixing is found from equation (12): N = 16 W, and the average LDPE droplet diameter in the PBB from equation (13): $d_k \approx$ 9·10^{-5} m.

The turbulent diffusion coefficient for apparatus with baffles is determined by equation (15) D_T = 0.05 m^2·s^{-1}, and the average velocity of droplet precipitation from equation (16) ω_{oc} = 6.7·10^{-6} m·s^{-1}. Substituting these values into equation (14), we find the Peclet number $Pe = 1.6 \cdot 10^{-5} < 0.3$, which satisfies the uniform distribution of the disperse phase along the apparatus height.

In Figure 1 there is LDPE distribution photo of the in bitumen obtained on a stereoscopic microscope MBS-10 with an increase of ×300. The average LDPE particle diameter on microphotographs is of the order $d_k \approx$ 1.1·10^{-4} m, which is in agreement with the calculated values.

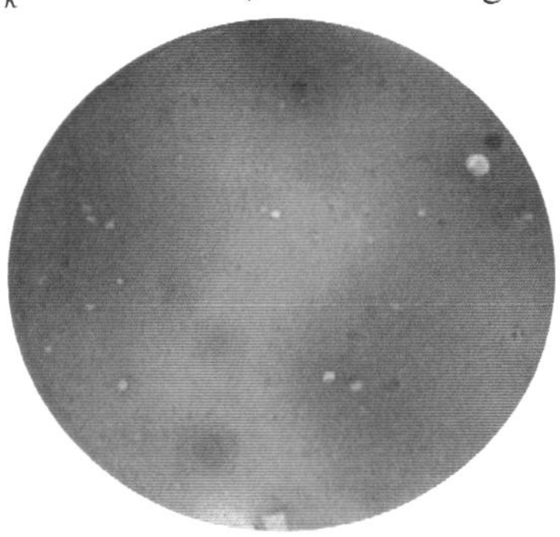

Figure 1. Photo of LDPE distribution in bitumen.

Verification the hypothesis on the determining effect of the LDPE particle size in achieving the specified qualitative PBB parameters was carried out by setting similar studies on a laboratory mixer with a volume of 5.0·10^{-3} m^3 with a full geometric similarity to a 3.0·10^{-3} m^3 mixer [8]. The conducted studies showed that the required PBB quality indicators are achieved at a rational value of the mixer rotational speed 450 rpm. Table 2 shows a comparison of the PBB quality index obtained in both mixers with a rational agitator blades speed value for each of them.

The LDPE average particle diameter, measured by using a stereoscopic microscope MBS-10, was of the order $d_k \approx$ 1.7·10^{-4} m. Using the presented mathematical model equations of LDPE dispersion in bitumen, we calculated the average droplet diameter: $d_k \approx$ 1.49·10^{-4} m. In this case the Reynolds number is equal to Re = 1161, which corresponds to the turbulent regime in the mixer. Thus, the presented data indicate the validity of using our mathematical model.

Table 2. PBB physical and mechanical indicators obtained in various volume mixtures at rational values of rotation frequency.

Mixer volume ($\times 10^3$ м^3)	Rotatio frequency (min^{-1})	Penetration at 25 °C, ($\times$ 10 mm)	Ductility, (mm)	Softening point, (°C)	Elasticity, (%)
3.0	800	62	705	73	83
5.0	450	65	745	71	80
initial bitumen	-	114	765	46	-

5. Discussion of the results

As a result of the conducted experimental studies we established that polymer-bitumen binder samples, obtained in different volume mixers with the regime parameters of mixing calculated according to the proposed mathematical model, have comparable quality and size indicators of the LDPE drops. Thus, the obtained results that testify the proposed mathematical model adequacy of the PBB producing process in which we combine petroleum bitumen with a complex modifier: thermoplastic elastomer, low density polyethylene and adhesive additive in devices with paddle stirrers. A confirmation of the hypothesis about the LDPE particle size determining influence in achieving the specified PBB qualitative parameters has been obtained.

6. Conclusions

Petroleum bitumen modification with a complex modifier composition (the thermoplastic elastomer, low density polyethylene and adhesive additive) can be carried out in vertical mixers with paddle stirrers at a temperature of 160 °C. To obtain a polymer-bitumen binder with the required parameters of penetration, softening point, ductility and elasticity, it is necessary to provide a turbulent mixing regime with a Reynolds number $Re > 1000$ and also have in the modifying process the average droplets of low density polyethylene not higher than 1.7×10^{-4} m. The Peclet number has to be $Pe < 0.3$. A mathematical model is presented for the process of obtaining a polymer-bitumen binder with a given composition, which allows to do optimal mixers design.

Acknowledgments

Applied scientific research was carried out with the financial support of the Ministry of Education and Science of the Russian Federation (the project unique identifier RFMEFI57716X0214).

References

[1] Zhang F and Yu J 2010 *Constr. Build. Mat.* **24** 410
[2] Sengoz B, Topal A and Isikyakar G 2009 *Constr. Build. Mat.* **23** 1986
[3] Belyaev V P, Belyaev P S, Klinkov A S and Polushkin D L 2012 *Global Sci. Potential* **10** 169
[4] Tayfur S, Ozen H and Aksoy A 2007 *Constr. Build. Mat.* **21** 328
[5] Belyaev V P, Malikov O G, Merkulov S A, Belyaev P S, Polushkin D L and Frolov V A 2013 *Components Sci. Technol. Progress* **1** 75
[6] Polacco G, Muscente A, Biondi D and Santini S 2006 *Eur. Pol. J.* **42** 1113
[7] Alataş D and Yilmaz M 2013 *Constr. Build. Mat.* **42** 161
[8] Belyaev V P, Belyaev P S and Polushkin D L 2012 *Sci. prospects* **5** 186
[9] Galooyak S S, Dabir B, Nazarbeygi A E and Moeini A 2010 *Constr. Build. Mat.* **24** 300
[10] Belyaev V P, Malikov O G, Merkulov S A, Polushkin D L, Frolov V A and Belyaev P S 2013 *Global Sci. Potential* **9** 29
[11] Scholten E J, Vonk W and Korenstra J 2010 *Int. J. Pavement Research and Technol.* **3** 216
[12] Belyaev P S, Mishchenko S V, Belyaev V P and Frolov V A 2017 *AIP Conf. Proc.*, **1876**, Issue 1 020096

[13] Strenk F 1975 *Mixing and Apparatus with Mixers* (Leningrad: Khimiya) p 384
[14] Karpushkin S V, Krasnyanskiy M N and Borisenko A B 2009 *Calculations and Selection of Mechanical Mixing Devices of Vertical Capacitive Apparatus* (Tambov: Tambov State Technical University Press) p 150
[15] Braginskii L N, Begachev V I and Barabash V M 1984 *Stirring in Liquid Media* (Leningrad: Khimiya) p 336
[16] Kalgin Y I 2006 *Road Bituminous Mineral Materials Based on Modified Bitumen* (Voronezh: Voronezh State University Press) p 272
[17] Rudenskaya I M and Rudenskiy A V 2010 *Organic Binders for Road Building Purposes* (Moscow: INFRA-M) p 256
[18] Belyaev P S, Klinkov A S, Minkin E. V, Malikov O G and Odnolko V G 2010 *Rheology of polymer systems. Selected chapters* (Moscow: Spectr Press) p 248

Development of Decision-Making Automated System for Optimal Placement of Physical Access Control System's Elements

Olga Danilova[1] and Zinaida Semenova[2]

[1] Omsk State Technical University, 11, Mira av., Omsk, 644050, Russia
[2]Siberian State Automobile and Highway University, 5, Mira av., Omsk, 644080, Russia

[1]E-mail: olga.danlot@yandex.ru,
[2]E-mail: semenova.z.v@gmail.com

Abstract. The objective of this study is a detailed analysis of physical protection systems development for information resources. The optimization theory and decision-making mathematical apparatus is used to formulate correctly and create an algorithm of selection procedure for security systems optimal configuration considering the location of the secured object's access point and zones. The result of this study is a software implementation scheme of decision-making system for optimal placement of the physical access control system's elements.

Keywords: system of physical protection, physical access control system, decision-making system, optimization problem.

1. Introduction

One of the most effective and civilized approaches for solving complex objects security problems is the use of physical protection systems (PPS), including a set of access control instruments (ACI), security television (ST) and fire-protection equipment (FPE). It should be noted that the problem of optimal placement of the PPS elements is at the stage of pre-project researches, and the decisions taken at this stage determine the quality and reliability of the PPS.

Formulation of the optimization problem and applied methods depend on the degree of parameters' uncertainty, the system operates. There are four types of tasks. The first ones are the problems with the determined parameters, and needed solutions are continuous or discrete [1]. Problems of the second type are the problem of decision-making under risk, characterized by the fact that the exact values are unknown for a number of parameters, but ranges of change are identified [2]. The third type suggests that the possible discrete values are given for each of the parameters, for which the performance coefficient values, associated with each of alternative solutions, are determined [3]. The fourth ones differ from above mentioned in that decision is based on analytics in the competitive environment of the opposing sides, and game model is used as a decision-making scheme [4]. The choice of optimization method depends on the criteria and limitations as well as the degree of discrete parameters and availability of information about their probability distribution laws.

Any optimal solution search algorithm for the problem of minimizing the function on a set are aimed at solving a particular class of optimization problems with certain properties of the functions and set. Therefore, a more universal algorithms aimed at solving a wide range of problems, yield specialized algorithms, using the specific properties of specific problem. So one of the first optimization methods is the exhaustive search method. Partial enumeration methods, methods of cuts, approximations methods, methods of local optimization are often used among the approximate methods [5].

The model testing includes an assessment of its consistency, sensitivity, relevance and performance. Consistency requires a logical analysis of the modeling results by varying the initial parameters. Optimization calculations' results are assessed in detail for the limit values of the parameters. The sensitivity analysis is based on changes in system's performance and efficiency at the highest variations of controlled parameters. One of the relevance testing method is to establish correspondence of modeling results to well-known special cases. The efficiency of the model is related to the resource estimate required for the collection of baseline and to conduct computer experiments [6].

The authors of this paper believe that it is possible to determine the rational distribution of the protection systems' elements based on the analysis of the access points and zones organization on the object. Integrated physical access control system (PACS) was chosen as the object of the study in the work.

2. Formulation of the problem

The developed decision-making system (DMS) has to provide the collection and processing of the input data; displaying, recording and further archiving information for later use. The main purpose of the DMS is the need to maintain a record of information about the object's state and dynamics. The accumulated data can be processed automatically for subsequent tactical and strategic analysis for management of decision-making process [7].

The initial data for the DMS development is the information about the elements and a set of minimum requirements for the characteristics of the security features on each element.

First, the system has to carry out a survey for the object's category determination, and after the actual configuration of the system is made.

Using the data a request to create an optimal system configuration for determined object's category is generated. At this stage we solve the problem of an optimal system configuration construction.

After determination of the optimal and actual system configuration it is necessary to perform a comparison to detect critical differences.

3. The theory

The problem of rational arrangement of elements in the security system is divided into three parts: the problem of choosing the optimal configuration; PACS elements location task on the access object and the problem of optimal allocation of resources between the parts of the system.

The problem of choosing the optimal configuration is formalized as follows: a finite set of all possible elements of the PACS F is considered divided into a number of G, Q, P subsets, each of which corresponds to a plurality of components of the system. The statement of the optimization problem is simple and natural. There are the optimization problem and the individual optimization problem. Let define the individual optimization problem for PACS is a pair ($G1$, c), where $G1$ is a set, the range of permissible values of identifiers subset, and c is the value of the function providing the mapping c: $G \rightarrow R^1$. It's needed to find a point $g1 \in G$, for which c $(g1) \leq$ c(y), for all y $\in G$. This point is a globally optimal solution for a given individual optimization problem. Mathematical formalization of individual tasks leads to the answer to the question: which of the *i-th* solutions is optimal one and does it meet the requirements of the relevant class system? The optimum element is an element met the conditions $c\,[i] \times x\,[i] \rightarrow min$.

Problems of an arrangement of elements on the object and resources distribution are solved using of provisions of the graph theory. Each object is represented as one enclosed zone i.e. including variants of other zones. Access points on the object are determined as vertices, and transitions between these points of access – as edges of the graph, i.e. overlaps of access zone. For each edge the category of access, it corresponds, is defined. While the treelike scheme building it is necessary, considering a

IOP Conf. Series: Journal of Physics: Conf. Series **998** (2018) 012006 doi:10.1088/1742-6596/998/1/012006

priority of categories of access (the general application, important, especially important), to define a category of access points on the object [8].

To solve the problem of resource allocation it is required to convert the edges weights of the weighted bipartite graph in the way, that each vertex has to become incident to at least one edge with zero weight, and weights of all the edges in this case weren't negative. Empty matching *M* is declared as current one. If the graph vertices are saturated with respect to the current matching, the current matching is optimal. If not, then it's needed to choose any free $b \in B$ vertex and look for M – alternating chain that begins at the b vertex, and consists only of zero-weight edges. Once such chain is constructed, the $M = M \oplus P$ operation is performed, and it is checked whether all the vertices are saturated with respect to this matching or not. If the chain is not built, for the $b \in B'$, $a \in A'$ vertices set, marked in the search process, $d = \{\min c(b,a) \mid b \in B', A \in A \setminus A'\}$ is determined and (B', d, A') operation is applied to the graph. From the $b \in B'$ vertices, which was incident to at least one zero-weight edge, search of *M*- alternating chain with only the zero-weight edges is resumed. If such a *P-chain* is built, the $M = M \oplus P$ operation is performed again and the saturation peaks is checked. And it continues until an optimum matching is found.

4. Experimental results

The solution of the above problems allows determining the approaches to the automation system design (AS) ensuring the collection and processing of information, used for decision-making on the compliance of the proposed PACS project with the real operating conditions of the informatization object.

Analysis of the structural scheme of access zones on the object is based on the formation of *c* access points of PACS and defined as a *C* set, which owns the ($c \in C$) access points. If the object has several structural units with different categories of access to each of them, considering the specifics of the functional features of the object, then the PACS has several subsystems with different categories of access zones, and consequently, different levels of user access (access subject). Then the *C* set is divided into *I* subsets of C_I with the c_I elements. $C_i \subset C$ subsets, *i = 1, ..., I* are subsets of *C* set. Subsets of C_i access points can be overlapping or nonoverlapping. This depends on whether there are the structural units of the common access zone and points or not. The totality of all *z* PACS access zones can be defined by a variety of *Z* to which the $z \in Z$ access points belong. In turn, the *Z* set, in general, divided into *J of Zj* subsets with *zj* elements. $Z_j \subset Z$ subsets, *j = 1, ..., J* is a proper subset of Z.

For formation of access zones it is necessary to submit the physical access control system on the object in the form of a graph. One of the main objectives of the placement is to create the best conditions for the implementation of access control functions. Each access point must be equipped with a certain set of elements, which by its characteristics must comply with this access category.

5. Discussion of results

The main module of the DMS is the databases for the correct design of which the corresponding infological model is developed. The basic structural elements of infological models are the entities, links between them and their properties. The entities in this case are the sets of the technical elements of the PACS, and the properties (attributes) - their technical characteristics. Each entity must define a key that is the minimum set of attributes from the values of which the desired entity instance can be unambiguously found. Then the corresponding database tables are formed, and a query structure is made on which during operation of the automated system temporary tables needed for the task are created. The block diagram of the considered DMS is shown in Figure 1.

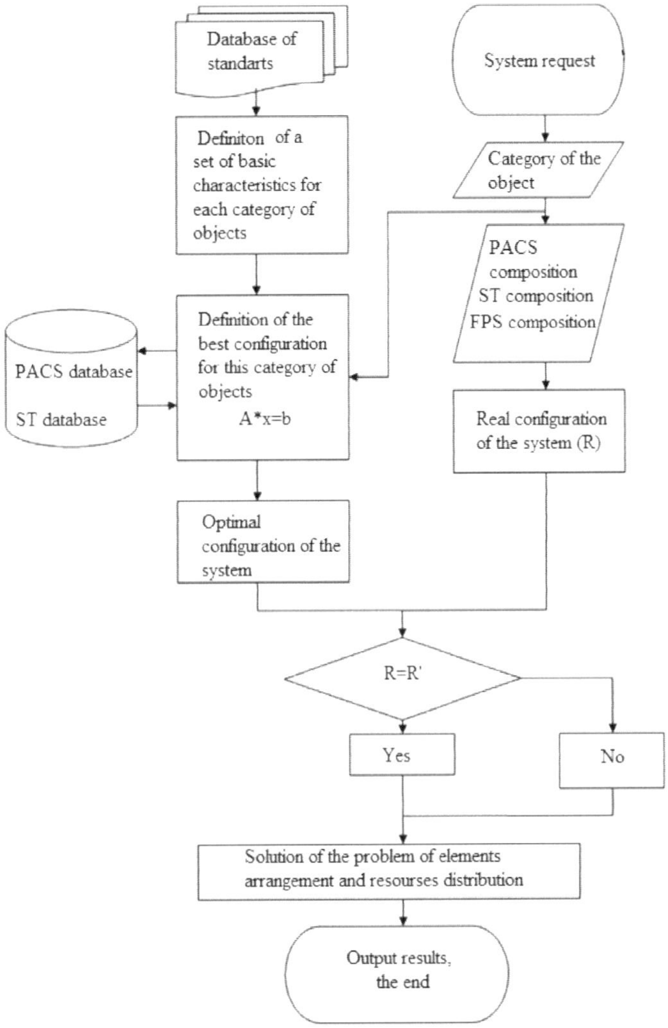

Figure 1. Block diagram of the work of decision-making system.

To make a decision it is necessary to perform the following operations:

1. The minimum requirements for the elements of the system are defined, considering the recommendations of the technical standards. Each solution is stored in a special database by categories: access control, ST, FPE, DMS. Under each configuration the specific solutions are selected.
2. The request to the category and structure of the object based on the presence of zones of different object types are made. Then, the DMS allows choosing the optimum configuration of the system and makes the distribution of elements on the object.
3. The selected configuration of PACS to meet the optimum one is analyzed.

6. Summary

As a result of the research the questions of design of new security systems and the analysis of existing design decisions have been considered Domain analysis allowed formulating requirements for security systems, defining the main blocks of the decision-making algorithm on the optimal placement of PACS elements.

The practical significance of the work is based on the fact that the results of access control systems design on the basis of the developed approaches will meet the necessary criteria for the significance of the object category, considering the cost of the planned system.

References

[1] Alekseev V Yu 1987 Complex application of methods of discrete optimization

[2] Aho A, Hopcroft J, UllmanJ and Aho M 1979 Construction and analysis of computational algorithms

[3] Emelyanov V V, Kureichik, V M 2003 Theory and practice of evolutionary

[4] Pankratiev E V, Chepovsky A M 2003 Algorithms and methods for solving scheduling problems and other extremum problems on large-scale graphs *Fundamental and Applied Mathematics* **9** **(1)** p 235-251

[5] Fedorov V V 1971 The theory of optimal experiment (planning of regression experiments)

[6] Balazs E 1969 Additive algorithm for solving linear programming problems with the variables taking values 0 and 1 *New episode* **6** p 217

[7] Berner L I, Garmash V B, Levin M Sh, Antropov M V 1985 Formation of valid options hierarchical control systems at the design stage. Problems and methods of decision-making: Coll. Works p 62-72

[8] Chen. B, Potts C N and Woeginger G J 1998 A review of machine scheduling: Complexity, algorithms and approximability. *Handbook of Combinatorial Optimization* **3**

Comparative Analysis of the Specifications on the Power Quality of the European Union and the Russian Federation

A V Ded, V N Maltsev, S P Sikorski

Omsk State Technical University, 11, Mira, av., Omsk, 644050, Russia

E-mail: ded_av@mail.ru

Abstract. Since July 2014 the interstate standard GOST 32144-2013 is the only document that defines standard requirements for the power quality in the territory of the Russian Federation. The new standard preamble specifies that this document considers the requirements of the European regional standard EN 50160-2010. However, GOST authors established the degree of standards conformity as nonequivalent. In connection with Russia's accession to the World Trade Organization (WTO) all requirements for goods including electric energy should correspond the international standard requirements. The article analyzes the above standard requirements and assesses the requirements for the power quality standards used in the European Union and in the Russian Federation.

Key words: power supply system, power quality, unified power quality index, power quality standards

1. Introduction

Membership in the World Trade Organization (WTO) obliges the participating State to implement fully the agreements reached by the organization member countries. For this, first and foremost, methods are to be brought in conformity with the established and approved rules for regulating the external economic activities of the WTO member countries. Regarding the issues of standardization, this requirement is to bring the existing rules and norms (standards) in line with the requirements of the Agreement on Technical Barriers to Trade [1, 2].

The field of meeting the standards (standardization) requirements has the main task to exclude the possibility of a regulatory document to influence the trade turnover between the WTO member countries in a technical barrier form. Countries should ensure that there is no difference in the requirements of national and interstate technical regulations and standards when developing, approving and applying them. So, in most cases, the only option is the international standards application. Thus, when developing any national technical regulation or standard, it is necessary to analyze the availability of a similar international standard (or its project) and to adopt it, in whole or in part, as the basis for the normative document being developed.

If national normative document requirements differ significantly from the international standards and thus can impede trading relations and trade turnover between countries who are WTO participants, the established norms differences from international standards are to be specified in the document.

In connection with Russia's accession to the WTO and necessity to balance Russian legislation with international community approaches, a new GOST 32144-2014 "Norms for the power quality in general-purpose power supply systems" was introduced in relation to the parameters of electric power quality indices [3,4].

This GOST 32144-2014 is developed on the basis of the earlier valid GOST 541149-2010 [5], it considers the provisions of the European Union regulations in the field of electromagnetic compatibility and the conceptual provisions principles of the European Union New and Global

Approach to the quality assurance problem [6]. According to the definition of GOST 32144-2014 authors, it considers the main normative provisions of the European standard EN 50160:2010 "Voltage characteristics of electricity supplied by public distribution networks" (Characteristics of the voltage supplied by public distribution networks) [7,8]. However, a preliminary analysis of the current and previous standards shows that only the structure of these documents is identical, but the norms differ significantly for certain power quality indices [4,5,8,9,10].

2. Statement of the problem

Electric power supply systems are to be designed and operated in various industries considering the implementation of existing norms and rules including power quality standards adopted at the state level.

The authors of this article analyze the regulatory requirements for the power quality indices established by the European standard EN 50160: 2010 and an intergovernmental standard GOST 32144-2013, to establish the degree of their correspondence or nonequivalence.

The findings can be applied in practice to evaluate measures aimed at maintaining the necessary working conditions and increasing the efficiency of the electrical equipment operation, considering the manufacturer country

3. Theory

Considering the current state of issues and tasks to ensure the power quality, the regulatory and methodological base in this direction has been substantially developed over the last ten years. The conditions necessary for providing power management and quality control were created. As a result, the required measuring instruments were developed and put into production; testing laboratories carried out power monitoring and quality control [11].

In 2013 the Eurasian Economic Community countries, due to the transition to the unified technical regulations and interstate standards, focused on the European Union requirements, re-designed and put into effect the relevant interstate standards to ensure the power quality requirements.

Since 01.06.2014 the Russian Federation introduced GOST 32144-2013 "Norms of power quality in general-purpose power supply systems" instead of the previous GOST R 54149-2010 [3, 5]. Nevertheless, the current GOST contains many indices from GOST 13109-97 [12 - 16]. In this regard, the GOST 32144-2013 standards on some indices are being criticized by scientific community and are to be revised [17, 18].

The current regulations are argued to lead to arbitration disputes between grid organizations and electricity consumers. In addition, the difference between Russian and European requirements for the power quality can lead to difficulties with import and export of electrical equipment and electricity [9].

4. Results of the experiments

As a result of the above-mentioned effective standards and their requirements to the power quality standards analysis, the main differences are determined for each standardized index measured during one week time interval.

In accordance with GOST 32144-2013 in synchronized systems, the frequency deviation should not exceed ± 0.2 Hz for 95% of the frequency measurement time interval and ± 0.4 Hz for 100% of the measurement time, see Table I.

Table 1. Requirements to frequency deviations in synchronized systems.

	GOST 32144-2013			EN 50160:2010		
Measuring interval	%	Δf, Hz	Range of change, Hz	%	Δf, Hz	Range of change, Hz
95%	± 0.4	± 0.2	49.8 – 50.2	–	–	–
99.5 %	–	–	–	± 1	± 0.5	49.5 – 50.5
100%	± 0.8	± 0.4	49.6 – 50.4	- 6 / +4	-3; +2	47.0 – 52.0

The standard EN 50160: 2010 determines that the frequency deviations should not exceed ± 0.5 Hz for 99.5% of time and should be in the range from + 2 Hz to - 3 Hz for 100% of time in synchronized systems.

GOST 32144-2013 is established for slow voltage changes, caused, as a rule, by changing the value of the network load, the interval within ± 10% of the nominal voltage level or the matched voltage within 100% of the measurement time. It should be noted that GOST 32144-2013 for four-wire three-phase systems takes the value of the rated voltage equal to 220 V [4].

At the same time, EN 50160: 2010 for analogous three-phase networks sets the nominal voltage between the phase conductor and the neutral at 230 V [8], and the limit of variation of the voltage deviation ± 10% of the nominal voltage level establishes at 95%, see table 2.

Table 2. Requirements to voltage deviations in synchronized systems.

	GOST 32144-2013			EN 50160:2010		
Measuring interval	%	δU(±), B	Range of change, B	%	δU(±), B	Range of change, B
95%	–	–	–	± 10	± 23	207 – 253
100%	± 10	± 22	198 – 242	–	–	–

For an indicator characterizing fluctuations in the voltage of the power supply lasting less than 1 minute - the dose of the flicker, the regulatory documents requirements of the European Union and Russia are also not the same.

GOST 32144-2013 specifies the norms of short-term PST and long-term dose of PLT flicker, which should not exceed the established limits for 100% of one week time interval. EN 50160: 2010 sets only the long-term dose rate of the PLT flicker, but at 95% of one week measurement time, see table 3.

Table 3. Requirements to flicker dose in synchronized systems.

	GOST 32144-2013		EN 50160:2010	
Measuring interval	P_{ST}	P_{LT}	P_{ST}	P_{LT}
95%	–	–	–	1.0
100%	1.38	1.0	–	–

For power energy quality indices that characterize the voltage non-sinusoidal - values of some coefficients of the harmonic components $K_{U(n)}$ and the total harmonic components K_U, the standards established by GOST 32144-2013 must be met within 95% of one week measurement time interval. During 100% of the measurement period, the same standards should not exceed the normative value of 95% of the observation period increased by 1.5 times. Document EN 50160: 2010 for non-sinusoidal voltage defines only requirements for values for 95% of one week measurements see eable 4.

Table 4. Requirements to the total coefficients of harmonic components.

	GOST 32144-2013 U_n, kV				EN 50160:2010 U_n, kV
Measuring interval					
	0.38	6-25	35	110-220	0.4-220
95%	8.0	5.0	4.0	2.0	8.0
100%	12.0	8.0	6.0	3.0	-

It should be noted that in the European Union where EN 50160: 2010 operates, unlike countries participating in the Interstate Council for Standardization, Metrology and Certification, which approved GOST 32144-2013, they are Armenia, Belarus, Kyrgyzstan, the Russian Federation,

Tajikistan and Uzbekistan, the norms for indicators KU (n) and KU are the same regardless of the voltage class (low, medium or high).

GOST 32144-2013 normalizes the requirements of the coefficients $K_{U(n)}$ and K_U for the harmonic components of the voltage up to the 40th order, determining the deviation in percent from the fundamental harmonic component of the direct sequence voltage U_1, see table 5 and table 6.

Table 5. Requirements to the coefficients of odd harmonic components.

Harmonic component number	GOST 32144-2013 U_n, kV				EN 50160:2010 U_n, kV
	0.38	6-25	35	110-220	0.4-220
n	Values of the coefficients K_U, %				
5	6.0	4.0	3.0	1.5	6.0
7	5.0	3.0	2.5	1.0	5.0
11	3.5	2.0	2.0	1.0	3.5
13	3.0	2.0	1.5	0.7	3.0
17	2.0	1.5	1.0	0.5	2.0
19	1.5	1.0	1.0	0.4	1.5
23	1.5	1.0	1.0	0.4	1.5
25	1.5	1.0	1.0	0.4	1.5
>25	1.5	1.0	1.0	0.4	1.5

Table 6. Requirements to the coefficients of odd harmonic components 3-fold.

	Electrical network voltage, kV				Electrical network voltage, kV
	Harmonic component number	GOST 32144-2013	EN 50160:2010	110-220	0.4-220
n	Values of the coefficients K_U, %				
3	5.0	3.0	3.0	1.5	5.0
9	1.5	1	1	0.4	1.5
15	0.3	0.3	0.3	0.2	0.5
21	0.2	0.2	0.2	0.2	0.5
>21	0.2	0.2	0.2	0.2	0.5

Document EN 50160: 2010 in this case is limited to the 24th harmonic component, motivating it by the fact that harmonics with higher order are much smaller in comparison with the fundamental harmonic component and may be less predictable due to the resonance effect.

Both standards for the power quality differentiate the limit values of the harmonic components coefficients for even (see Table 7) and odd harmonics, dividing the latter by multiples and not multiples of three, respectively.

One of the main differences between GOST 32144-2013 and the European standard EN 50160: 2010 is the absence of the maximum permissible value for an important index in the operated Russian networks as the coefficient of voltages unbalance in the zero sequence K_{0U}. EN 50160: 2010 rationing the values of the unbalance factor of the voltage by the negative sequence K_{2U}, but only by 95% time interval, see Table 8.

Table 7. Requirements to the coefficients of even harmonic components.

Harmonic component number	GOST 32144-2013 Electrical network voltage, kV				EN 50160:2010 Electrical network voltage, kV
	0.38	6-25	35	110-220	0.4-220
n	Значения коэффициентов K_U, %				
2	2.0	1.5	1	0.5	2.0
4	1.0	0.7	0.5	0.3	1.0
6	0.5	0.3	0.3	0.2	0.5
8	0.5	0.3	0.3	0.2	0.5
10	0.5	0.3	0.3	0.2	0.5
12	0.2	0.2	0.2	0.2	0.5
>12	0.2	0.2	0.2	0.2	0.5

Table 8. Requirements for non-symmetry voltages.

	GOST 32144-2013		EN 50160:2010	
Measuring interval	K_{2U}	K_{0U}	K_{2U}	K_{0U}
95%	2.0	2.0	2.0	-
100%	4.0	4.0	–	–

5. Discussion of the results

The narrower scope of the requirements for the frequency deviation in GOST 32144-2013 compared to EN 50160: 2010 may lead to difficulties in connecting to synchronized power systems of cogeneration plants and distributed generation facilities, their development is indicated as a priority task of the Energy Strategy of Russia till 2030 [19].

At the same time, the methods adopted in EN 50106: 2010 and applied to the establishment of the norms of the total coefficients of harmonic components K_U and the harmonic components $K_{U(n)}$, are justified. This assumption is due to higher harmonics having the greatest negative impact on low voltage network consumers [9].

The current interstate standard GOST 32144-2013, the norms for medium and high voltage networks define a more stringent framework for changing quality indicators compared to EN 50160: 2010. For low voltage networks, GOST 32144-2013, by introducing the zero-sequence unbalance coefficient, considers the presence of consumers in the networks and its load is distributed unevenly.

At the same time, the need to separate the norms of the power quality in terms of voltage classes in GOST 32144-2013 is primarily related to the need to meet the requirement of sharing responsibility for ensuring the voltage quality between electric energy suppliers (backbone networks) and consumers (territorial network organizations) [4].

The fulfillment of GOST 32144-2013 requirements obliges not only the network organizations to maintain individually the required power quality in the distribution networks of consumers. The requirements of this GOST stimulate the manufacturers of electrical equipment and end-users, who buy it, to be responsible for the absence in the networks of inadmissible conductive electromagnetic interference generated during the operation of this equipment and installations.

6. Conclusion

Thus, in all key parameters GOST 32144-2013 does not comply with the European standard EN 50106: 2010. Therefore, the indication, at present, of information on the non-equivalent degree of correspondence on it must be considered a necessary condition.

It is worth noting that the structure of GOST 32144-2013 is formed in the established international practice of testing the power quality. In this connection, GOST 32144-2013 is more similar in its structure to EN 50160: 2010 than the previous regulatory framework [5, 12]. To this end, the requirements for the power quality indicators [4], the methods of their measurement [20], as well as the requirements for measuring instruments [21] have been singled out into separate standards.

The requirements of the European standard EN 50160: 2010 are primarily designed for distribution networks of countries that have different requirements for both the design of electrical networks and different, in comparison with the level of the countries of the Eurasian Union, including the Russian, technical condition of the networks

References

[1] Official website of the World Trade Organization, *Access mode: http:www.wto.org*
[2] Russia and the WTO. Russian version of the official website of the World Trade Organization, available at: *http://www.wto.ru*
[3] On the introduction of the interstate standard. Order of the Federal Agency for Technical Regulation and Metrology (Rosstandart) July 22 2013 N 400-st, available at: *http://docs.cntd.ru/document/499061116*
[4] GOST 32144-2013 Electric Energy. Compatibility of technical means is electromagnetic. Norms of power quality in general-purpose power supply systems. 2013 (Moscow). *Standartinform*
[5] GOST R 54149-2010. Norms of quality of electrical energy in general-purpose power supply systems 2012 (Moscow) *Standardinform*
[6] Guidelines for the implementation of the directives based on the principles of the new approach and the global approach available at: *http://www.ctec.lv/userfiles/files/ New% 20Approach% 20and% 20European% 20standardisation.pdf*
[7] Nikiforov V V 2015 *Business magazine Neftegaz* **9** 42-47
[8] EN 50160-2010. Voltage characteristics of the electricity supplied by public distribution networks
[9] Vagin G Ya 2013 *Industrial energy* **1** 39-43
[10] Zimin R Yu, Abramovich B N and Sychov Yu A 2017 available at: *http: //www.lib.tpu.ru/fulltext/c/2014/C82/011.pdf.*
[11] Nikiforov V V 2011 *News of electrical engineering* **3** 69
[12] GOST 13109-97 Electric Energy. Compatibility of technical means is electromagnetic. Norms of quality of electrical energy in general-purpose power supply systems 1998 (Moscow) *Publishing house of standards* **32**
[13] Vagin G Ya 2003 *Industrial Electric Power Engineering and Electrical Engineering*. 3
[14] Vagin G Ya and Sevostyanov A A 2004 *Industrial energy* **9** 39-43
[15] Vagin G Ya and Sevostyanov A A 2010 *Industrial Power* **11** 45-48
[16] Sheiko P A and Zhelezko Yu S 1999 *Materials of the scientific-technical seminar* (Moscow) AO VNIIE
[17] Sudnova V V, Kartashev I I, Tulskiy V N and Kozlov V V 2013 *News of electrical engineering* **4** 82
[18] Sudnova V V, Kartashev I I, Tulsky V N and Kozlov V V 2014 *News of electrical engineering* **2** 86
[19] Decree of the Russian Federation Government November 13, 2009, N 1715-p available at: *http://www.garant.ru/products/ipo/prime/doc/96681*
[20] GOST 30804.4.30-2013 Electric Energy. Compatibility of technical means is electromagnetic. Methods for measuring the quality of electrical energy 2014 (Moscow) *StandardInform* 51
[21] GOST 30804.4.7-2013 Compatibility of technical means is electromagnetic. General guidance on measuring instruments and measurements of harmonics and interharmonics for power supply systems and associated equipment 2014 (Moscow) *Standardinform* 33

SENSORS FOR METERING HEAT FLUX AREA DENSITY AND METROLOGICAL EQUIPMENT FOR THE HEAT FLUX DENSITY MEASUREMENT

D O Doronin[1]

[1]Research and Production Company Etalon Joint-Stock Company, 175, Lermontov str., Omsk, 644009, Russia

E-mail: doronin@etalon.lan

Abstract. The demand in measuring and studies of heat conduction of various media is very urgent now. This article considers the problem of heat conduction monitoring and measurement in various media and materials in any industries and branches of science as well as metrological support of the heat flux measurement equipment. The main study objects are both the sensors manufactured and facilities onto which these sensors will be installed: different cladding structures of the buildings, awnings, rocket fairings, boiler units, internal combustion engines. The Company develops and manufactures different types of heat flux sensors: thermocouple, thin-film, heterogeneous gradient as well as metrological equipment for the gauging calibration of the heat flux density measurement. The calibration shall be performed using both referencing method in the unit and by fixed setting of the heat flux in the unit. To manufacture heterogeneous heat flux gradient sensors (HHFGS) the Company developed and designed a number of units: diffusion welding unit, HHFGS cutting unit. Rather good quality HHFGS prototypes were obtained. At this stage the factory tests on the equipment for the heat flux density measurement equipment are planned. A high-sensitivity heat flux sensor was produced, now it is tested at the Construction Physics Research Institute (Moscow). It became possible to create thin-film heat flux sensors with the sensitivity not worse than that of the sensors manufactured by Captec Company (France). The Company has sufficient premises to supply the market with a wide range of sensors, to master new sensor manufacture technologies which will enable their application range.

Keywords – heat flux sensors, heat flux measurements, heat conduction, calibration equipment

1. Introduction

Heat flux density measurement is of great importance both for the experimental research and for the commercial production. There are numerous instruments which assist in solving the heat flux density measurement problems. These problems include measurement of local and integral heat flux to determine heat losses, measurement of the material heat-protection and heat-transport properties, measurement of the objects' heat-emission [1,2,3].

There are different methods of obtaining information of heat flux:

1) Analytical methods based on the temperature measurement and numerical solution of the relevant equation describing the process. Solution of the direct or inverse heat conduction problem.
2) Experimental methods based on the heat vs. physical value relationship.

A wide-spread method for the heat flux measurement is thermoelectric method. It is based on Seebeck thermoelectric effect which consists in the fact that heating of two heterogeneous conductors' junction produces thermal electromotive force (TEMF) causing electrical current in the circuit.

The main heat-measurement objective is the development of such metering instruments which would ensure the measurements locality, low heat resistance, high sensitivity, good result reproducibility and having reliable metrological support [1,2,3].

IOP Conf. Series: Journal of Physics: Conf. Series **998** (2018) 012008 doi:10.1088/1742-6596/998/1/012008

Any metering instrument is based on the primary metering transducer which is the first element to take up the physical value in question. In our case this is primary heat-flux transducer (HFT) – heat-flux sensor (HFS). The heat flux in question is converted into the temperature difference ; the temperature difference is converted into the thermoelectrical potential difference. The most popular HFT is auxiliary wall HFT. The auxiliary wall HFT operation principle consists in the measurement of the temperature difference arising when the heat flux passes through a finite thickness wall located on the heat flux route on the normal line to its direction, or the values of the temperature values of two parallel sections of the wall. An HFT has two mandatory elements: intrinsic wall where the temperature difference occurs and this difference meter (thermoelectric battery) [1, 2, 3].

2. Problem Description

The Company key objectives are:

- Supply a wide range of heat flux sensors to the consumer market.
- Development and manufacture of heat-flux sensors based on the new technologies.
- Holding internal factory acceptance tests of the sensors manufactured.
- Design of the units for the heat-flux sensor manufacture and metrological support.

3. Theory

Manufacture of Thermocouple Heat-Flux Sensors

Thermocouple heat-flux sensors are thermoelectric transducers operating on "auxiliary wall" principle. They consist of a battery of identical voltaic thermoelements (Figure 1) connected in parallel in terms of the heat flux being measured and in series in terms of the electrical signal generated. The sensor tightness in rigid or flexible version is ensured by the insulating sealant compound. These sensors are classified as lengthwise heat flux sensors [1, 2, 3].

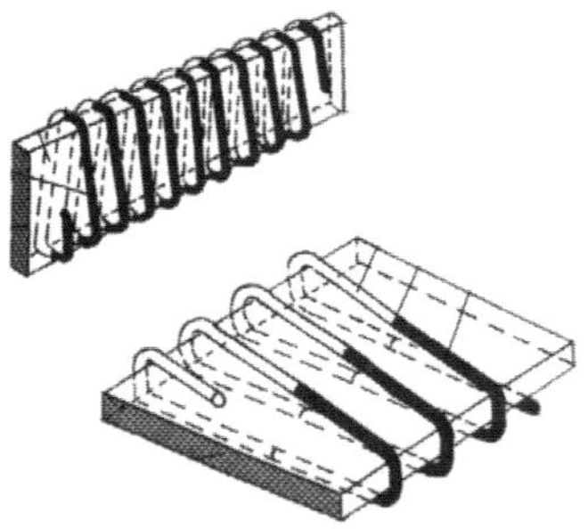

Figure 1. Thermoelectric Battery Schematic Diagram.

Our company manufactures HFS 0924, HFS 0925 (reference) heat-flux sensors intended as shop instrument for measuring thermal flux density during the test of various construction items and elements both in laboratory (using climatic chamber) and in environment conditions. The main thermocouple wire is made of constantan, bimetallic wire (main wire section with plating) is represented as copper electroplating [1,2,3].

The Company also designed and manufactured an experimental high-sensitivity heat-flux sensor for measuring low heat-flux densities (up to 10 W/m^2).

Manufacture of Thin-Film Heat-Flux Sensors

Thin-film heat-flux sensor is a thin vacuum-metallized substrate. To provide temperature gradient overlaps made of low heat-conduction material are used, currently this material is POLI-FIX 330 double-sided glue tape. Temperature-leveling surfaces are represented as foil-clad polyimide or copper foil (Figure 2)

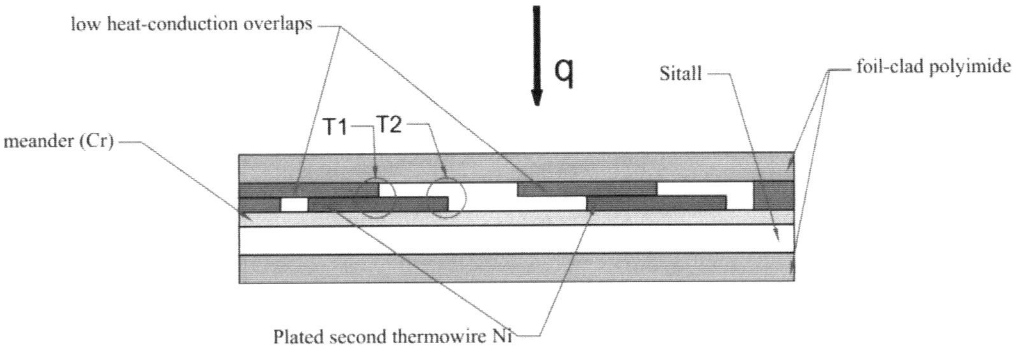

Figure 2. Schematic Diagram of the Thin-Film Heat-Flux Sensor.

The sensor operation consists in the fact that there are contacts with poor and good heat conduction. Heat flux line distribution is not symmetrical relative to the junction centre, the electrical charge is distributed along the heat flux line parallel to the poor-conduction substrate surface. The sensor is not based on the study of the temperature difference along the sensor thickness, it detects electrical potential tangential gradients produced by the heat flux [4].

The metal plating was performed using EVA-600 vacuum deposition unit as per magnetron deposition method. Substrate was made of sitall plate. Thermowires were made of chromium and nickel.

The advantages of this sensor compared with thermocouple sensors are: small dimensions, wider operating temperature and heat-flux density ranges.

Manufacture of Heterogeneous Heat-Flux Gradient Sensors

A Heterogeneous Heat-Flux Gradient Sensor (HHFGS) is a cross-type heat-flux sensor (Figure 3).

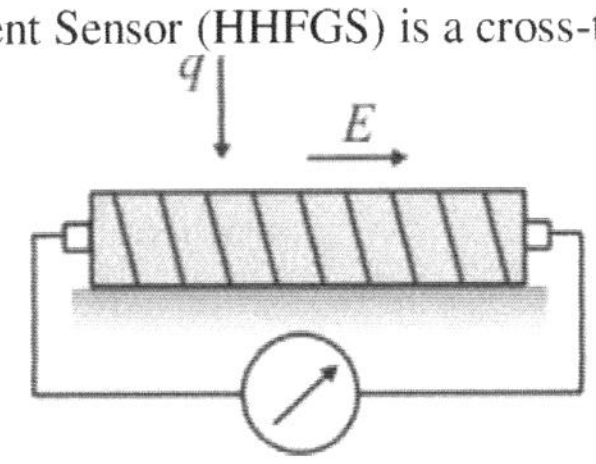

Figure 3. Cross-Type HFS.

The HHFGS operation principle is based on cross-wise Seebeck thermoelectric effect – occurrence of thermal EMF with a field vector normal to the heat-flux vector, in the heat-conduction anisotropy, electric conductivity and thermoelectric coefficient. Transverse thermal EMF arises when the heat-flux vector does not match the crystal main axis with heat-conduction anisotropy and thermoelectric coefficients [5,6].

To manufacture the sensor, our Company designed and fabricated diffusion welding unit. The vacuum attained in the welding chamber is 16 Pa, with the maximum pressure acting on the workpiece to be welded equal to 10 tons, maximum temperature attained in the chamber of 597 °C, maximum briquette size of 35x35 mm.

To produce ready-made gradient heat-flux sensors the structure was cut on a special cutter into individual elements at 45° to the main crystal axes.

The Company produced a silicon- (Si) and aluminum- (Al) based GHFS (Fig. 4).

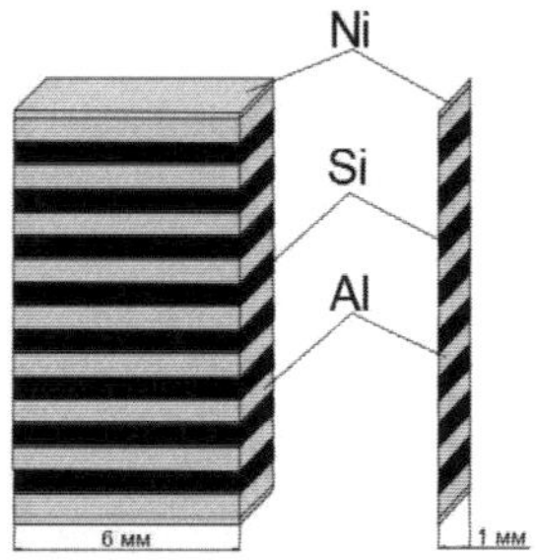

Figure 4. Schematic Diagram of the HHFGS Produced at *NPP "Etalon" Company.*

HHFGS has the following advantages: very high sensitivity of 300...500 mV/W, abnormally low time response $10^{-8}...10^{-9}$ s making them virtually inertia-free for the study of most heat processes, operability under electromagnetic impacts and insensitivity to pressure drops, high thermal stability (up to 400 °C or more) which makes them an important diagnostic element at thermal power-engineering facilities [5,6].

Metrological Equipment for Heat-Flux Density Meter Calibration

The Company manufactures a range of metering instruments for the calibration of heat-flux density metering:

- UTM–1 heat-metering unit is designed for the reproduction of heat-flux area density during the calibration of heat-flux area density meters by means of referencing in fixed-temperature mode. The unit provides calibration of heat-flux area density meter having different design (thermoelectric, voltaic, semiconductor etc.), different shapes and dimensions (within the working zone diameter of 300 mm), including those with the sensors with the design and shape different from the sensors against which they are referenced. The heat flux area density range produced in the heat-metering chamber is 20...2000 W/m^2.
- UTM–2 heat-metering unit is designed for the transmission of the heat-flux area density value using direct measurement of the heat-flux area density in fixed-temperature mode. The unit provides calibration of heat-flux area density meter having different design (thermoelectric, voltaic, semiconductor etc.), different shapes and dimensions (within the working zone diameter of 150 mm), including those with the sensors with the design and shape different from the sensors against which they are referenced. The heat flux area density range produced in the heat-metering chamber is 1...3000 W/m^2
- UTM–P heat-metering unit is designed for the transmission of the heat-flux area density value using direct measurement of the heat-flux area density in fixed-temperature mode. The unit provides calibration of heat-flux area density meter having different design (thermoelectric, voltaic, semiconductor etc.), different shapes and dimensions (within the working zone diameter of 200 mm), including those with the sensors with the design and shape different from the reference sensors. The heat flux area density range produced in the heat-metering chamber is 100...10000 W/m^2.
- To gage HHFGS our Company developed and designed a special unit (Fig. 5).

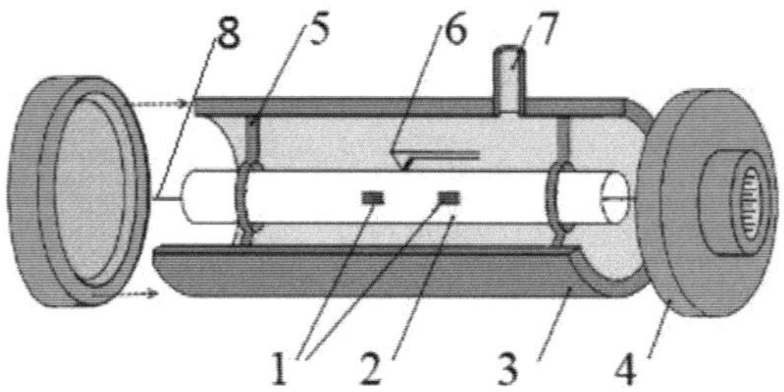

Figure 5. Schematic Diagram of the HHFGS Gauging Unit.

Legend: 1 – HHFGS to be calibrated; 2 – tube; 3 – housing; 4 – cover; 5 – holder; 6 – thermocouple; 7 – vacuum branch; 8 – heater.

- The HHFGS to be gauged are installed onto the tube with the thermocouple. The cover is tightly closed and the working volume is vacuumed to eliminate convection heat-transfer between its elements. In the cover tight connections for the power-supply to the heater and signal reception from the gradient heat-flux sensor and thermocouple. The heat flux from the heater is transferred by меаты ща emission radial to the outer walls. The thermocouple is used to determine the reference temperature [3,4]. The range of the heat-flux area density in the chamber is approximately 100…30000 W/m^2.

4. Experiment Results

High-Sensitivity Heat-Flux Sensor

The high-sensitivity heat-flux sensor was tested on UTM-1 heat-metering unit and is further tested at Construction Physics Research Institute (NIISF RAASN), Moscow.

The sensor thermocouple materials are constantan-copper. Filling compound MG CHEMICALS 832HT. Sensor dimensions: 40x80 mm, thickness 2.8 mm.

The tests conducted on UTM-1 unit gave good sensitivity and reproducibility results, see Table 1.

Table 1. HFS sensitivity as function of the heat-flux density

Heat Flux Density, [W/m^2]	Sensitivity, [mV/(W/m^2)]
92.54	0.242
188.39	0.241
290.62	0.240
392.68	0.239

Based on the calculations made the sensitivity measurement error was less than 0.7 %. The sensor conversion ratio was 4.163 (W/m^2)/mV.

Based on the preliminary tests conducted at the Construction Physics Research Institute the sensor we manufactured demonstrates a more stable behavior than its equivalents.

Thin-Film Heat Flux Sensor

Thin-film heat-flux sensors are manufactured on two substrate types: sitall (thickness 0.5 mm, foil-plated textile laminate 0.3 mm). The main thermowire is chromium (Cr) meander with the metallization thickness of 0.5 µm, as the second thermowire – nickel (Ni) with the metallization thickness of 1 µm. The sensitive element location area on the substrate was 40x50 mm (Fig.6).

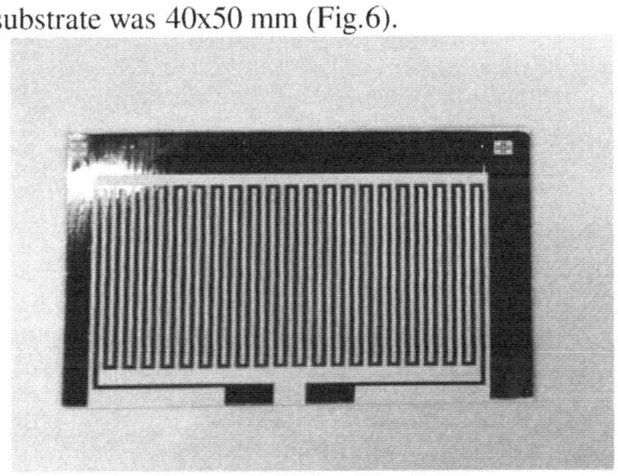

Figure 6. Thin-Film HFS.

As the overlaps to create the temperature gradient synthetic-rubber-based POLI-FIX 330 double-sided glue tape was used. Thickness 0.08 mm, heat conduction : $\lambda \leq 0.035$ W/(m·K) at 0 °C. As temperature-leveling coatings 80 µm foil-clad polyimide was used, copper thickness 35 µm.

The tests were conducted on UTM-1 heat-metering unit. The sensor made on the textile laminate substrate demonstrated unsatisfactory results, with the sensitivity of 0.00158 mV/ (W/m^2). The sensor made of sitall substrate gave better results: its sensitivity made 0.00338 mV/ (W/m^2). These sensitivity readings are approximately equal to the sensitivity specifications of similar sensors manufactured by Captec Company (France) [4].

Heterogeneous Heat-Flux Gradient Sensor

Si-Al-based heat-flux gradient sensor was manufactured using a diffusion welding unit at NPP Etalon Company.

The following materials were used for the sensor manufacture:

- КДБ-12 silicon plates, diameter 100 mm, thickness 500 μm, orientation (100), with two-sided polishing
- A5 grade aluminum foil, thickness 0.2 mm, T-hard

Si and Al plates were cut into rectangular plates 35x34 mm. Then they were alternately laid into the vacuum chamber for diffusion welding.

The diffusion welding mode:

The pressure onto the workpiece was about 30 atm., chamber pressure – 10 Pa, initial heating was performed up to 558 °C, the entire workpiece residence time at this temperature was 10 minutes, then the workpiece was heated to 577 °C. After the workpiece shrinkage to 20 μm the heating was stopped. The heating time was 52 minutes, cool-down time – 54 minutes. When the temperature in the vacuum chamber was 150 °C, air was injected and the welded workpiece was removed.

For subsequent soldering of the leads to the sensitive elements the workpiece was coated with chemical-grade nickel.

To produce finished heat-flux gradient sensors the structure was cut into individual elements at 45 °C to the main Si crystal axes, for cutting a special tool was used. As a result sensors shown in Fig. 7 were produced. The sensor width was 6 mm, thickness 1 mm.

Figure 7. HHFGS Pilot Samples.

Preliminary tests at UTM-1 heat-metering unit showed the dependence of the thermal EMF on the heat-flux density. For further analysis of the HHFGS's produced tests at large heat-flux density values at HHFGS calibration units will be held.

5. Discussion of Results

The high-sensitivity heat-flux at the initial research stage came up to expectations, particularly: it enabled attaining the sensor high sensitivity with minor error and the linearity of its signal as function of time. Its disadvantage is large dimensions which may be an issue in case of its use under certain conditions.

During the design and manufacture of the thin-film sensor at this stage it became possible to obtain optimum structure of the sensitive element and thermowire metallization modes. The studies conducted demonstrated that the substrate material is very important for the sensor. In the future the manufacture of a thin-film HFS on a flexible basis is planned which will enable making it a more universal metering instrument. The sensor sensitivity improvement at the expense of applying other metals for thermowires and contact heat-insulation is also planned [5].

At the Unit developed for HHFGS diffusion welding, rather good HHFGS pilot samples were used [5]. The Unit enables rather fine tuning of all the parameters for diffusion welding as well as monitoring of the welding process. The welding modes will have to be worked out. The materials pre-welding treatment is critical: depending on their treatment degree and quality the welding modes and workpiece quality will change, the diffusion does not cover the entire surface of the elements [5].

6. Conclusions

1. High-sensitivity heat-flux sensor after improvements (dimension reduction, sensitivity raising) together with the metrological equipment will enable reliable check of the cladding structure heat conduction as well as monitoring of low heat fluxes which is necessary for the structures' commissioning. This sensor is

different from its peers by the low output data error, it will enable on-site metering of heat fluxes with high accuracy.

2. Further upgrade and studies of the HHFGS gauging unit. The Unit advantages are stable maintaining of the heat flux in the wide temperature range which enables the sensor gauging in severe conditions.

3. Further upgrade and research of fine-film heat-flux sensors will enable creating flexible sensors with rather a good sensitivity for their use in the laboratories and in various industries where heat fluxes need to be measured in non-conventional conditions and to measure radiation heat flux.

4. HHFGS diffusion welding unit enables producing high-quality HHFGS which will allow their mass-scale production. The following activities are planned: HHFGS manufacture from other silicon and aluminum grades, sensor encapsulation, performance characteristics studies in HHFGS gauging unit. The sensor high sensitivity will enable their application without auxiliary signal amplification equipment. Due to a very low thermal inertia indicator it would be possible to instantaneously read the heat flux value in the sensor location area. The wide operating temperature range will enable its use in extreme conditions which is impossible for the most sensors produced worldwide. Due to their characteristics the sensors will find their application in high-energy industries as well as in aerospace industry.

References

[1] Grishchenko T G , Dekusha L V , Vorobyev L I , Burova Z A, Kovtun S I and Dekusha O L *2017 Heat Metering: Theory, Metrology, Practice.* Monograph in three books. Book 1: Methods and Equipment for Heat Flux Measurement (Kiev) Institute of Engineering Thermal Physics of the National Academy of Sciences of Ukraine

[2] Brodnikov A F, Vikhareva N A , Cherepanov V Ya *2017Measurements and Reference Standards of the Thermal Values* Textbook (Novosibirsk: NF ASMS)

[3] Dekusha LV 2015 *Heat Metering Equipment Based on the Heat Flux Thermoelectric Transducers* Thesis Institute of Engineering Thermal Physics of the National Academy of Sciences of Ukraine

[4] *Captec scientific catalogue* Available at: http://www.captec.fr

[5] Sapozhnikov S Z, Mityakov Yu and Mityakov A V 2012 *Gradient Heat Metering Foundations* (St. Petersburg. Polytechnic University Publishing House)

[6] Sapozhnikov S Z, Mityakov Yu,Mityakov A V., Gusakov A A and Zaynullina E R 2017 *Thermal Physics Experiment Metrology*: Textbook (St. Petersburg. Polytechnic University Publishing House)

IOP Conf. Series: Journal of Physics: Conf. Series **998** (2018) 012009 doi:10.1088/1742-6596/998/1/012009

Development of the disable software reporting system on the basis of the neural network

S Gavrylenko[1], O Babenko[2] and E Ignatova[3]

[1,2] National technical university «Kharkiv Polytechnical Institute», Kharkiv, Ukraine
[3] Intertek Ltd, Kharkiv, Ukraine

Abstract. The PE structure of malicious and secure software is analyzed, features are highlighted, binary sign vectors are obtained and used as inputs for training the neural network. A software model for detecting malware based on the ART-1 neural network was developed, optimal similarity coefficients were found, and testing was performed. The obtained research results showed the possibility of using the developed system of identifying malicious software in computer systems protection systems

Key words: PE-file structure, antivirus programs, computer systems, malicious software, signature method, heuristic method, ART-1 neural network

1. Introduction

The most notable event of the first summer month of 2017 was the Trojan.Encoder.12544 worm epidemic, which is mentioned in the media as Petya. A or Mbr locker 256. It is the first time in the history of Ukraine, that banks, gas stations, stores, sites of government agencies were struck in a few hours due to hacker's attack. Even the sites of the Cabinet of Ministers and some of the largest media outlets were paralyzed. This malware infected computers of many organizations and individuals in 60 countries around the world. The damage from the virus attack is estimated at $ 8 million.

It is known that over a year, viruses cause damage to hundreds of billions of dollars, and about the same amount is indirect damage associated with the development of software and other measures to protect against viruses. It should be noted that this problem is compounded by the dynamic growth in the number of mobile devices, the general switch to cloud technology and the spread of the Internet technologies, which leads to an increase of the number of malicious software.

It becomes obvious that the analysis of such a number of malicious code and the formation of the entire spectrum of virus signature is an almost unrealized task. That is why the actual topic is the development of effective methods and technologies for counteracting computer viruses based on heuristic methods.

Analysis of literature [1-4], as well as research on the methods of heuristic analysis in antivirus programs showed a high variety of existing approaches and methods of heuristic analysis: intelligent subsystems based on the theory of artificial intelligence, methods of fuzzy logic, cluster analysis, the theory of neural networks, genetic algorithms and other. The main disadvantage of the heuristic method is the high frequency of false positives.

Statistical methods based on control cards (e/g, Shukhart control maps, CUSUM maps, EWMA cards, etc.) can also be used to solve the set tasks [5]. In addition, methods for statistical data processing, for example, BDS testing [6], can be used for further refine of obtained results.

These methods are based on the assumption that for a computer system (CS) there is a template for normal behavior and any significant deviations from it may be due to the influence of intruders. That is why the very important task is to select or form a template, which would reproduce the functional portrait of the CS and record its abnormal behavior with the given accuracy. At the same time, the more input data is analyzed, the more accurate the result of the evaluation. Meanwhile, if the model or evaluation criterion is chosen incorrectly, parametric methods lose their basic authority, which can lead to an increase in false positives.

The conducted studies have shown that the main way of eliminating these shortcomings is to improve the models of information technology and the reasoned choice of criteria for evaluating abnormal behavior of computer systems.

2. Results of the development and research

The conducted researches have shown that one of the perspective directions of heuristic analysis of computer viruses is the use of neural networks [7-16]. On the one hand neural networks, can behave as a deterministic machine, on the other as a fuzzy system, evaluating new data that did not participate in the formation of the neural network. Such a result is achieved by learning networks [17] and not by the formation of reaction rules, as is done in classical approaches. Ability of the networks to educate by examples makes them more attractive in comparison to systems that function according to a defined system of rules formulated by experts. The education process can be considered the architectural determination of the network and finding the coefficients of the connections between neurons. The neural network adjusts the weight of the connections depending on the existing training set and is a parallel computing device, since it is based on a set of simple computing elements - the parallel functioning neurons. Such a system is resistant to damage, that is, the network will work even if part of the neurons fails. In the education process, the neural network is capable to detect complex interdependencies between incoming data and output, as well as generalization. This means that in case of successful training, the network will be able to return the correct result based on the data that was missing in the training sample, as well as incomplete or partially distorted data.

It was decided to develop a heuristic analyzer based on the neural network, namely ART-1.

ART Network is a vector classifier. The input vector is classified depending on which of the many images, previously remembered, it is similar to. The ART network classifies its classification in the form of excitation of one of the neurons from recognition layer. If the input vector does not match any of the memorized images, a new category is created by memorizing the image, which is identical to the new input vector. If it is determined that the input vector is similar to previously remembered vectors in terms of a certain similarity criterion, the remembered vector will change (learn) under the influence of the new one in such a way as to become more similar to this input vector.

The remembered image will not change if the current input vector doesn't look similar to it. Thus the dilemma of stability-plasticity is solving. The new image may create additional classification categories, however, it cannot force the existing memory to change itself. Several types of neural networks have been developed on the basis of adaptive resonance theory, in particular, the ART-1 network is designed to work with binary input images or vectors.

The architecture of the ART-1 neural network is shown in figure 1.

It has control elements G1 and G2 that provide control over the operation process [18]. In Figure 1, due to the large number of links between the Z and Y layers of the elements, only one generalized pair of weights, the connections between the interface neurons and the recognition neurons are indicated. Most of the links shown in Figure 1. are excitatory: from the input layer of the Fa elements to the neurons R, G1, G2 and Fb, from the neurons G1, G2 according to the neurons of the Fb and Y layers. Braking signals transmit only the plurality of bonds from the interface elements to the R- A neuron, from the R neuron to the Y neurons and from the Y neurons to the G1 element. All ART-1 networks communicate only binary signals 0 or 1. Each element in the interface or Y-layer of the ART-1 network has three sources of input signals. An arbitrary interface element can receive signals from elements of the input layer S, and elements of the Y-layer and from the neuron G1. Similarly, an element can receive signals from the interface elements, neurons R and G2. In order to translate the neurons of the interface or recognition layers into an active single state, the presence of incoming excitatory signals from two sources is required [20].

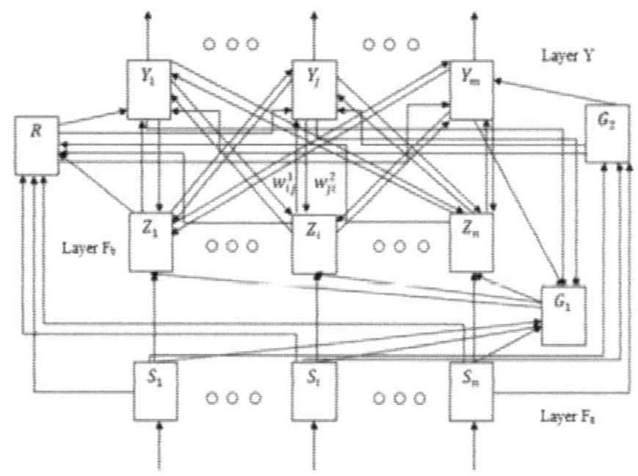

Figure 1. An architecture of the neural network APT−1.

It has control elements G1 and G2 that provide control over the operation process [18]. In Figure 1, due to the large number of links between the Z and Y layers of the elements, only one generalized pair of weights, the connections between the interface neurons and the recognition neurons are indicated. Most of the links shown in Figure 1 are excitatory: from the input layer of the Fa elements to the neurons R, G1, G2 and Fb, from the neurons G1, G2 according to the neurons of the Fb and Y layers. Braking signals transmit only the plurality of bonds from the interface elements to the R- A neuron, from the R neuron to the Y neurons and from the Y neurons to the G1 element. All ART-1 networks communicate only binary signals 0 or 1. Each element in the interface or Y-layer of the ART-1 network has three sources of input signals. An arbitrary interface element can receive signals from elements of the input layer S, and elements of the Y-layer and from the neuron G1. Similarly, an element can receive signals from the interface elements, neurons R and G2. In order to translate the neurons of the interface or recognition layers into an active single state, the presence of incoming excitatory signals from two sources is required [19].

If a prototype found with a certain accuracy specified by a special similarity parameter matches the input process, then it is modified to become more similar to the presented process. When the input process is not sufficiently similar to any of the existing prototypes, a new class is created on its basis. This is possible due to the fact that the network has a large number of redundant or unallocated elements that are not used until there is no need for it (if there are no undocumented neurons, then the input image does not cause network reactions) [19].

ART Neural networks - dynamic objects, which are described by systems of ordinary differential equations, so their training in the general case is rather laborious. However, ART network models can be simplified if we assume that the change in the output signals of the neurons occurs much faster than the change in the weight vectors of their bonds. Therefore, in the neural networks of the theory of adaptive resonance we can assume that after selecting for the learning of an acceptable Y-element (the resonance between the present and stored in the memory image), the output signals of all neurons remain unchanged for a long period of time, during which there are changes in connections` weights. The algorithm for training the neural network is given in [20].

The input data for training the neural network are based on the analysis of the PE structure of the file.

In figure 2 it an example of PE structure of the file is shown and the areas for the further analysis are emphasized.

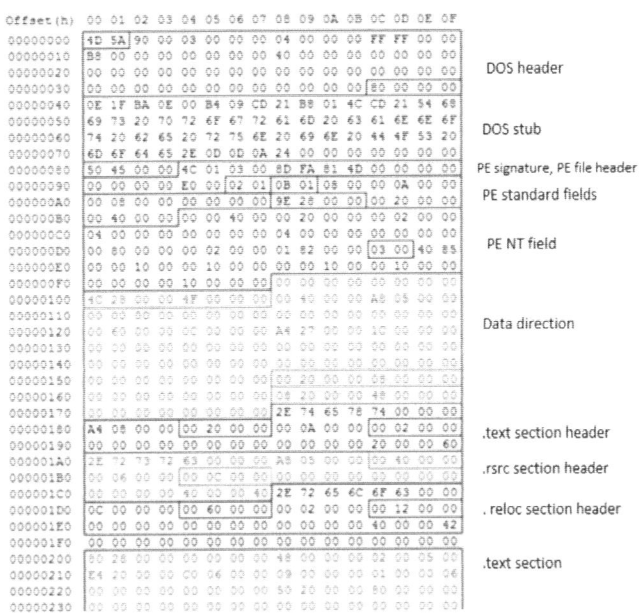

Figure 2. An example of PE file structure.

The PE structure of harmful and secure software was analyzed, namely:

- 290 Worm type files
- 1050 files such as Trojan,
- 1153 files of type Backdoor
- 1000 secure files

As a result of the analysis of the PE structure of the investigated files received:

- a table with API-functions and libraries in which they are included. There were 24945 records received.
- a table with strings. Found in total 175651 rows, their length varied from 6 to 70 characters.

The analysis of the data received from malicious and secure software allowed to highlight functions and rows, and to form the signs (Table 1), which are inherent to the considered viruses and to form a table of attributes. For further analysis it was decided to use 49 signs. These signs were later used as bitmaps for file analysis. The binaries of malicious files like Worm, Backdoor, Trojan and secure software were obtained as a result of searching for selected attributes in files (Figure. 3).

For the correct functioning of the state identification system, an optimal coefficient of similarity has been selected experimentally.

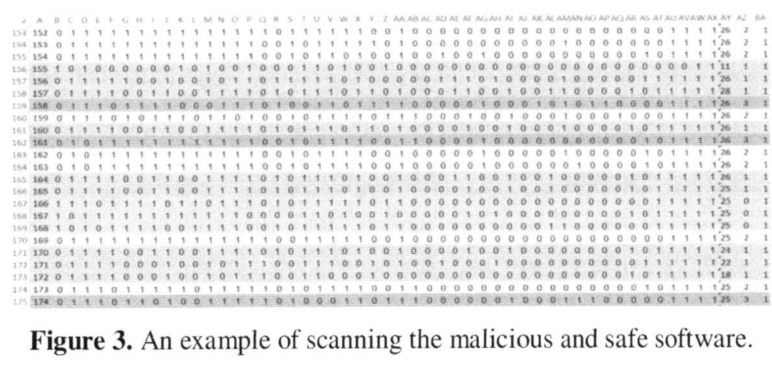

Figure 3. An example of scanning the malicious and safe software.

Table 1. Features malicious signatures

	Functions/lines	Var 1	Var 2	Quantity in the viruses
1	callnexthookex	-	-	42
2	copyfile	CopyFileA	CopyFileW	132
3	createfile	CreateFileA	CreateFileW	178
4	enumcalendarinfo	EnumCalendarInfoA	-	43
5	findfirstfile	FindFirstFileA	FindFirstFileW	127
6	getcurrentprocessid	-	-	70
		...		
37	host	-	-	10
38	admin	-	-	-
39	hostname	-	-	9
40	localhost	-	-	19
41	sendmail	-	-	28
42	userhost	-	-	9
43	wininet	-	-	35
44	wnet	WNetEnumResourceA	WNetOpenEnumA	55
45	getmodulefilename	GetModuleFileNameA	GetModuleFileNameW	206
46	getmodulehandle	GetModuleHandleA	-	247
47	loadlibrary	LoadLibraryA	-	135
48	setfilepointer	SetFilePointerEx	-	158
49	virtualalloc	VirtualAllocEx	-	158

At the first stage, 100 signatures of the type Backdoor are taken for basic knowledge of the neural network. The initial similarity factor is 0.6, since it is the minimum allowable for the neural network. Signatures of 100 secure files were submitted to the input of the neural network for recognition. As a result of the program's operation, some of the input signatures were displayed on the console. Console-derived signatures were attributed to signatures of the type Backdoor, which was a mistake (Figure. 4) and required an increase of the coefficient of similarity to the optimal. The optimal coefficient is considered to be the coefficient for which there are no false recognitions. The results of the experimental selection of the optimal similarity coefficient for malicious software of the type Backdoor are given in Table. 2.

Table 2. Experimental match of the optimal coefficient of similarity for the viruses type Backdoor

Coefficient of similarity	Quantity of false alerts for the safe signatures
0.6	18
0.7	10
0.8	1
0.84	0

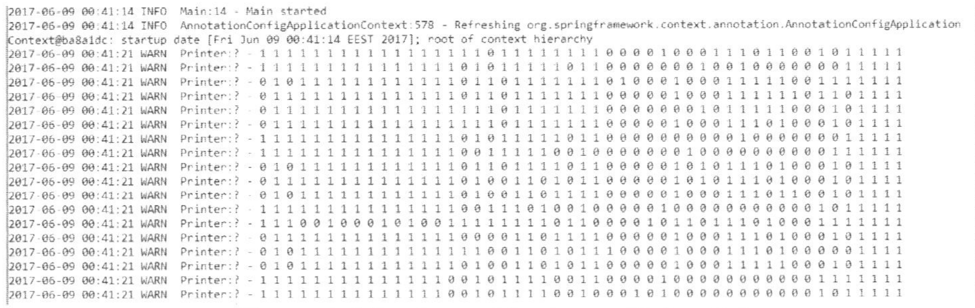

Figure 4. Mistakes making by coefficient 0.6.

Thus, the absence of false positives for malicious software such as Backdoor managed to achieve with a similarity factor of 0.84.

At the second and third stage, 100 signatures of malicious software such as Trojan and Worm are taken for basic knowledge of the neural network. The results of the experimental selection of the optimal coefficient of similarity are given in Table 3 and Table 4.

Table 3. Experimental match of the optimal coefficient of similarity for the viruses type Trojan

Coefficient of similarity	Quantity of false alerts for the safe signatures
0.6	10
0.65	3
0.7	0

Figure 5 describes optimal coefficients of the similarity grouped by malicious signature types

Table 4. Experimental match of the optimal coefficient of similarity for the viruses type Worm

Coefficient of similarity	Quantity of false alerts for the safe signatures
0.6	29
0.7	14
0.8	17
0.9	10
0.97	0

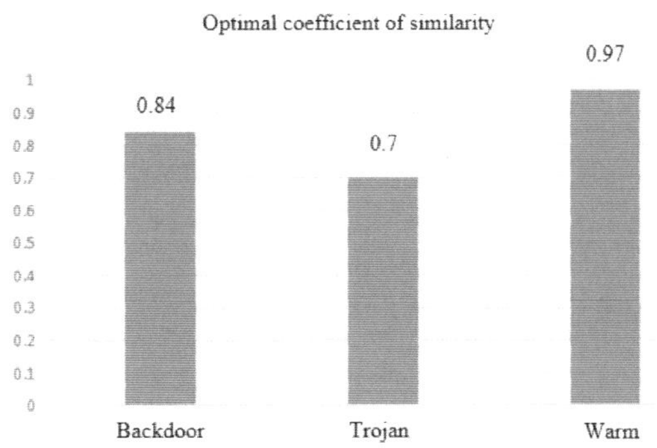

Figure 5. Optimal coefficients of the similarity.

IOP Conf. Series: Journal of Physics: Conf. Series **998** (2018) 012009 doi:10.1088/1742-6596/998/1/012009

In the future, 3 experiments with different input data and different similarity coefficients were carried out. The first one uses 100 signatures of the type Backdoor as the base of initial knowledge, 400 signatures are presented to the recognition input, which represent a mixture of all possible signatures (100 Backdoor signatures, 100 Trojan signatures, 100 Worm signatures, 100 signature files of secure files). The second one is the base of 100 signatures of type Trojan, the input is 400 signatures. The third is the base of 100 signatures of type Worm, the input is 400 signatures.

At the initial base of Backdoor malicious signatures, an additional group of malicious Trojan 28% signature was detected.

Thus, the absence of false positives for malicious software such as Backdoor managed to achieve with a similarity factor of 0.84.

At the second and third stage, 100 signatures of malicious software such as Trojan and Worm are taken for basic knowledge of the neural network. The results of the experimental selection of the optimal coefficient of similarity are given in Table 4 and Table 5.

At the initial base of Warm malicious signatures, only signatures from this group were 100%. At the initial base of Trojan malware, Trojan 100%, Backdoor -46%, Worm - 1% were detected.

The results of the identification system showed that when training the neural system by the Backdoor sample, the system also begins to identify harmful Trojan (28%) signatures, since these types of signatures have a high similarity coefficient, since they perform similar actions from the point of view of the operating system.

When learning a worm-patterned system, only a signature of this type is detected, due to a relatively high optimal similarity factor of 0.97, the tokens of this type are abandoned from Signature Backdoor, Trojan, and secure software.

Also, when training a neural system with Trojan sampling, in addition to this sample, the system recognizes backdoor signatures (46%) and insignificant number of worm-signatures (1%).

This percentage of signature recognition (Backdoor - Trojan and vice versa) is due to the fact that these types of viruses have similar actions from the point of view of the operating system namely harm to it: the desire to obtain unauthorized access to data or remote control of the operating system and the computer as a whole; collection of information and its transmission to the attacker, its destruction or malicious change, the disruption of the computer; the use of computer resources for other purposes.

3. Conclusions.

In this article, the methods of constructing antivirus programs are analyzed. The PE structure of harmful and secure software has been analyzed, API functions and rows are found, inherent to these files, and selected part of them for further analysis. The result of the search for highlighted features in files is binary vectors of malicious software such as Worm, Backdoor, Trojan and secure software, and used as inputs for training the neural network.

The software model of the heuristic analyzer on the basis of the ART-1 neural network has been developed, the optimal similarity coefficients were found, and the tested computer virus detection system was tested.

The test results showed that when training the neural system with the Backdoor sample, the system also begins to identify 28% of the malicious Trojan type signatures, since these types of signatures perform such actions from the point of view of the operating system. When learning the system, the worm sample is only susceptible to virus of this type.

The obtained research results showed the possibility of using the developed system of identifying malicious software as an auxiliary method in information security systems of the COP

References

[1] Semenov S, Gavrylenko S and Davydov V 2014 *Data Protection in Computer-Aided Control Systems* ("LAP Lambert Academic Publishing" Germany)

[2] Lukatsky A 2001 *Attack Detection* (St. Petersburg: VHV-Petersburg)

[3] Shelukhin O I, Sakalema D Zh and Filinov A S 2013. *Intrusion Detection into Computer Networks* (Moscow: Hot line-Telecom) p 220

[4] Gavrylenko S and Semenov S 2015 Approximating computer system operation technologies under external action through the brusselator model with perturbation in the form of dynamic chaos *Industrial Engineering Journal* **1** (44) p 36-40 Romania

[5] Gavrylenko S, Chelak V and Hornostal O 2016 *Metrology and metrology assurance*" pp 342-347 Bulgaria

[6] Gavrylenko S, Semenov S and Chelak V 2016 *Bulletin of NTU "KhPI". Informatics and modeling* vol 21 p 18-125 Ukraine

[7] Grishin A V 2011 *Information technologies and computing systems* vol **1** p 53-64

[8] Yemelyanova Yu G 2011 *Software systems: theory and applications* vol **3** (7) p 3-15

[9] Izrailov K E 2012 *Vestnik Inzekona. Series: Engineering* **8** (59) p 150-153

[10] Nesteruk F G, Nesteruk L G and Nesteruk G F 2009 systems *Automation and Remote Control* **70** p 491-501

[11] Kochetkova A S 2007 Monitoring *Bulletin of Volgograd State University Series 9: Research of Young Scientists* 6 p 163-166

[12] Magnitsky N A 2008 *Dynamics of Non-uniform Systems* p 200-205

[13] Marchenko O O, Matvienko S V and Nesteruk F G 2007 *Scientific and Technical Bulletin of Information Technologies, Mechanics and Optics* 39 pp 83-93

[14] Skorik F A and Sayenko I B 2013 *Innovations in science* (materials of XVI international correspondence scientific-practical conference) p 151-156

[15] Tumoyan Y P 2009 Development of the method of simulation of network attacks on the basis of neural networks and probabilistic graphs, News of Southern federal university. *Technical sciences* **90** p 148-153

[16] Novikov O O and Krasnopevtsev A A 2012 *Safety of information technologies* vol 1 p 47-50

[17] Bezobrazov S V and Golovko V A 2010 *Neuroinformatics* vol 7 p 273-288

[18] Dmitrienko V D, Zakovrotny A Yu, Noskov V I and Mezentsev M V 2014 *Fundamentals of neurocomputing: educational and methodical manual for practical classes* p 140 Ukraine

[19] Ossovsky S 2002 *Neural Networks for Information Processing* (Moscow: Finance and Statistics) p 344

[20] Gavrylenko S Yu, Semenov S G and Babenko O S 2015 *Systems of information processing* **10** (135) p 126-129 Ukraine

IOP Conf. Series: Journal of Physics: Conf. Series **998** (2018) 012010 doi:10.1088/1742-6596/998/1/012010

The mathematical model accuracy estimation of the oil storage tank foundation soil moistening

M I Gildebrandt[1], R N Ivanov[1], AV Gruzin[1], L B Antropova[1] and S A Kononov[2]

[1]Omsk State Technical University, 11, Mira av., Omsk, 644050, Russian Federation
[2]The Siberian Automobile and Highway University, 5, Mira av., Omsk, 644080, Russian Federation

E-mail: r.bantik@mail.ru

Abstract. The oil storage tanks foundations preparation technologies improvement is the relevant objective which achievement will make possible to reduce the material costs and spent time for the foundation preparing while providing the required operational reliability. The laboratory research revealed the nature of sandy soil layer watering with a given amount of water. The obtained data made possible developing the sandy soil layer moistening mathematical model. The performed estimation of the oil storage tank foundation soil moistening mathematical model accuracy showed the experimental and theoretical results acceptable convergence.

Key-words: soil foundation, moistening, mathematical model, accuracy, tank

1. Introduction

Quality preparation during the soil foundation construction is one of the key factors to ensure the oil its refinery products storage vertical steel tank sustainability during the operating period [1]. The earlier researches revealed the soil moisture significant impact on its deformation properties [2]. The existing technologies for the tank soil foundation preparation include premoistening before compaction [3]. For choosing the compaction rational process parameters there is a need to develop the mathematical model of the tank foundation soil moistening with a limited amount of water. Obviously, the accepted mathematical model accuracy is required to be estimated for confirming its correctness.

2. Study subject

The study subject is the accuracy of the finite thickness sandy soil moistening with a given amount of water describing by the proposed mathematical model.

3. Experimental

Initially, the moistening laboratory studies of the medium sized sandy soil layer weighing 16 kg, placed into the specialized stand presented in figure 1A were carried out. Sandy soil layer thickness h=0.4 m is defined by the vertical steel tanks soil foundations construction existing technologies. The amount of water, applied for watering the sandy soil selected from the top, was calculated on the basis of the assumption of achieving the optimum moisture content of 11% and constituted 1.7 kg. The specialized stand design made possible the periodic soil sampling with a mass of 2÷3 g from the predetermined depths for the current humidity control. Soil samples humidity control was carried out by using the halogen moisture analyser HC103, in this case the soil samples mass measurement accuracy was 0.001 g. The obtained experimental results allowed estimating the medium sized sandy soil moistening character both over time and in the sample depth (figure 1B).

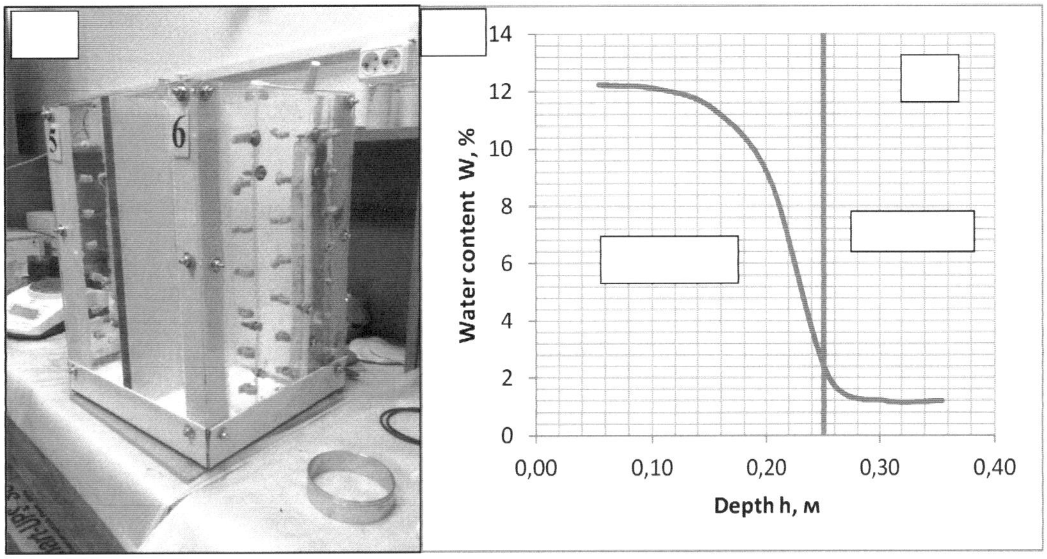

Figure 1. The specialized stand for the experiment (A); the sandy soil sample water content change in depth, 10 minutes after its moistening (B).

4. Theoretical studies

At the stage of theoretical studies, the obtained laboratory data analysis made it possible to identify two specific sections of the soil water content dependence *w* on the depth *h*: the wetted section (section I: depths are from 0 to h\` (t), m) and the section with the initial water conent (section II: depths are from h\` (t) to 0.04, m), where h\` (t) is the humidity boundary depth variation over time. For section I approximation, the third degree polynomial is proposed to be used:

$$w = a_0 + a_1 * h + a_2 * h^2 + a_3 * h^3 \tag{1}$$

where *w* is the actual soil water content at the given depth *h*, u.f.; *h* is the given depth, m; a_0, a_1, a_2, a_3 are the coefficients defined in accordance with the experimental results.

The proposed mathematical model coefficients of the previously performed laboratory studies were determined by the least squares method (table 1).

Table 1. The laboratory data processing.

Depth h, m	Experimental moisture w, %	Mathematical model coefficients	
0.055	12.2	a_0	14.704
0.105	12.1	a_1	-808.393
0.155	11.3	a_2	80961.430
0.205	8.8	a_3	-2726667
0.255	1.5		

5. Results discussion

At the final stage of the studies for confirming the proposed mathematical model correctness (1), its accuracy was estimated. The obtained laboratory data statistical analysis has shown that the carried out measurements error is properly described by the normal distribution law. The humidity experimental values normal distribution hypothesis was tested according to the Pearson chi-square test with 95% confidence level, the Pearson test calculated value is $\chi_p^2 = 17.9$ which is less than the accepted theoretical value of $\chi_m^2 = 25.0$ [4]. Thus, the finite thickness sandy soil layer wetting with a given amount of water is proposed to be described by the following dependence:

$$w = 14.704 - 808.393 * h + 80961.430 * h^2 - 272667 * h^3,\ at\ h < h`. \quad (2)$$

For identifying the water content determination absolute error, the mathematical model adequacy dispersion S_{ad}^2 of the oil storage tank foundation soil watering is necessary to be calculated. The mathematical model adequacy dispersion is determined by the formula (3) [6]:

$$S_{ad}^2 = \frac{m}{N-l} \cdot \sum_{i=1}^{N} (\overline{y_i} - \hat{y}_i)^2 \quad (3)$$

Where m is the concurrent experiments quantity;
N is the number of units;
$\overline{y_i}$ is the average value of the output quantity;
$\hat{y}_i$ is the investigated object regression equation value including the statistically significant coefficients;
l is the significant coefficients number defined as a result of the conducted experiments N.
The soil moisture specifying absolute error can be determined in accordance with the proposed mathematical model by using the following equation [5, 6]:

$$\Delta = 2 \cdot \sqrt{\frac{S_{ad}^2}{F_T}} = 2 \cdot \sqrt{\frac{0.038}{7.71}} = 0.13 \quad (4)$$

where the theoretical F-test F_T with 95 % confidence level for the third-degree polynomial is F_T=7.71.

6. Conclusions

The increased requirements for the oil storage tanks sandy foundation preparation quality needs the the soil premoistening technology improvement. The tank foundation preparation quality is influenced by factors such as the moistened layer thickness, required water amount and time needed for achieving the optimum water content by the sandy soil layer. The preparation process rational parameters choice could reduce the material costs and spent time for the planned work performance. To solve this problem, the oil storage tank foundation sandy soil finite layer wetting mathematical model with the given amount of water to the final water content required values is suggested to be used. The conducted laboratory studies showed that the sandy soil moistening with a given amount of water has a complex, non-linear nature. Based on the results of laboratory data processing, the sandy soil moistening mathematical model was proposed and suggested mathematical model using error was estimated.
The oil storage tank foundation soil moistening proposed mathematical model presented by the third degree polynomial makes it possible to calculate the humidity dependence on the soil depth at the soil depth of less than 0.3 m, with the absolute error of 0.13% and 95% confidence level.

7. References

[1] Gruzin A V, Gruzin V V and Shalay V V 2016 Theoretical researches of rammer's operating element dynamics in a soil foundation of oil and oil products storage tank *Procedia Engineering* **152** 182-189

[2] Gruzin A V, Tokarev V V, Shalai V V and Logunova Yu V 2015 The Artificial Additives Effect to Soil Deformation Characteristics of Oil and Oil Products Storage Tanks Foundation *Procedia Engineering* **113** 158-168

[3] Set of Rules 2011 22.13330.2011 *Bases of buildings and structures Updated version of SNiP* 2.02.01-83 (Moscow: CPP)

[4] Korn G and Korn T 1978 *Handbook of Mathematics (for scientists and engineers)* (Moscow: Nauka)

[5] Lepjavko A P 2012 *Metrological basis of thermotechnical measurements* (Moscow: ASMS)

[6] Volodarskij E T Malinovskij B N and Tuz Ju M 1987 *Planning and organization of the measurement experiment* (Kiev:Vysshaja shkola)

Two-point method uncertainty during control and measurement of cylindrical element diameters

V I Glukhov[1], V V Shalay[1] and H. Radev[2]

[1]Omsk State Technical University, 11 Mira av., Omsk, 644050, Russia
[2]Mechanical Engineering Faculty, Technical University of Sofia, Sofia, Bulgaria

E-mail:mips@omgtu.ru

Abstract. The topic of the article is devoted to the urgent problem of the reliability of technical products geometric specifications measurements. The purpose of the article is to improve the quality of parts linear sizes control by the two-point measurement method. The article task is to investigate methodical extended uncertainties in measuring cylindrical element linear sizes. The investigation method is a geometric modeling of the element surfaces shape and location deviations in a rectangular coordinate system. The studies were carried out for elements of various service use, taking into account their informativeness, corresponding to the kinematic pairs classes in theoretical mechanics and the number of constrained degrees of freedom in the datum element function. Cylindrical elements with informativity of 4, 2, 1 and θ (zero) were investigated. The uncertainties estimation of in two-point measurements was made by comparing the results of of linear dimensions measurements with the functional diameters maximum and minimum of the element material. Methodical uncertainty is formed when cylindrical elements with maximum informativeness have shape deviations of the cut and the curvature types. Methodical uncertainty is formed by measuring the element average size for all types of shape deviations. The two-point measurement method cannot take into account the location deviations of a dimensional element, so its use for elements with informativeness less than the maximum creates unacceptable methodical uncertainties in measurements of the maximum, minimum and medium linear dimensions. Similar methodical uncertainties also exist in the arbitration control of the linear dimensions of the cylindrical elements by limiting two-point gauges.

Keywords. Technical measurements, linear sizes, element informativeness, methodical uncertainty.

1. Introduction

The International Organization for Standardization technical Committee TC213 "Dimensional and geometrical product specifications and their control" updated all terminology to linear and angular values at the level of international standards [1-5] in the first decade of the 21st century without convincing justification. As a result, millions of engineers – users of standards in all processes of the product life cycle – designers, technologists, metrologists, control masters, university and college teachers found themselves in a difficult situation. New terms and definitions did not specify or improve old terms that served for 25-50 years. The new engineering terms abolished the old ones. Users began to note the complexity of the new terminology in their publications [6].

Between the new and old terminology there must be continuity in the development and improvement of terms, including on the following unresolved problems:

- there is no classification of geometric specifications for a functional purpose, coordinating dimensions of the elements of parts are not introduced;

- coordinate systems are not used to read the geometrical characteristics of elements, parts and assemblies;
- the engineering principles for ensuring the quality of products in terms of the geometric characteristics accuracy are not used: the principle of inversion, the principle of the unity of design, technological and measuring bases, the principle of linear-angular two-dimensionality of geometric characteristics, and the principle of geometric elements informativeness [7];
- systematic functional deviations of the component surfaces shape parts are considered random variables and are estimated by statistical characteristics.

2. Statement of the problem

The possibility of reliable measurements of the component parts dimensions and shape is a necessary condition for manufacturers of precision products [8]. Know-how – to know how to measure – is a basis of knowledge about the measurements required by a designer, a technologist and a metrologist [9]. The evaluation of the uncertainty in measuring geometric characteristics is a prerequisite for products quality control of [10].

Technical measurements make it possible to evaluate the part quality by the accuracy of geometric characteristics in the production process. The quality evaluation is carried out by controlling the dimensional and geometric accuracy of the component parts to meet the requirements of the design project. The transition of international GPS standards to the evaluaion of the dimensional accuracy of parts based on the average size of elements using the two-point method requires estimation of the measurements uncertainty, which is the aim of this work.

The specific problem of evaluating the influence of form and an joints deviations of cylindrical dimensional elements surfaces of various informative content is stated and solved on the expanded uncertainty of measurements of average, maximum and minimum linear sizes of elements by the two-point method.

3. Theory of elements functional linear sizes

Any workpiece of a technical product consists of geometric elements, which are parts of the material of the workpiece, limited by one or several geometric surfaces, primarily flat and cylindrical. Each element of the part performs certain functions [7]. The set of main design datums performs the functions of positioning the part in the product and identifying its generalized coordinate system. In sliding joints, the set of main datums performs the third function - it provides the kinematics of rotation and (or) the translational movement of the product part. The set of auxiliary design datums performs the functions of positioning the attached part to the considered one and identifying the auxiliary coordinate system for the attached part. The actuating elements perform the working functions of the part. Dimensional elements determine the overall dimensions of the part, and free elements with dimensions without tolerances perform the functions of connecting elements that unite all the geometric elements into a single solid body of the part.

The datum of a geometric element can limit a part by one to six degrees of freedom (up to three linear "t" and up to three angular "r"), which is a characteristic of the information "c" of the element and coincides with the class of kinematic pairs in theoretical mechanics. The maximum informativeness of the cylindrical element datum axis is equal to four and consists of two linear constraints along the normal to the axis in two perpendicular directions and two angular constraints around the same directions ($4c = 2t + 2r$).

Datums are combined into sets of datums to constraint the detail of the six degrees of freedom ($6c = 3t + 3r$) and form a generalized or auxiliary coordinate system in which the dimensions and coordinates of the elements should be measured. The parts are joined on sets of datums in order to combine the generalized coordinate system of the attached part with the auxiliary coordinate system of the base part.

The surfaces of all geometric elements have shape deviations in the form of alternating smooth harmonic protrusions and cavities. In addition, elements with informative content less than the maximum have protrusions and cavities of surfaces in the form of location deviations. Surface protrusions form the datum of the elements. Fits in the joints of the shafts and holes are formed by protrusions of the elements real surfaces. Consequently, the dimensions of the elements along the projections of real surfaces are the functional maximum material dimensions of the elements.

The cavities of the elements real surfaces constraint the shape and location deviations of the, and together with the micro-roughness, they are the containers for the joint lubrication. Therefore, the size of the element along the cavities of the real surface is the functional minimum material size, determining the geometric accuracy of the element.

Thus, each real dimensional geometric element has two boundary functional sizes: the maximum material size and the minimum material size. By definition, these are the sizes of two regular (ideal) elements of the same type as the dimension element, nominally located relative to the coordinate system of the datums set, in which the dimensional element fulfills its service purpose, and covering the real surfaces of the dimension element with a tangent: the first - outside the material element - forms the maximum material size of the; the second - from the material of the element - forms the minimum material size. The dimensions of the maximum and minimum material are constrainted by the upper and lower limits of the element size. Their reliable measurement ensures the quality of the joints during assembly.

The average size of the element should be calculated as a half-sum of the maximum and minimum material sizes. The connection with the gap S_m between the average dimensions of hole D_m and the shaft d_m (Fig. 1, *a*) can provide positive differences in the size of the minimum material hole D_{LM} with the size of the material maximum of shaft d_{MM} and the size of the maximum material of the hole D_{MM} with the material minimum size of the shaft D_{LM} only for one of the details translational motion. Rotational movement of the shaft in such joint is impossible (figure 1, *b*) as the difference in the maximum sizes of hole D_{MM} and shaft d_{MM} form the tension N_{MM}.

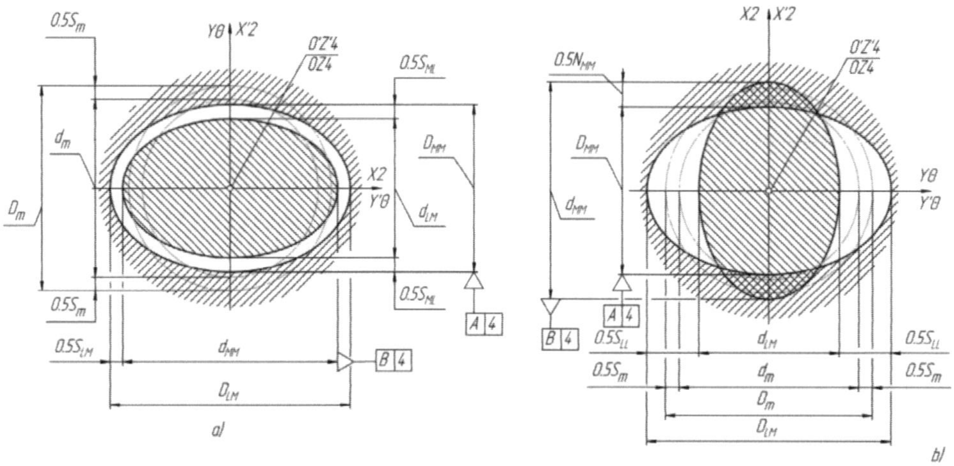

Figure 1. The model of the interference formation along the diameters of the maximum material in the fit with gap on the average hole and shaft diameters with informativeness 4.

4.The two-point method uncertainty during control and measurement of diameters

Arbitrary means of reliable control of the outer and inner cylindrical elements diameters are the limiting gauges. American engineer Taylor F W suggested the principles of designing limiting gauges more than a century ago. To control the maximum material size, the go gauge should have a geometrical shape of the mating part-the ring for a shaft and the plug for a hole. Controlling the size minimum material is necessary to be carried out by a no-go two-point gauge - a bracket for a shaft and a two-point internal caliper for a hole.

The control by two different limit gauges of the suitability of the maximum and minimum material sizes confirmed the facts of different service functions of the element two dimensions and the influence of shape deviations on the formation of two boundary dimensions of the element. Limit gauges are designed to control the validity of cylindrical elements with informativeness *4*, which have location deviations.

The complex method of controlling the maximum size of the cylindrical elements material by methodical go gauges does not create a methodical uncertainty of measurements.

In connection with the standardization of the term "two-point size" [1], the two-point method was used to control and measure the maximum material size, and the minimum material size of cylindrical elements. When monitoring using two-point gauges - the diameters of the maximum d_{MM} and the minimum d_{LM} of the outer cylindrical element material (fig. 2), methodical uncertainties do not arise for the shape of a cylindrical surface profile such as the ovality *EFV* (fig. 2, *a*) and the conicity *EFC* (figure 2, *c*), as well as for deviations of the saddle shape type *EFS* (Fig.3, *c*) or barrel-type *EFT* (figure 3, *b*). The control should be carried out in the two extreme and middle sections of the longitudinal profile and at least six sections of the transverse profile. In this case, the average diameter *dm* is not controlled by the limit gauges.

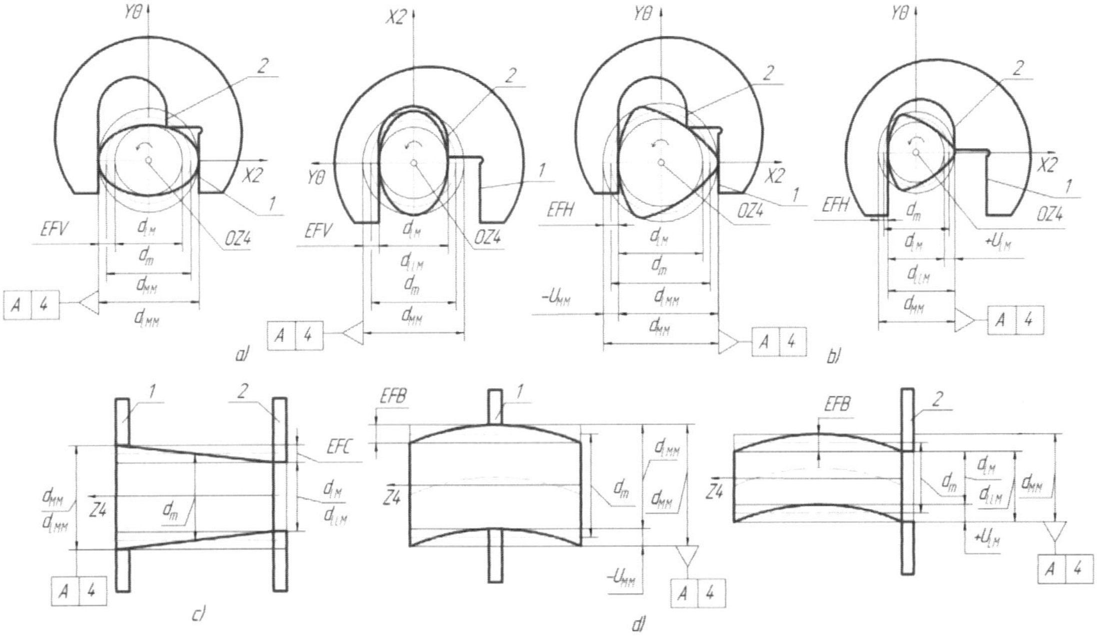

1- go gauge to maximum material diameter; *2* – no – go gauge to minimum material diameter.

Figure 2. The two-point method uncertainties models of controlling diameters of maximum and minimum material of the outer cylindrical element with informativeness 4 using limit gauges.

The methodical uncertainties of U_{MM} and U_{LM} occur when controlling the diameters of the maximum d_{MM} and the minimum d_{LM} material of cylindrical elements with the shape deviations of the cross-sectional profile with the facet pattern *EFH* (figure 2, *b*) and with deviations in the longitudinal section shape with the curvature type *EFB* (straightness of the element axis) (Fig. 2, *d*). Essentially, the control of the element serviceability is carried out on the average diameters instead of the maximum and minimum material diameters. The methodological uncertainties of the two-point control method are equal to the numerical values of the facet *EFH* and the curvature *EFB* profiles of the transverse and longitudinal sections and will cause the same parts actual dimensions falling beyond the tolerance interval. Uncertainty of control should be excluded in arbitration control: the diameter of the maximum material should be controlled by a go gauge ring by a complex method; in order to estimate the diameter of the minimum material, the cut and curvature values of the element axis should be additionally measured, corrections should be made to the results of the control, or the diameter of the minimum material should be measured by universal means.

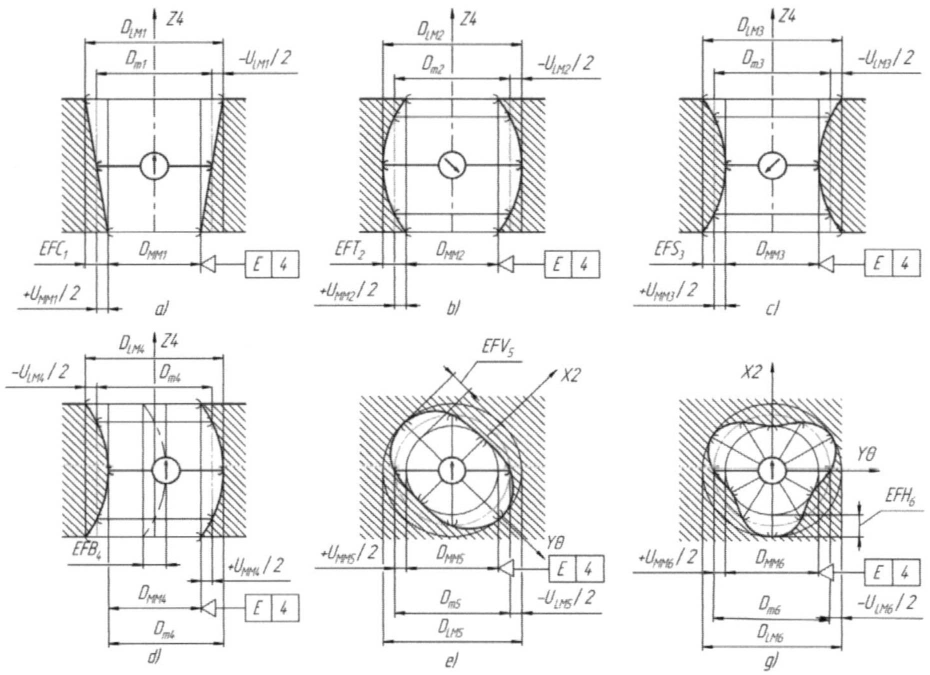

Figure 3. Models of uncertainties in measurements by the two-point method of the average diameter of an inner cylindrical element with informativeness 4.

When measuring two diameters of the maximum D_{MM} and minimum material D_{LM} of cylindrical elements with informativeness 4 using digital or scale measuring means by the two-point method (figure 3), methodical uncertainties are not formed by the profile shape deviations of the longitudinal section of the cone-shaped *EFC* (figure 3, a), barrel-like *EFT* (Fig.3, *b*), saddle-shaped *EFS* types (figure. 3, *c*) and the shape deviation of the cross-section profile of ovality type *EFV* (figure 3, *e*). However, if the element average diameter D_m is taken as the measurement result, its methodological uncertainty will be determined by the differences in the average diameter D_m and with the maximum material size D_{MM} and with the minimum material size D_{LM}. If the element has profile shape deviations of the longitudinal section of curvature type *EFB* (figure 3, *d*) and the cross-sectional profile of the facet type *EFH* (figure 3, *g*), the methodological uncertainties of the U_{MM} and U_{LM} measurements of the maximum D_{MM} and minimum D_{LM} material of the are numerically equal to the shape deviations values *EFB* and *EFH*. This means that in any section of the element only the average diameter D_m is measured with the same uncertainty. Consequently, methodical uncertainties of the cylindrical elements diameters when measuring (figure 3) and control (figure 2) by the two-point method coincide.

A cylindrical element can have information content of less than four constraints (figure 4). With informativeness of 2 or 1 (figure 4, *a*), the element performs a datum function that constraints a detail of two translational (*2t*) or one rotational (*1r*) degrees of freedom, in a system with a flat datum having the informativeness *3* (*3c* = *1t* + *2r*). An element with informativeness θ (zero) performs the function of actuator, which should not have any point of contact with the attached part (figure 4, b), i.e. ensure the clearance fit.

The sets of *A3B2S1* component datums have a total informativeness value 6, which allows them to materialize the generalized coordinate systems *OX4Y2Z0*. Unused by the holes, the constraints of degrees of freedom, due to reduced informativeness, form angular *EPE* and linear *EX, EY* in the of axes and the centers of the holes location deviations. These location deviations reduce the diameters of the maximum material D_{MM} and increase the diameters of the minimum material D_{LM}. According to the principle of two-dimensionality of sizes [7], these diameters are formed by two regular cylinders nominally located in coordinate system *OX4Y2Z*θ and tangentially covering real surfaces of holes outside the material and from the parts material.

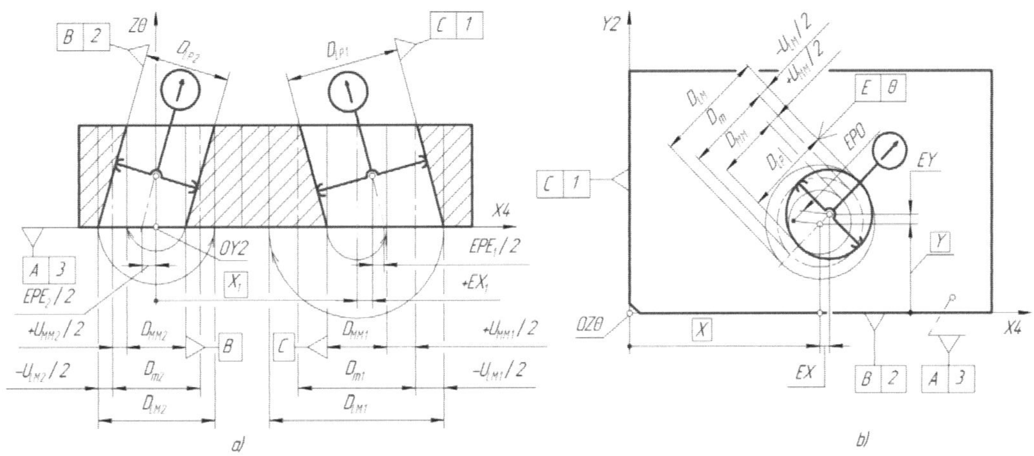

Figure 4. Uncertainties models of measurements of internal cylindrical elements average diameters of with informativeness*2*, *1*, and *0* by the two-point method.

Therefore, the two-point method cannot measure the diameters of the maximum and minimum of the material and estimate their average diameter D_m, because the method does not take into account the location deviation of the cylindrical elements centers and axes and can be used for elements with informativeness*4* that have only shape deviations (figure 3). The total uncertainty of the accepted value of the mean diameter will be a sum of unaccounted shape deviations of the surfaces and unaccounted location deviations of the elements centers and axes in the generalized coordinate system of the part.

5. Discussion of results

As a result of the of linear sizes two-point measurements accuracy studies, it is established that methodical uncertainties do not arise only in the maximum and minimum sizes with the maximum informativeness of elements having deviations of ovality, conicity, barrel shape, or saddle shape for cylindrical elements. Methodical uncertainty is formed when measuring elements with maximum informativeness for shape deviations of the cut type and the type of curvature of the cylindrical elements. Methodical uncertainty is formed by measuring the average size of the elements for all types of shape deviations. The two-point measurement method cannot take into account deviations in the location of a dimensional element, so its use for elements with informativeness less than the maximum creates unacceptable methodical uncertainties in measurements of the maximum, minimum and average linear dimensions. Similar methodical uncertainties also exist in the arbitration control of the linear sizes of the part elements by limiting two-point gauges.

6. Conclusion

According to the research results, the following conclusion can be made:

1. Each dimensional geometric element limits by a real surface has two linear sizes: two permissible size covers for the normalization of the accuracy class and two functional sizes for the operation of the part in the product: the maximum material size and the minimum material size.

2. According to the principle of sizes two-dimensionality, the size of the material maximum and the size of the material minimum he of the dimensional element are the dimensions of two regular (ideal) elements of the same type as the dimension element, nominally located relative to the coordinate system in which the dimensional element fulfills the service purpose, with the tangentially covering the real surface of the dimensional element: the first - outside the material of the element - materializes the size of the material maximum, the second - from the material of the element - materializes the size of the material minimum.

IOP Conf. Series: Journal of Physics: Conf. Series **998** (2018) 012011 doi:10.1088/1742-6596/998/1/012011

3. To implement these definitions, when measuring two sizes of an element, we need a coordinate system in which the dimensional element operates, and knowing the base informativeness of an element, i.e. the number of degrees of freedom that the ISO does not use.

4. Measurement of local linear sizes according to ISO standards by the two-point method is carried out by the shortest distance between opposite points of a dimensional element real surface in a set of pairs of points. By the processing of the measurement results the maximum size, the minimum size and the average size, which is preferred as an ordered one and having the biggest methodical uncertainty, are found.

5. The evaluation of the methodical extended uncertainty of two-point measurements of cylindrical elements linear sizes, made by comparison with the dimensions of the maximum and minimum material, showed that measurement uncertainty is not formed only in certain particular cases, when the maximum and minimum dimensions of the elements are located at opposite points of real surfaces.

6. The greatest influence on the uncertainty of two-point measurements of linear sizes has the deviation of a cut-type (roundness with an odd number of faces) and curvature (straightness) of the elements.

7. The uncertainty of two-point measurements of the linear sizes of elements that have information content less than the maximum is influences by location deviations such as positional deviation, perpendicularity, symmetry and alignment of the bases of the measured elements, which leads to unacceptably large values of uncertainties and goes beyond acceptable limits.

8. There is only one way to solve the described problem: the transition of ISO standards to a functional approach with geometric products specifications based on the two dimensionality principle with the classification of elements by informativeness, with the introduction of coordinate systems for reading characteristics, the development of mathematical models for the maximum and minimum material sizes and software for their control.

References

[1] ISO 8015:2011, Geometrical product specifications (GPS) – Fundamentals–Concepts, principles and rules

[2] ISO 286-1:2010, Geometrical product specifications (GPS) – ISO code system for tolerances on linear sizes – Part 1: Basis of tolerances, deviations and fits

[3] ISO 17450-1:2011, Geometrical product specifications (GPS) – General concepts- Part 1: Model for geometrical specifications and verification

[4] ISO 14405-1:2010, Geometrical product specifications (GPS) – Dimensional tolerancing. Part 1: Linear sizes

[5] Nielsen H S *2013Proceedings of the Institution of Mechanical Engineers*, Part B. J. Eng. Manufact **227** 643-649.

[6] Wiesner H2015 *Journal SKF Evolution* **1** 27-30

[7] Glukhov V I 2014 Dynamics of Systems, Mechanisms and Machines (Dynamics) (IEEE Conf. Pub.) p 1-9

[8] Saunders P, Gludice S and Swart J 2014 *Int. J. Metrol. Qual. Eng.* **5** p 203-212

[9] Maropoulos P, Ceglarek D, 2010 *CIRP Annal. Manufast. Technol.* **59** p 740-759

[10] Vetturi D, Lancini M, Bodini I, and Pasinetti S 2013 *Int. J.Metrol. Qual. Eng.* **4** p 35-39

IOP Conf. Series: Journal of Physics: Conf. Series 998(2018) 012012 doi:10.1088/1742-6596/998/1/01:

The linear sizes tolerances and fits system modernization

V I Glukhov, VA Grinevich, and V V Shalay

Omsk State Technical University, 11 Mira av., Omsk, 644050, Russia

e-mail: mips@omgtu.ru

Abstract The study is carried out on the urgent topic for technical products quality providing in the tolerancing process of the component parts. The aim of the paper is to develop alternatives for improving the system linear sizes tolerances and dimensional fits in the international standard ISO 286-1. The tasks of the work are, firstly, to classify as linear sizes the elements additionally linear coordinating sizes that determine the detail elements location and, secondly, to justify the basic deviation of the tolerance interval for the element's linear size. The geometrical modeling method of real details elements, the analytical and experimental methods are used in the research. It is shown that the linear coordinates are the dimensional basis of the elements linear sizes. To standardize the accuracy of linear coordinating sizes in all accuracy classes, it is sufficient to select in the standardized tolerance system only one tolerance interval with symmetrical deviations: Js for internal dimensional elements (holes) and js for external elements (shafts). The main deviation of this coordinating tolerance is the average zero deviation, which coincides with the nominal value of the coordinating size. Other intervals of the tolerance system are remained for normalizing the accuracy of the elements linear sizes with a fundamental change in the basic deviation of all tolerance intervals is the maximum deviation corresponding to the limit of the element material: EI is the lower tolerance for the of the internal elements (holes) sizes and es is the upper tolerance deviation for the outer elements (shafts) sizes. It is the sizes of the material maximum that are involved in the of the dimensional elements mating of the shafts and holes and determine the fits type.

Keywords – Technical products, size elements, linear coordinating sizes, tolerance intervals, basic limit deviations.

1. Introduction

The most common dimensional elements, which comprise the details of technical products, are cylindrical and plane-parallel prismatic elements (elements of the "cylinder" type and the "two parallel planes" type, according to the international ISO standards for the geometrical product specification GPS). The sizes of these elements are expressed in units of length and are standardized as linear sizes [1]. The accuracy of linear sizes is normalized by the International System of limits and fits [2, 3]. High efficiency of the system is confirmed by almost a century of its application without significant changes for the technical products design. Based on the numerical values of linear sizes tolerances and F W Taylor principles, half a century ago, a system of limit gauges was created to control the accuracy of dimensional elements [4] and is still used today [5].

The location of the dimensional elements of the part in the GPS standards is determined by the linear distances [6], which do not belong to linear sizes and do not have a tolerance system. Therefore, linear distances are used as theoretically exact dimensions (TED), and their accuracy is normalized towards the part datums [7] using positional tolerances [8], which do not have standardized numerical values either and are set by the project designer. The control of design documentation shows that the projects quality is mainly reduced by inconsistencies in the accuracy standardization of the linear

IOP Conf. Series: Journal of Physics: Conf. Series **998** (2018) 012012 doi:10.1088/1742-6596/998/1/012012

distances between dimensional elements. A lot of researchers face this problem when calculating dimensional chains and estimating the cost of assemblies [9], analyzing size tolerances in technological processes [10] and assembling products [11], investigating the influence of the location tolerances on the maximum material size of dimensional elements in automotive and aircraft industry [12,13], developing new algorithms for setting dimensions and geometric tolerances [14], creating software packages for computer-aided design [15], developing the models for three-dimensional linear sizes tolerances and geometrical tolerances [16], including those which take into account the forces of gravity and friction in assembling units [17].

2. Statement of the problem

Modern automated design of technical products and technological processes is carried out in a rectangular coordinate system. However, in the ISO standards for geometrical products specification GPS, coordinate systems are not applied. The reason is simple: in the matrix GPS model [18] there are no coordinate systems for reading the parts geometrical specification.

When designing products, the position of each geometric element in the generalized coordinate system of the workpiece is specified using the linear coordinates x, y, z.

Consequently, the first task of the study is the transition from the sizes of linear distances to the mathematical linear coordinates of the parts elements and the development of a tolerances system for the normalization of coordinating sizes accuracy.

The second task is to refine the functional purpose of the concept of "basic deviation" of linear sizes tolerance intervals for dimensional elements in the system of tolerances and fits ISO 286-1.

3. Accuracy of the parts linear coordinating sizes

Each part has a generalized *OXYZ* coordinate system, which is identified by a set of main design datums that constrain the detail of six degrees of freedom - three linear and three angular. The coordinate planes of the rectangular *OXYZ* coordinate system have different informativeness: 3, 2, and 1. The coordinates axes of the rectangular *OXYZ* coordinate system also have different informativeness: 4, 2 and θ (zero) [19]. The coordinating dimensions of the part geometric elements should be specified with respect to the coordinate planes and axes of coordinates with the maximum informativeness. The more such coordinates the part has, the higher will be its quality in positioning accuracy of geometric elements, including size elements.

To solve the problem of normalizing the accuracy of linear coordinating sizes, we consider structural models of linear elemental sizes of two dimensional elements: prismatic plane-parallel with informativeness 3 (Fig.1, a) and cylindrical with informativeness 4 (Fig.1, b)

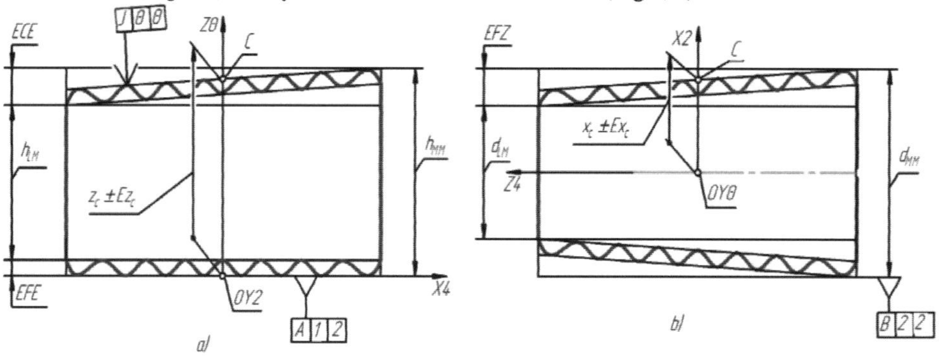

Figure 1. Structural models of linear element sizes of two dimensional elements:
a) plane-parallel prismatic; *b*) cylindrical.

Both dimensional elements form their own rectangular coordinate systems *OX4Y2Zθ* and *OX2YθZ4* thanks to their maximum informativeness of the datums: flat datum A3 with informativeness 3 for the prism element and the reference axis B4 with informativeness 4 for the cylindrical element. The analysis of the models shows that the dimensional basis of the elements linear sizes is the linear coordinates $z_c \pm Ez_c$ and $x_c \pm Ex_c$ from the datums to the center C of the opposite plane for the prism element and to the center C of the opposing straight line forming the surface of the cylindrical element while its rotating about the datums axis. Therefore, if the nominal sizes are equal, the tolerance of the linear coordinating size is a part of the linear elemental size tolerance. Taking into account that up to 60% of the linear size tolerance is taken by deviations in the shape and location of its real surfaces [20], the tolerance of the linear coordinating size should be one grade exact than the accuracy of the linear size and depend only on the length of the normalized coordinate. This will allow to normalize the accuracy of the linear coordinating sizes according to the standards ISO 286-1 and ISO 286-2 for the system of tolerances and fits of linear sizes, because both sizes are varieties of one linear physical quantity "length", have a common measurement unit and comparable tolerances.

Three linear coordinates in the part generalized coordinate system specify the position of one point of the origin of the auxiliary coordinate system for the attached part or one point of the executive geometric element center that performs the work functions of the part. Linear coordinating sizes of x, y, z points are the distances from the coordinates origin of the part generalized coordinate system to the projections of the positioning point on the linear scales of the X, Y, Z coordinate axes [19]. Every linear coordinate has only one value, is one-dimensional and has a normal distribution, symmetric with respect to its nominal size. The accuracy of coordinating sizes is higher the more informativeness the axes of coordinates has; i.e.4 and 2. Therefore, linear coordinating sizes measured along the coordinate axes with zero informativeness *(θ)* should be avoided in the part design, or their nominal values should be reduced to zero *(θ)*. The principle of the shortest dimensional chains requires the reduction of the values of all linear coordinating sizes in order to increase their accuracy.

The symmetry of the normal distribution law of linear coordinating sizes with respect to their nominal values requires, for the accuracy normalization, only one symmetric tolerance interval ***Js*** for holes and ***js*** for shafts that have the same limit deviations $\pm IT/2$ equal in absolute values to half the tolerance IT in each accuracy grade (Fig. 5). The basic deviation of the linear coordinating size is the zero deviation *(θ)*, which coincides with the nominal size value.

The zero nominal value of the linear coordinating size is its distinguishing feature from the linear size of the element. Therefore, the first interval of sizes in the system of tolerances ***Js*** and ***js*** should start from zero: *θ ... 3mm.*

The second distinguishing feature of the linear coordinating size is its grouping around the nominal size. Therefore, the most probable values of the coordinating size are located in the interval covering the nominal size. In this case, the value of the linear coordinating size may differ from the normal linear sizes.

Finally, linear coordinating sizes may have negative values if the direction of their reference from the origin is opposite to the positive direction of the coordinate axis.

4. Accuracy characteristics of linear element sizen

The accuracy in the linear coordinating dimensions, which belong to the structure of elements linear sizes (figure 1), determines the dimensional basis of the element's accuracy. The geometric accuracy of the size element is determined by the deviations of the shape and location of the element real surfaces. These deviations make the linear dimension of each element variable, limited by two functional boundary dimensions. The main functional element dimension - the maximum material size - is formed by the protrusions fit points of the real surface of the hole D_{MM} or the shaft d_{MM} (figure 2). participates in mating, determines the nature and forms an an envelope element that serves as a design datum.

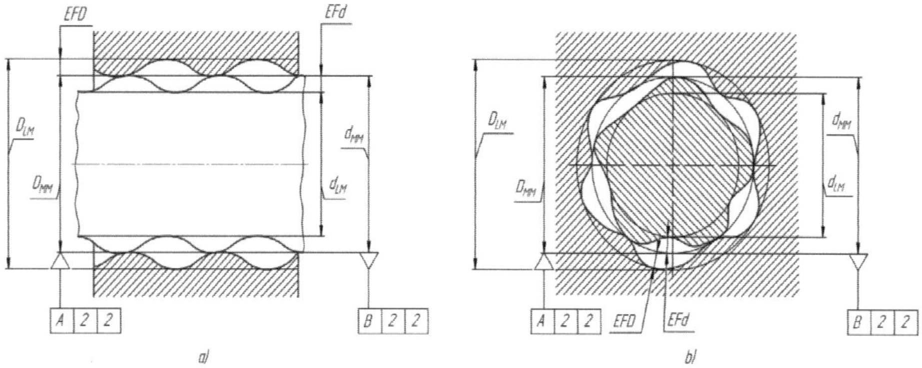

Figure 2. The influence of surface shape deviations on different functional purpose of linear sizes of material maximum and material minimum in the mating of real cylindrical elements
a) mating model in longitudinal section; *b)* the cross-sectional mating model.

The valleys of the real surface form the second functional dimension of the element - the minimum material size of hole D_{LM} or shaft d_{LM},, which limits the deviation of the shape and location of the element surfaces.

In accordance with the principle of two-dimensionality of linear sizes [19], the maximum material size and the minimum material size of the real dimensional element are the sizes of two regular (ideal) elements of the same type as the dimensional element, nominally located relative to the coordinate system in which the dimensional element fulfills its functional purpose, and tangentially covering the real surface of the dimensional element: the first - outside the element material - forms the maximum material size, the second - from the element material – forms the minimum material size of the dimensional element.

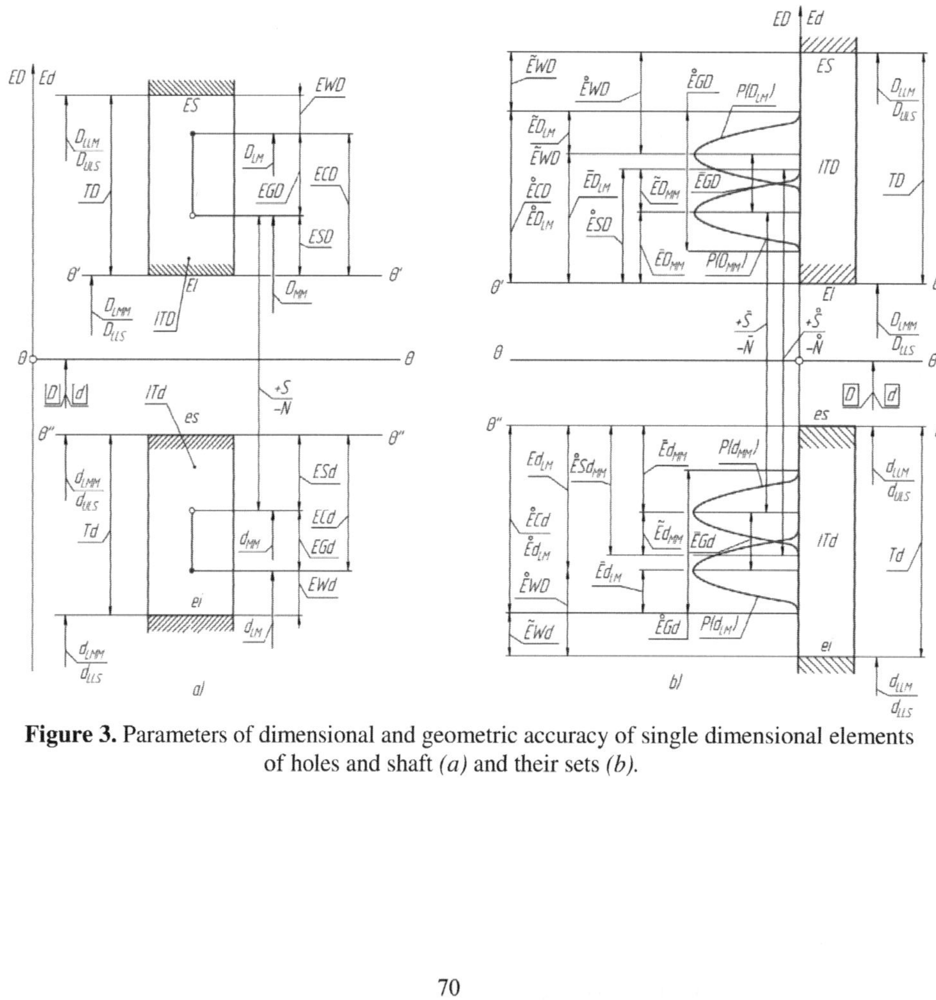

Figure 3. Parameters of dimensional and geometric accuracy of single dimensional elements of holes and shaft *(a)* and their sets *(b)*.

Thus, the linear size of the element has two boundary values that are counted in the part coordinate system and are constrainted by to two maximum and minimum material limits of the element according to the tolerance and fit system ISO 286-1 and ISO 286-2.

Two values of the linear size of a single real element make it possible to evaluate a number of accuracy characteristics that expand the functionality of every boundary size (figure 3*a*):

0′- 0′ and *0″– 0″* are the reference systems for the indices of dimensional elements accuracy forming the limits of the material maximum D_{LMM} for hole and d_{LMM} for shaft;

ESD and *ESd* are the indices of dimensional accuracy for hole and shaft;

$ESD=|D_{LMM} - D_{MM}|$, $ESd= d_{LMM} - d_{MM}$;

EGD and *EGd* are the indices of geometrical accuracy for hole and shaft:

$EGD =|D_{MM} - D_{LM}|$, $EGd= d_{MM} - d_{LM}$;

ECD and *ECd* are the complex indices of total, dimensional and geometrical accuracy for hole and shaft:

$ECD=|D_{LMM} - D_{LM}|$, $ECd= d_{LMM} - d_{MM}$;

EWD and *EWd* are the values of accuracy margins for material wear for hole and shaft dimensional elements:

$EWD =D_{LLM} - D_{LM}$, $EWd= |d_{LLM} - d_{LM}|$,

where D_{MM} and d_{MM} are the maximum material dimensions for hole and shaft;

D_{LM} and d_{LM} are the minimum material dimensions for hole and shaft;

D_{LLM} and d_{LLM} are the minimum material sizes for hole and shaft.

For a single dimensional element, all the accuracy indices are systematic quantities with known uncertainties in the measurements. For a set of dimensional elements of a batch of parts, the accuracy characteristics become probabilistic quantities, consisting of systematic and random components (figure3,b).

Since the element accuracy reference point is the maximum material limit, the basic deviation in the linear sizes system of tolerances and fits must be the deviation of the element maximum material limit: the lower deviation ***EI*** for all tolerance intervals of the holes ***A ... ZC*** and the upper deviation ***es*** for tolerance intervals ***a ... zc.*** This requirement in existing ISO 286 standards is maintained mating in the hole system and shaft system only for tolerance intervals from *A* to *H* holes and from *a* to *h* shafts. It should also be extended to the remaining tolerance intervals from *J* to *ZC* for holes and from *j* to *zc* for shafts. The main deviation, which is now the closest to the zero line, has no physical justification.

The symmetric tolerance interval *Js* refers to the linear coordinating dimensions and should be taken out in a separate matrix. Then the system of tolerances on linear sizes will indeed become a system without numerous corrections in it. The choice of fits for mating will be made only on the maximum material limits, i.e. on the basic deviations of all tolerances intervals for holes and shafts.

5. Experimental results

An experimental verification of the linear sizes accuracy of cylindrical elements was carried out at the production conditions on a circular grinding technological operation in a batch processing sequence N = 50 shafts (figure 4).

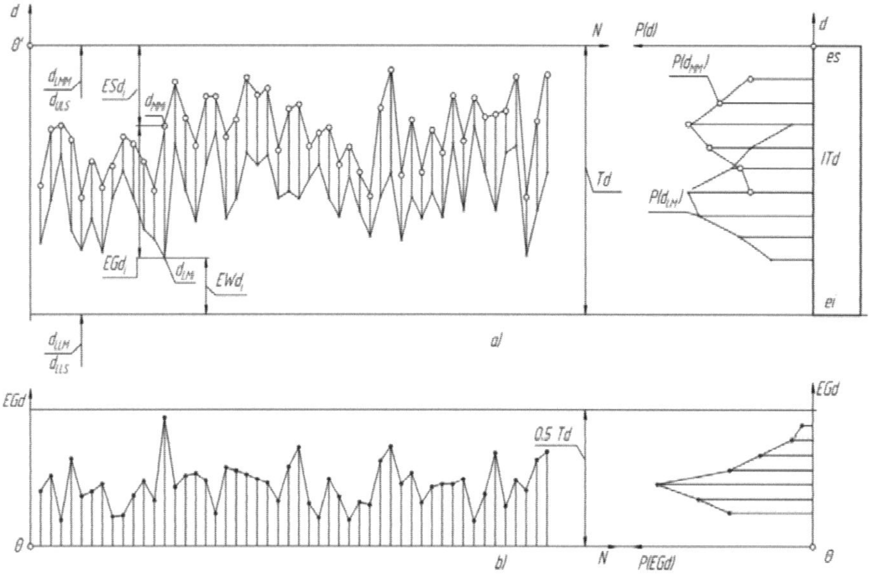

Figure 4. Dot diagrams and histograms of dimensional accuracy distribution of maximum material diameters, minimum material diameters and geometric accuracy of a cylindrical element for a batch of 50 shafts on a circular grinding operation.

The results of the experiments have fully confirmed the following theoretical conclusions of the research:

• the dimensional element of every part has two boundary linear sizes - the maximum material size d_{MM} (the maximum diameter) and the minimum material size d_{LM} (the minimum diameter);

• both the element boundary sizes have their own distribution laws within the tolerance interval; their grouping centers are displaced from each other by the average deviation of the cylindrical surface shape in terms of diameter;

• the geometrical accuracy of the *EGd* element is determined by the difference in the maximum and minimum material size and its dispersion area occupies 40% of the diameter tolerance interval;

• the guaranteed accuracy margin for the wear of the element material is about 20% of the diameter tolerance interval.

6. Discussion

As a result of the work, two suggestions to improve the International System of Tolerances and Fits ISO 286 [2, 3] were updated.

The first suggestion relates to the introduction of the concept of a linear coordinating size and its accuracy normalization with the help of the tolerance interval ***Js*** symmetric deviation for hole and ***js*** for shaft in order to indicate the location of the parts geometrical elements. To do this, it is recommended to exclude this deviations ***Js*** and ***js*** as a separate matrix (figure 5) for linear coordinating element sizes from the tolerances and fits general system of ISO 286.

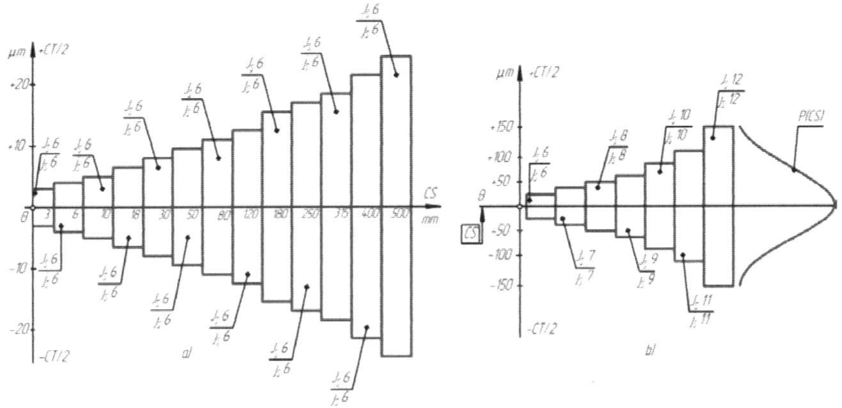

Figure 5. Symmetric tolerance intervals for linear coordinating sizes of external and internal geometrical elements of parts:
a) for the 6 accuracy tolerance grade of the coordination sizes range CS = θ ... 500mm;
b) for 6 ... 12 accuracy tolerance grades of the interval of coordinating sizes CS = 50 ... 80 mm.

Symmetrical deviations ***Js*** holes and ***js*** for shafts have the same numerical values, in their tolerance grades and equal to half the standard international tolerance ± *IT / 2*. Therefore, for the tolerances of the linear coordinating sizes of holes and shafts, a new general designation, for example ± *CT / 2* (Coordinate Tolerance), could be assigned.

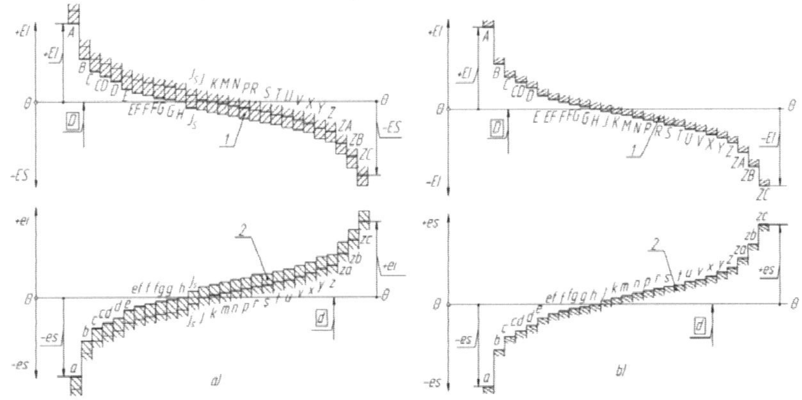

Figure 6. Models of two systems of basic deviations of elements linear sizes tolerance intervals of 1-hole material; 2 - shaft material:
a) system of nearest deviations from the zero line;
b) system of functional deviations of the maximum material limit;

The second proposal relates to a change in the content of the "basic deviation" concept as the deviation of the maximum material limit in place of the deviation nearest to the zero line of the nominal size (figure 6).

The proposed model (b) does not contain symmetric deviations of the tolerance intervals ***Js*** and ***js.***

7. Conclusion

The conclusion include the following findings:

1. In geometry, linear coordinates determine the position of a point in three-dimensional space. Likewise, linear coordinating dimensions are necessary for specifying the location of geometric elements in the space of the part material.

2. Linear coordinating dimensions are the dimensional basis of the geometric elements linear sizes, and their accuracy can be normalized by International system of linear sizes tolerances and fits ISO 286.

3. Linear coordinating dimensions have a normal symmetrical distribution with respect to the nominal size value. Therefore, for normalizing the accuracy of coordinating sizes, only one symmetric tolerance interval ***Js*** or ***js*** from the ISO 286 tolerance system is required. The nominal values of linear coordinating sizes may differ from the normal values of linear sizes.

4. The accuracy tolerance grade of the linear coordinating size should be one grade more precise than the grade of the positioned element; at the same time, the tolerance should correspond to the nominal length of the coordinating size. The nominal length of the linear coordinating size can be equal to zero, so the first size interval in the tolerance system should start from zero: *θ ... 3mm.*

5. In order to improve the international system of tolerances and fits of linear sizes ISO 286, it is recommended to consider the symmetric tolerance interval ***Js*** and ***js*** in a separate matrix as a functional tolerance for linear coordinating sizes.

6. For the same purpose, it is recommended the deviation of the material maximum limit *EI* for holes and *es* for shafts to consider the basic functional deviation for international system of linear sizes tolerances and fits *ISO286*, instead of the deviation that is nearest to the zero line as unreasonable.

7. The introduction of coordinating sizes will require a chain of changes in GPS standards, including a masterplan, coordinate systems for reading the dimensional and geometric specifications of parts, the informativeness of datums and datums systems, and the classification of dimensions according to a functional feature.

References

[1] ISO 14405-1:2010, Geometrical product specifications (GPS)– Dimensional tolerancing – Part 1: Linear sizes
[2] ISO 286-1:2010, Geometrical product specifications (GPS)– ISO code system for tolerances on linear sizes – Part 1: Basis of tolerances, deviations and fits
[3] ISO 286-2:2010, Geometrical product specifications (GPS) – ISO code system for tolerances on linear sizes – Part 2: Tables of standard tolerance classes and limit deviations for holes and shafts
[4] ISO/R 1938:1971, ISO system of limits and fits - Part 1: Inspection of plain workpieces
[5] ISO 1938-1:2015, Geometrical product specifications (GPS) – Dimensional measuring equipment - Part 1: Plain limit gauges of linear size
[6] ISO 14405-2:2011, Geometrical product specifications (GPS) – Dimensional tolerancing – Part 2: Sizes other than linear sizes
[7] ISO 5459: 2011, Geometrical product specifications (GPS) – Geometrical tolerancing – Datums and datum systems
[8] ISO 1101: 2012, Geometrical Product Specifications (GPS) – Geometrical tolerancing – Tolerancing of form, orientation. location and run out
[9] Heling B, Aschenbrenner A, Walter M S J and Wartzack S 2016 Procedia CIRP **43** p 262 – 267
[10] Muholzoev A and Masyagin V 2016 Procedia Engineering **150** p 959 – 962
[11] Franciosa P, Gerbino S and Patalano S 2013 Procedia CIRP **10** p 169 – 177
[12] Humienny Z, Berta M 2015 Procedia CIRP **27** p 264 – 269
[13] Leonard P, Pairel E, and G ordano M 2013 Procedia CIRP **10** p 30 – 36
[14] Fengxia Zhao, Kunpeng Zhang, Linna Zhang and Peng Zheng 2015 Procedia CIRP **27** pp 254 – 259
[15] Zhang Jin and Qiao Lihong 2013 Procedia CIRP **10** p 259 – 266
[16] Humienny Z and Berta M 2013 Procedia CIRP **10** p 312 – 316
[17] Corrado A, Polini W, Moronib G and Petrob S 2016 Procedia CIRP **43** p 130 – 135
[18] ISO/TR 14638:1995, Geometrical product specification (GPS) – Masterplan
[19] Glukhov V I 2014 Dynamics of Systems, Mechanisms and Machines (Dynamics) p 1-9
[20] Glukhov V I, Zlatkina O Y and Ivleva I A 2016 IOP Conf: Materials Science and Engineering, **124** p 1-8

IOP Conf. Series: Journal of Physics: Conf. Series 998(2018) 012013 doi:10.1088/1742-6596/998/1/0120

Automation is an Effective Way to Improve Quality of Verification (Calibration) of Measuring Instruments

M Golobokov[1] **and S Danilevich**[2,3]

[1] State Regional Center of Standardization, Metrology and Testing in the Novosibirsk Region, Federal Budgetary Institution (FBU), 2/1, Dzerzhinsky av., Novosibirsk, 630112, Russia
[2] Novosibirsk Branch of the Academy of Standardization, Metrology and Certification (Training), Federal State Autonomous Educational Institution of Supplementary Professional Education (FGAOU DPO), 36, Revolution str., Novosibirsk, 630004, Russia
[3] Novosibirsk State Technical University (NSTU), 20, Karl Marx av., Novosibirsk, 630073, Russia

E-mail: malachit_40@inbox.ru

Abstract. In order to assess calibration reliability and automate such assessment, procedures for data collection and simulation study of thermal imager calibration procedure have been elaborated. The existing calibration techniques do not always provide high reliability. A new method for analyzing the existing calibration techniques and developing new efficient ones has been suggested and tested. A type of software has been studied that allows generating instrument calibration reports automatically, monitoring their proper configuration, processing measurement results and assessing instrument validity. The use of such software allows reducing man-hours spent on finalization of calibration data 2 to 5 times and eliminating a whole set of typical operator errors.

1. Introduction

Instrument calibration is the key method of governmental control to ensure uniformity of measurements. The purpose of such calibration is to establish whether a specific instrument does or does not comply with a given set of parameters. In order to ensure calibration efficiency and quality, two tasks need to be resolved [1,2].

First, efficient calibration techniques for specific instruments need to be developed that ensure the required level of calibration reliability at a minimal cost. Such task normally comes down to selecting reference templates and fine-tuning the calibration procedure (namely, selecting the metrological characteristics to be monitored during calibration), provided calibration reliability is high. Second, calibration technicians must be highly qualified and productive.

2. Goal setting

Calibration method quality is defined by its reliability, i.e. the probability of false recognition of a defective instrument to be valid (second type calibration error) or false recognition of a valid instrument to be invalid (first type calibration error). However, calibration reliability indicators are unknown for most existing methods. There are methods that vary greatly across groups of instruments with similar properties. For example, calibration of Fluke 572 infrared thermometers is executed in 22 points within the testing range, whereas that of Fluke 572-2 with identical metrological characteristics is executed in 5 points. In a number of cases, the same method is applied to verify instruments with different technical characteristics. The goal is to study the possibility to improve the efficiency of instrument calibration by substantiating the number of calibration points, standardizing test conditions and accuracy requirements applied to reference samples. High quality and labor productivity of calibration technicians are provided by automating test result processing.

3. Achievement methods

The goal of assessing calibration reliability can be resolved by simulation modeling of the calibration procedure [1,2]. Calibration reliability criteria include: manufacturer risk R_p, customer risk R_z, likelihood of first P_1 and second P_2 type error, likelihood of any incorrect decision P_n. In order to assess reliability, one needs to build a stochastic instrument tolerance model, assess accuracy of measurements taken during calibration, carry out modeling, and analyze the results. Generation of a representative sample requires a few hundred same-type instruments to be available and significant costs to arrange field research. Therefore, using the available instrument calibration data is feasible.

Available EDM systems would not ensure the required level of calibration data processing automation. For this reason, proprietary instrument calibration data processing software has been developed. The software stores data on metrological characteristics of more than 3,500 types of instruments and allows generating reports in accordance with calibration method specifications, selecting reference samples and monitoring their on-schedule calibration. Also allows monitoring compliance of calibration specifications with regulatory standards, making sure reports are filled out completely, performing calculations, assessing compliance of calibration results with established requirements, generating output documents, preparing necessary reports on the tasks completed, storing and recording the results of all the tasks completed.

The software represents a set of 23 OpenOffice.Calc 3.2.1 files developed based on [3,4]. Each file contains data on a specific instrument type and has a universal structure comprising a basic package of work sheets.

'Catalogue' sheet stores details on instruments, standard values of influence quantities, the list of control points, and the calibration standards. 'Modules' sheet presents data on metrological characteristics of additional measuring channels. 'OSI' sheet provides details on the reference samples applied.

'Report' sheet is the key work sheet. A report structure is set up as the name of instrument and the type of calibration are selected, all necessary details in 'Catalogue' being translated to 'Report'. Necessary calculations are executed automatically as the report is being filled out. If any fields are left blank or filled out incorrectly or whenever the actual metrological characteristics do not comply with the established standards, a notification window shows up. A conclusion on instrument compliance with the established standards is prepared based on the calibration results.

'Rep_1', 'Rep_2', 'Not_1', 'Not_2', 'Cert_1' and 'Cert_2' sheets are used for printing both sides of calibration report, non-validity report and calibration certificate. The sheets are filled out automatically based on data entered in the 'Report' sheet. The other side of the calibration certificate shows calibration data and its uncertainty. As soon as processing is completed, all data is exported to 'Archive'.

'Archive' sheet is used to store data and generate in-house reports: the summary of used calibration marks and output documentation templates and the metrological task record log. 'Catalogue' sheet is used for recording available calibration methods.

A uniform data storage format allows analyzing data promptly in order to assess calibration data reliability. Selection of theoretical distributions, verification of their compliance with observations and sorting out of errors are executed using statistical methods.

Calibration method research findings

Integration of the software allowed automating instrument calibration data processing, namely the measurement uncertainty calculation process. It allowed ensuring uniformity of how output documentation is finalized, storing calibration results in a space effective manner, and generating in-house reports. Application of the software allowed assessing reliability of pyrometer and thermal imager calibration data and developing recommendations on improving how the calibration methods.

The derived stochastic pyrometer and thermal imager error models have shown a strong positive correlation for two groups of calibration points (below and above ambient temperature). The results correlate with the non-contact temperature measurement principle and provide justification for revising the common rule that calibration points are distributed evenly within the test range. It appears feasible to calibrate these instruments at temperatures near the bottom measurement limit, close to ambient temperature, at the point of range relevant to a modified method of tolerance standardization (100 °C) and at temperatures close to the upper measurement limit.

Table 1 shows calibration data reliability indicators obtained by calibration procedure simulation modeling [2]. These have been calculated for the calibration of Testo 875 – 882 thermal imagers in six and four points of the test range. A first-category blackbody model was used as a reference template. The number of realizations during calculations was 4.8 x 10^6.

Table 1. Testo thermal imager calibration reliability indicator values.

Reliability indicator	Six calibration points (minus 20, plus 30, 100, 200, 280 and 350°C)	Four calibration points (minus 20, plus 30, 100 and 350°C)
Rp [%]	6.3	6.1
Rz [%]	4.1	3.9
P_1 [%]	8.3	7.9
P_2 [%]	12.1	12.8
P_n [%]	9.3	9.0

Table 1 reveals that the likelihood of false decisions during calibration P_n drops insignificantly (by 0.3%) as the number of calibration points decreases from 6 to 4. The results obtained prove that the number of calibration points can be reduced with negligent effect on calibration data reliability.

The results of field research on how the size-of-source effect impacts calibration reliability is provided in [5]. Values of the size-of-source effect in absolute units and in percentages of the main permissible error limit Δ_{max} for Testo thermal imagers can be found in Table 2. Unless action is taken to eliminate the size-of-source effect, calibration reliability indicator values are as follows: R_p = 31.9%, R_z = 4.2%, P_1 = 42.2%, P_2 = 7.8%, P_n = 33.8%. Obviously, the resulting risk of false decisions during calibration is too high, hence unacceptable.

Table 2. Size-of-source effect for Testo thermal imagers.

T [°C]	-20	50	100	200	280	350
SSE [°C]	1.5	0.4	1.0	2.2	3.4	4.2
SSE [% Δ_{max}]	75	20	50	55	61	60

In order to eliminate the size-of-source effect impact, its distance to radiator and its aperture in calibration methods need to be standardized.

The impact of atmospheric transmission is dramatic for devices which spectral range includes strong water vapor absorption bands. During calibration of AKIP 9301 – 9303 infrared thermometers with spectral range of 5 to 14 μm, calibration data reliability indicators show the following values: R_p = 11.5%, R_z = 4.6%, P_1 = 13.8%; P_2 = 22%, P_n = 15.1%. Reduction of calibration outcome to nominal values of temperature and relative air humidity provides for the following values: R_p = 5.4%, R_z = 2.9%, P_1 = 6.5%, P_2 = 14.6%, P_n = 7.8%, which basically match the reliability indicators obtained using first-category reference samples. The suggested algorithm of bringing the measuring outcome to nominal values of influence quantities has been integrated in the software described.

The likelihood of any false decision for various types of infrared thermometers and thermal imagers varies from 8% to 24% [2]. Such values can hardly be accepted as satisfactory although they are formally not standardized in any of the calibration methods.

Expanded uncertainty of temperature reproduction by a first-category blackbody within the range from minus 50 to plus 200°C varies from 0.6°C to 0.9°C. Correlation of maximum permissible error limits and reference blackbody uncertainty within the specified range is from 1/1.3 to 1/2.2 for various types of verified instruments. This results in a high likelihood of control errors. Details on liquid thermostat blackbody models with temperature reproduction uncertainty within the specified range of 0.13°C to 0.34°C are provided in [6].

Inserts in the form of a blackbody model have been fabricated for fluid thermostats available at Novosibirsky CSM. The inside of the inserts is covered with paint of high emissivity (over 0.9). Effective emissivity has been calculated based on the diffuse interreflection theory [7, 8] using a trial version of SHEET 321 software [9]. The temperature difference of the coolant and the insert inside have been determined experimentally and considered in the temperature reproduction uncertainty budget.

Pre-trial outcome allows assessing the temperature reproduction uncertainty within the range from minus 50°C to 200°C with a value from 0.05°C to 0.36°C, which is 2.5 to 12 times lower than the first-category reference template uncertainty. In this case, Testo thermal imager calibration data reliability indicators will be R_p = 1.9%, R_z = 1.4%, P_1 = 2.5%, P_2 = 4.3%, P_n = 2.9% (calibration in six points). It is 2.8 to 3.3 times less than the first-category reference template reliability indicators.

4. Conclusion

Integration of the software has allowed collecting research materials and reducing man-hours required to finalize the calibration data 2 to 5 times. In total, over 50 thousand of calibration reports have been prepared from 2010 through 2017, with over 8 thousand man-hours saved. The software cost efficiency has been evaluated at RUB 4.8 mln.

The research completed has allowed assessing calibration data reliability based on the existing methods. The possibility of reducing the number of calibration points within the test range has been justified for a few types of infrared thermometers and thermal imagers with hardly any higher risk of false calibration decisions. The impact of the size-of-source effect and atmospheric transmission on reliability indicators has been studied.

Liquid thermostat blackbody reference templates have been designed and fabricated. Recommendations have been provided to improve calibration methods for some types of heating instruments.

References

[1] Danilevich S, Kolesnikov S and Palchun Y 2011 Use of Simulation Modelling for Checking Monitoring and Testing Procedures *Measurement Techniques* **7** p 846-850

[2] Golobokov M and Danilevich S 2017 Determining the Sighting Indicators and the Size-of-Source Effect in Calibration of Infrared thermometers and Thermal Imagers *Kontrol. Diagnostika* **6** p 42-49

[3] Golobokov M and Danilevich S 2016 Simulation Model Procedures for the Calibration of Measuring Instruments *Kompetentnost* **4** p 40-47

[4] Liebmann F and Kolat T Traceability and Quality Control in a Radiation Thermometry Laboratory (http://www.minervaipm.com/downloads/product/230/papers_articles /Traceability_and_Quality_Control_in_a_Radiation_Thermometry_Laboratory_52527_KB.pdf)

[5] Pitonyak A 2004 OpenOffice.org Macros Explained (translated by D. Chernov). Hentzenwerke Publishing(http://websprav.admin-smolensk.ru/freesoft/freesoft/ OpenOffice.org%20Macros%20Explained.Master.pdf)

[6] Pitonyak A Useful Macro Information for OpenOffice.org (http://www.pitonyak.org/AndrewMacro.odt)

[7] Prokhorov A. and Hanssen L 2004 Effective Emissivity of a Cylindrical Cavity with an Inclined Bottom: I. Isothermal Cavity *Metrologia* **41** p 421-431

[8] Prokhorov A and Hanssen L 2010 Effective Emissivity of a Cylindrical Cavity with an Inclined Bottom: II. Non-Isothermal Cavity *Metrologia* **47** p 33-46

[9] Information on http://www.virial.com/steep321.html

IOP Conf. Series: Journal of Physics: Conf. Series **998** (2018) 012014 doi:10.1088/1742-6596/998/1/012014

Software and Hardware System for Fast Processes Study When Preparing Foundation Beds of Oil and Gas Facilities

A V Gruzin[1], V V Gruzin[2] and V V Shalay[1]

[1]Omsk State Technical University, 11, Mira av., Omsk, 644050, Russia
[2]Kazakh Agro Technical University named after S. Seifullin, 62, Pobedy av., Astana-city, 010011, Kazakhstan

E-mail: polyot-m@mail.ru

Abstract. Analysis of existing technologies for preparing foundation beds of oil and gas buildings and structures has revealed the lack of reasoned recommendations on the selection of rational technical and technological parameters of compaction. To study the nature of the dynamics of fast processes during compaction of foundation beds of oil and gas facilities, a specialized software and hardware system was developed. The method of calculating the basic technical parameters of the equipment for recording fast processes is presented, as well as the algorithm for processing the experimental data. The performed preliminary studies confirmed the accuracy of the decisions made and the calculations performed.

1. Introduction

The expansion of existing and development of green fields located at considerable distance from the places of material and technical supply, in the complicated engineering, geological and climatic conditions, actualizes the development of existing and search for new technical and technological solutions in the design, construction and operation of oil and gas buildings and structures [1-4]. Analysis of the current normative documentation [5-6], regulating technical and technological approaches in the design and maintenance of buildings and structures foundation beds showed the lack of adequate technologies that meet modern challenges. Currently, this work is underway aimed at solving a complex problem, as the rational use of materials and mineral resources, the reduction of planned facilities commissioning terms while ensuring the stability and reliability of this new unit during its lifetime. An example of such modern approach to solving this complex problem is the development of technology for direct compaction. The technology idea consists in forming a field of vertical compressive stresses of required spatial configuration in the soil half-space [7-8], thus a reduction is expected to be achieved in installed power requirements for the rammer equipment used compared with the currently used. It is obvious that the study of the specific interaction dynamics of building machine working bodies and foundation beds will determine both as rational technical as rational technological parameters of the buildings and constructions bases of oil and gas in difficult geological and climatic conditions. The specific recommendations availability for the selection of rational parameters would reduce material and time costs for the foundation beds preparation for oil and gas facilities.

In the course of analytical studies earlier performed, dependencies that describe the dynamics of the rammer element working body in the foundation bed of a vertical steel tank for oil storage were obtained [9]. To assess the obtained ratios accuracy, it was decided on simulating the interaction of the rammer mechanism working body and disconnected particulate soil, typical for clay base of the tank for storing oil under laboratory conditions. As a rule, most speed and displacement sensors are the contact type and introduce an additional disturbing effect on the measurement results. Therefore, the decisive factor in choosing the method for recording the process under study was the condition to minimize the influence of the recording device. Modern photo- video- and computer technology development level allows remote, contactless nature of such measurements. The main advantage of recording with high-speed photo and video shooting is the ability to visualize fast processes, contactless and remotely data receiving in the most informative form, namely - photo and video. The absence of any influence on the process under study excludes the introduction of additional errors in the data obtained. It is important for an accurate and unambiguous interpretation of the objects under study dynamics. Furthermore, the complex use of photo-, video- and computing simplifies and accelerates both recording and subsequent processing of the results [10].

To make laboratory research of the construction machines working bodies' influence dynamics on the foundation bed, a specialized software and hardware system was developed, including a laboratory stand, recording equipment and a software product for video frames processing.

2. Laboratory bench

To determine the laboratory bench dimensions it is necessary to know the nature and magnitude of the vertical compressive stresses spatial distribution from the acting external dynamic load in the soil. In the course of laboratory research, models made of steel with mass up to 0.3 kg were used as ramming machine working bodies. The models planned impact rate on the soil half-space should not exceed 1.7 m/s. The lack of unambiguous and reliable data on the spatial distribution of vertical compressive stresses in the ground, resulting from impact on the ground, facilitated the adoption of two assumptions. Due to the small masses and impact velocities, it is proposed: to use the model of spatial distribution of vertical compressive stresses for the static load case, to limit the considered area of vertical compressive stresses distribution with area 10 % of the stress from the actual load. In accordance with accepted assumptions, the dimensions of the bench will be determined with the magnitude of the vertical compressive stresses spatial distribution area which is 10 % of the acting load:

$$\begin{cases} H_{st} \geq 4k_i H \\ B_{st} \geq 6k_i H \end{cases} \quad (1)$$

where H_{st} is the bench width, m; B_{st} is the bench height, m; H is the linear dimension of the model in a direction perpendicular to the direction of its movement, m; k_i is impact coefficient considering the dynamic nature of the soil loading (assumed to be one).

Thus, in accordance with (1) a box construction with width H_{st} = 0.7 m and height B_{st} = 0.75 m was selected as a bench for laboratory research (figure 1 A).

3. Recording equipment

The recording equipment for conducting laboratory studies includes: a high-speed video camera (1), a control computer (2) and an uninterruptible power supply (3) (figure 2 B).

Initially, to select the required technical characteristics of the recording equipment, it is necessary to determine the speed of the process under study recording and its duration, which is provided with technical capabilities of the fastest video camera itself. For existing machines and mechanisms, as well as technologies used for the foundation bed compaction of typical buildings and structures, according to various sources, the duration of the soil compacting process as a rule is in the range from 10…95 ms [11-13]. At the same time, the researchers do not give unambiguous recommendations on the necessary number of shots and time discreteness between them for qualitative and quantitative assessment of the model impact on the ground.

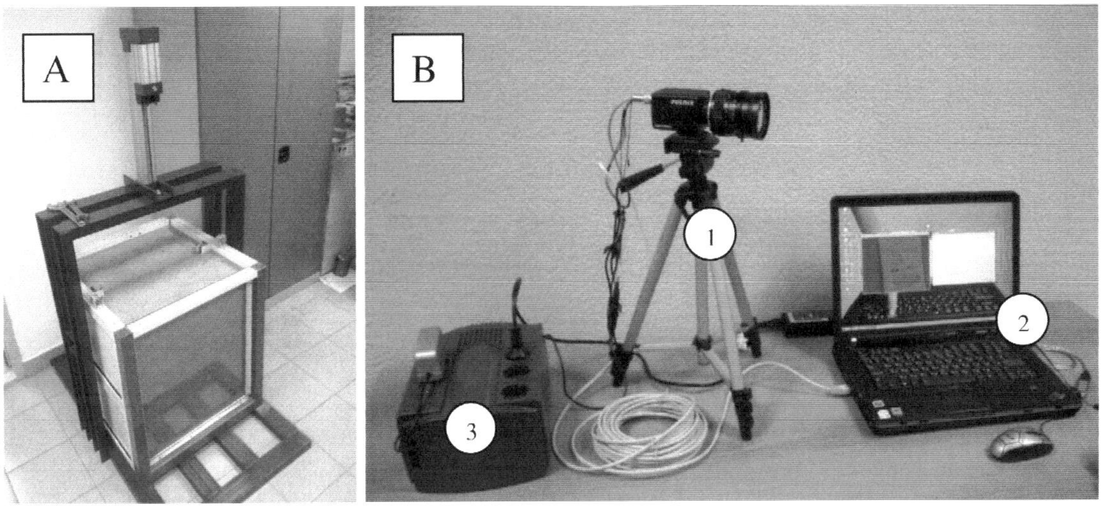

Figure 1. Laboratory bench (A); recording equipment (B).

It was decided to use high-speed video, providing process under study recording with discreteness between the frames taken an order less than its duration. That is, to study the dynamics of the model in the ground, it is proposed to perform high-speed video shooting with duration of at least ten frames. In this case, the temporal discreteness between individual frames should not exceed 0.001 s. Such temporal discreteness between frames corresponds to a shooting frequency 1000 frame/s. These requirements correspond to the high-speed video camera TMC-6740GE with maximum shooting frequency 1250 frame/s (figure 2 B), thereby fully meeting the terms of time resolution.

In turn, the recording equipment required spatial resolution is determined with characteristics of the speed camera lens and the relative spatial position of the recording equipment and the area under study. The angular resolution of the high-speed video camera photosensitive element cannot be considered, as it is much higher in most modern cameras than angular resolution of the used regular lenses. If a specific video lenses model has been selected than condition for obtaining high-quality images of the object allows specifying the requirements for the mutual spatial position of the recording equipment and the study area. This requirement observance shall be ensured with the corresponding depth of field (DOF) Δp. DOF Δp is the distance along the optical axis of the shooting lens, within its limits the objects located are displayed on the photosensitive element with a sufficient sharpness degree. Since the recording of the model impact on the ground is supposed to be carried out from a distance that excludes damage to the lens and video camera with ground particles and ensures the minimum vibrational impact on the study, the DOF value Δp is proposed to be calculated with formula [14]:

$$\Delta p = 2\frac{l_{obs}^2 \cdot d_{perm}}{D \cdot f'}. \tag{2}$$

where l_{obs} is observation distance from the camera lens front surface to the subject, m; d_{perm} is the value of the permissible confusion diameter circle, m; D is diameter of the entrance pupil, m; f' is back focal length of the lens, m. Obviously, formula (2) also allows to select the necessary objective lens for the given requirements to the spatial resolution of the researches and/or the relative spatial position of the recording equipment and the area under study.

In the experimental studies, sandy soil of medium size with an average particle size 0.38 mm was used. Recording of soil particles movement under the influence of an external force was intended to perform up to d_{perm} = 0.1 mm. The DOF value was assumed equal to the maximum longitudinal (along the axis of the lens) size of the rammer mechanism working body Δp = 45 mm. Using lens Navitar DO-5095, having focal length f' = 50 mm, f-number $K_г$ = 0.95…16 and the angular field in the space of objects 2ω (H × V) = 14°36' × 11°00', allowed to calculate the technical parameters of the shooting. The minimum f-number corresponds to the fully open aperture; the maximum value corresponds to the fully closed aperture. The maximum observation distance for the extreme f-number will be:

$$\begin{cases} l_{obs}^{\max}\left(K_г = 0.95\right) = \sqrt{\dfrac{\Delta p \cdot f'^2}{2 \cdot d_{perm} \cdot K_г}} = \sqrt{\dfrac{45 \cdot 50^2}{2 \cdot 0.1 \cdot 0.95}} \cong 770\,mm = 0.77\,m \\ l_{obs}^{\max}\left(K_г = 16\right) = \sqrt{\dfrac{\Delta p \cdot f'^2}{2 \cdot d_{perm} \cdot K_г}} = \sqrt{\dfrac{45 \cdot 50^2}{2 \cdot 0.1 \cdot 16}} \cong 190\,mm = 0.19\,m \end{cases}, \tag{3}$$

The linear dimensions of the observation field in horizontal $H_г$ and vertical planes $V_г$ for the extreme f-number $K_г$ will be:

- for $K_г$ = 0.95:

$$\begin{cases} Hr = 2l_{obs}\, tg\omega_H = 2 \cdot 0.77 \cdot tg7°18' \cong 0.197\,m \\ Vr = 2l_{obs}\, tg\omega_V = 2 \cdot 0.77 \cdot tg5°30' \cong 0.148\,m \end{cases} \tag{4}$$

- for $K_г$ = 16:

$$\begin{cases} Hr = 2l_{obs}\, tg\omega_H = 2 \cdot 0.19 \cdot tg7°18' \cong 0.049\,m \\ Vr = 2l_{obs}\, tg\omega_V = 2 \cdot 0.19 \cdot tg5°30' \cong 0.037\,m \end{cases} \tag{5}$$

The maximum size of the model in the plane under study is L = 65 mm. Angular field of the selected lens in the object space in a vertical plane equals to 2ω(V) = 11°00'. Thus, the maximum f-number $K_г$ of the lens in the planned laboratory tests will be:

$$K_г = \frac{2\Delta p\, tg^2\omega\, f'^2}{L^2\, d_{perm}} = \frac{2 \cdot 45 \cdot tg^2 5°30' \cdot 50^2}{65^2 \cdot 0.1} \cong 4.9 \tag{6}$$

4. Software product for processing video frames

Special software (SS) "Single-frame analyzer" was designed to handle high-speed frame video footage to obtain data on the objects dynamics (figure 2).

The following requirements have been presented to the developed SS:

- automatic loading of SS video frames with time reference depending on the shooting speed;
- determination of the image true scale;
- the possibility of scaling the video frame;
- ability to determine the current values of both the linear displacement and the speed of a given point of the object, and its vertical and horizontal components;
- possibility of changing the frame exposure rate during its processing.

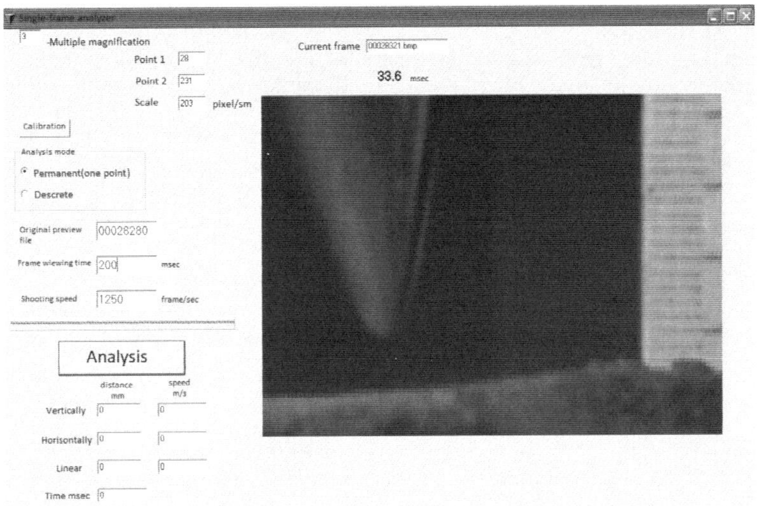

Figure 2. SS working window "Single-frame analyzer".

The process of video frames processing consists of two stages: the preparatory stage and the stage of the frame processing (figure 3).

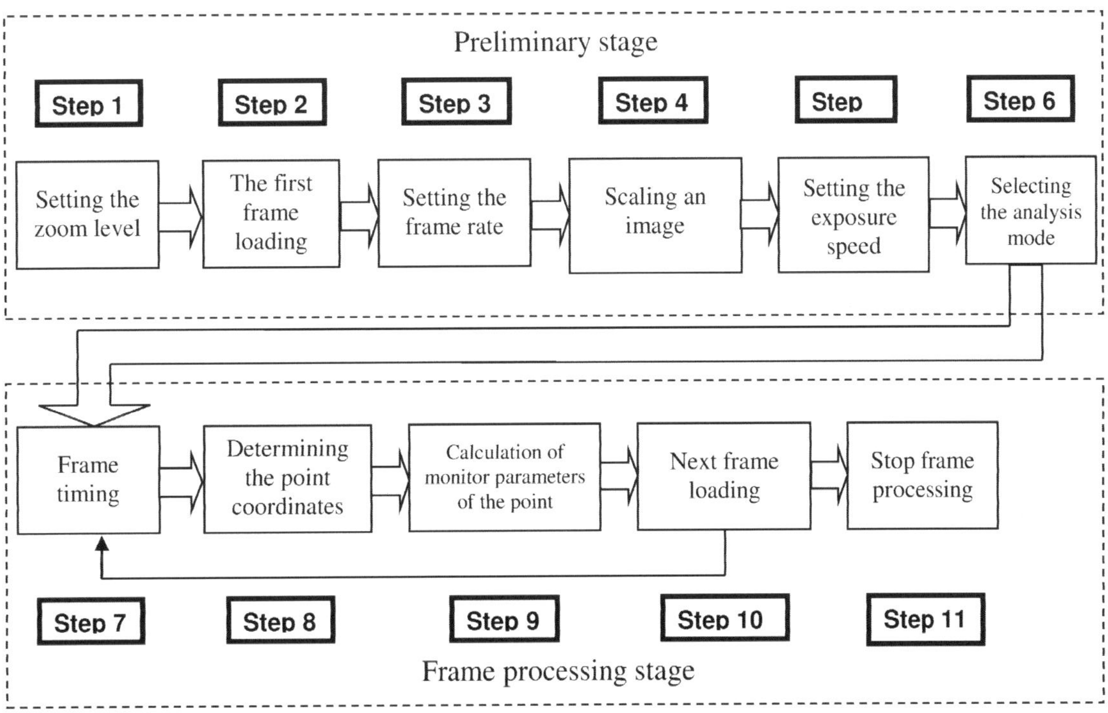

Figure 3. The sequence of processing video frames.

The preparatory stage is intended for setting parameters for viewing and processing high-speed video shooting. The first step in the preparatory stage is to set the viewing scale of the video frames. This step reduces the effect of the screen pixel resolution the frames are viewed on to the processing results. The next step is to load the first frame of high-speed video. For the SS operation all the shooting is to be sequentially numbered. Since the high-speed TMC-6740GE video camera used in the laboratory studies sequentially numbered each video frame, then at this step the frame with the smallest number in the series is loaded.

IOP Conf. Series: Journal of Physics: Conf. Series **998** (2018) 012014 doi:10.1088/1742-6596/998/1/012014

After downloading the frame number is displayed on the SS working panel in the "Source Viewer" and "Current Frame" windows. Simultaneously for this frame, the current video time is set to zero, and it is also displayed on the SS working panel above the frame itself (figure 2). In the case of the standard frame size, the SS automatically detects the video speed and displays its value in the "Rate of shooting" field. In addition, it is possible to enter manually the value of the shooting speed in this field. The value of the video speed allows calculating automatically the current time of each frame uploaded. For correct calculation of the moving and speed values of the analyzed object, in the fourth step of the preparatory stage, the scale of the displayed image is determined. To do this, two points must be marked in the field of the exposed frame; the distance between them is 1 cm vertically on the real object. The setting of the frames exposure rate is performed in the fifth stage. Entered in the field "Frame scan duration" number allows to set the time convenient for processing the automatic change of the captured video frames. At the final stage of the preparatory phase, the analysis mode is set: constant or discrete. Constant analysis mode allows monitoring sequentially the dynamics of one selected point, determining its movement and speed by consecutive clicks on each analyzed frame. Discrete mode allows monitoring the dynamics of different points on every two consecutive frames.

At the stage of processing frames, the dynamics of the objects under study are analyzed directly. Frame processing begins with the time reference of the loaded frame. The time reference of the frame is calculated by the frame number in the series. Clicks on the point chosen for analyzing the dynamics on each subsequent frame allow calculating the speed and movement of the point from frame to frame. At the next processing stage, the calculated data is displayed automatically in the SS lower left corner. To ease the analysis calculated numerical values are represented as orthogonal components (horizontal and vertical components of the displacement and speed), as linear values. At the end of the set exposure time of the current frame, the next frame is loaded automatically, and the processing of the loaded frame is resumed. After processing of the last video frame is completed, the first frame is automatically loaded, the time counter is reset and the initial repetition of the stage of the video frames selected series processing is possible.

5. Discussion of the results

In preliminary laboratory studies the designed hardware and software system was tested during the motion study under the influence of the gravitational force of the metal cone (figure 4 A) and water drops (figure 4 B).

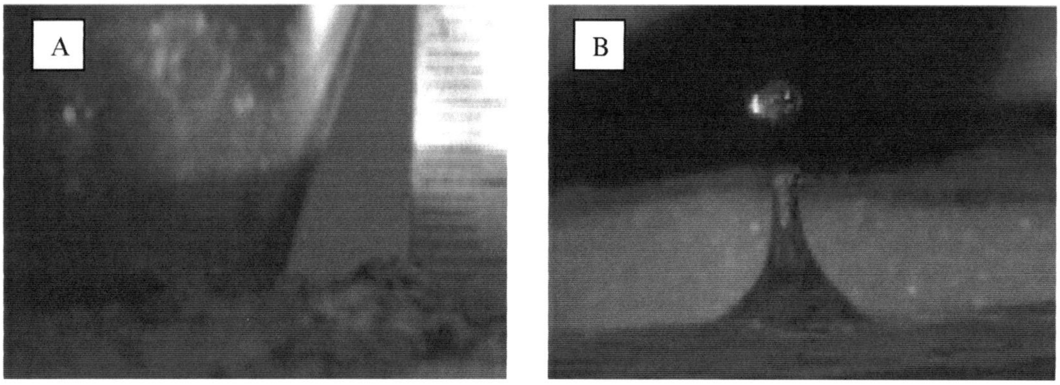

Figure 4. Video frames of the fall of the conical model (A) and the formation of a splash of water (B).

Video processing of free fall and subsequent impact of the conical model on sandy ground via SS "Single-frame analyzer" allowed to determine the nature of the model velocity change over time (figure 5). The preliminary data allowed identifying in the dynamics of the conical portion two characteristic parts: section I "Acceleration" and section II "Breaking". The interface portions is shown with dashed red line in Fig. 5. Section I "Acceleration" is characterized with a constant increase in the speed of the model. It is noteworthy that the duration exceeds the available models fall under the gravitational force. Sometime after the contact with ground the model continues the accelerated motion. The beginning of the model impact on the ground is marked with a blue line. Section II "Breaking" is characterized with a sharp decrease in the speed of the model with its maximum value to zero.

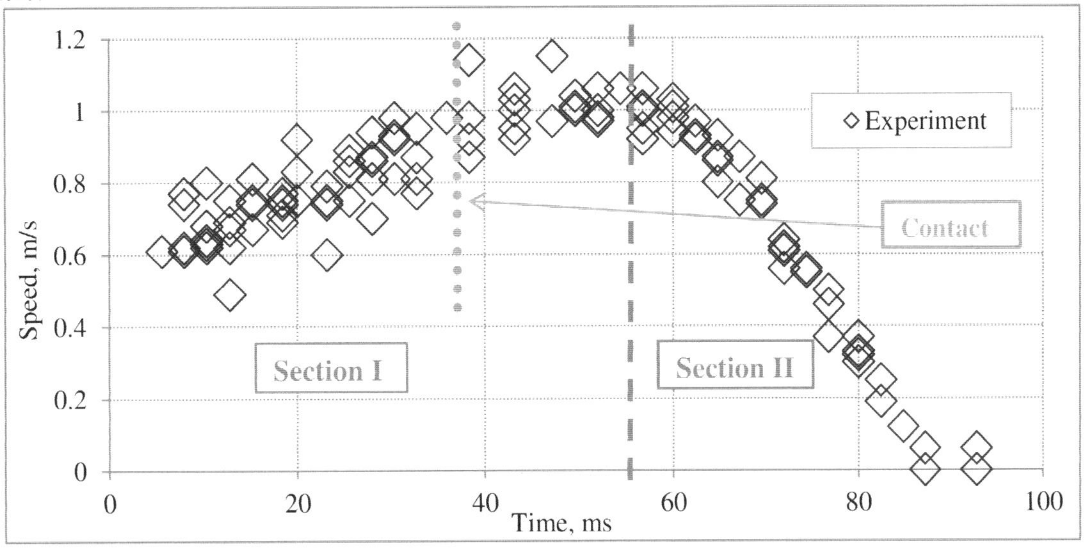

Figure 5. The model speed change during the free fall and the subsequent impact on the ground.

The preliminary data also made possible to give an evaluation characteristic of the integral precision of the hardware and software system and the proposed methods for studying fast processes in the preparation of foundation beds of oil and gas facilities.

Linear approximation of discrete data to changes in the model free fall speed until the hit into the sandy ground has allowed to establish a functional dependency of model change speed under gravitational force and compare it with the known one (figure 6).

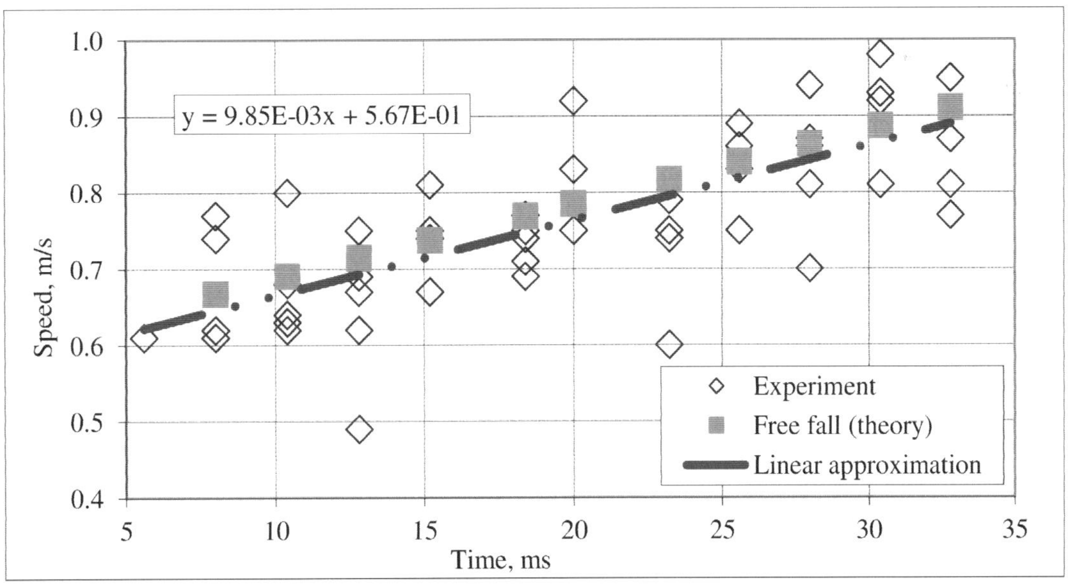

Figure 6. Change in velocity of the model free fall.

The calculated value of the acceleration relative error of free fall via the used techniques of data recording and processing was 0.4 %. In view of the stochastic nature of the physical and mechanical properties of sandy soil the result obtained is satisfactory.

6. Conclusions

In accordance with the proposed methodology there has been developed software and hardware system for studying rapid processes in the preparation of foundation beds of oil and gas facilities. The layout scheme and dimensional characteristics of the laboratory bench were defined. A high-speed video camera, controlled by PC was used as recording equipment. Specialized software product "Single-frame analyzer" was designed to handle video. Equipment used and program product enable fast processes recording with time increments up to 1 ms, with spatial discreteness up to 0.1 mm. The relative error in determining the dynamic characteristics of the processes under study using the methods for recording and processing data is 0.4 %.

References

[1] Gruzin A V, Tokarev V V, Shalai V V and Logunova Yu V 2015 *Proc.Eng.* **113** 158

[2] Gruzin A V, Ermakov V S, Belyaev N M and Katunin A V Pat. 2592929 RU, IPC E02D 27/35 (2006.01). Fundam container with improved thermal insulating properties No 2015122634/03; appl. 10/06/15; publ. 07.27.16, Bull. No 21

[3] Gruzin A V, Antropova L B, Rusanova A D, Katunin A V, Gildebrandt M I and Sirotin A D Pat. 2621799 RU, IPC E02D 1/02 (2006.01). A method for controlling the deformation properties disconnected particulate soil № 2016101450; appl. 31/05/16; publ. 7.6.17, Bull. № 16

[4] Gruzin A V, Tokarev V V, Shalai V V and Belyaev N M Pat. 2625498 RU, IPC B09B 3/00 (2006.01). A method of recycling of sludge as a soil vertical vessel base. No 2016112254; appl. 03/31/16; publ. 07.14.17, Bull. No 20

[5] SP 45.13330.2017. Earthworks, bases and foundations. The updated edition of SNiP 3.02.01-87. Intr . 08.28.2017. M .: Ministry of Construction of Russia, 176

[6] SP 22.13330.2011. Foundations of buildings and structures. The updated edition of SNiP 2.02.01-83. Intr . 2011-05-20. M.: Regional Development of Russia, 166

[7] Gruzin A V, Gruzin V V and Kucherenko M V 2012 *LAP LAMBERT Acad. Publ.* 136

[8] Abramenkov D E, Gruzin A V and Gruzin V V 2012 *Palm. Acad. Publ.* 336

[9] Gruzin A V, Gruzin V V and Shalay V V 2016 *Proc. Eng.* **152** 182

[10] Gruzin A V, Gruzin V V, Kucherenko M V and Katunin A V Pat. 2518018 RU, MPK G01P 3/68 (2006.01)A method and a device for contactless speed and displacement measurement of the object № 2012136314/28; zajavl. 22.08.12; opubl. 27.02.14, Bul. № 6

[11] Kharkhuta N Ya 1973 *Machines for soil compaction. Theory, calculation and design* (Moscow: Mechanical engineering) p 176

[12] Fedulov A I, Ivanov R A and Puchkov V V 1983 *Shock soil compaction* (Novosibirsk: IGD AN SSSR) p 118

[13] Tarasov V N, Boyarkina I V, Kovalenko M V, Kuznetsov S V and Schlegel I F 2006 *The theory of the impact in the construction and mechanical engineering* (Moscow: Ed. ACB) p 336

[14] Gruzin A V, Gruzin V V and Shalay V V 2017 *AIP Conf. Proc.* Vol. 1876. DOI: 10.1063/1.4998865

Asessment of adequacy of the monitoring method in the activity of a verification laboratory

R N Ivanov, V A Grinevich, A A Popov, V V Shalay and L D Malaja

Omsk State Technical University, 11, Mira av., Omsk, 644050, Russia

E-mail: irnsoft@mail.ru

Abstract. Questions of assessing adequacy of a risk monitoring technique for a verification laboratory operation concerning the conformity to the accreditation criteria, and aimed at decision-making on advisability of a verification laboratory activities in the declared area of accreditation are considered.
Key words - risk assessment, laboratory, accreditation criteria.

1. Introduction
In modern conditions of market relations, an urgent task is to assess and identify risks in the activities of laboratories. Risks sssessment and identification is one of the main requirements in the quality management system in accordance with Russian state Standard GOST R ISO 9001-2015 [1]. The need for identifying and assessing risks in the verification laboratory should be met with a view to improving the quality of the laboratory activities.

The main risks in the verification laboratory are: confirmation of the competence of the accredited party, the risk of obtaining unreliable results, the risk of "not being" certified.

At present, a lot of methods for assessing and identifying risks are developed [2]. One of the most promising of them is the method of constructing and analyzing the risks probability tree. To assess the risks of not satisfying the criteria, in particular accreditation criteria in the verification laboratory, expert methods can be used if it is impossible to find the frequency of not satisfying the accreditation criteria.

2. Statement of the problem
For the successful operation of the calibration laboratory, it is necessary to monitor the risks of not satisfying the accreditation criteria in the declared area of the verification laboratory accreditation. One of the main tasks is to check the adequacy of the methodology for monitoring risks in the laboratory. To verify the adequacy of the methodology for monitoring risks in the verification laboratory, it is necessary to check the adequacy of the received risk assessments for non-compliance with the accreditation criteria.

3. Theory
Order No. 326 of the Ministry of Economic Development of Russia on May 30, 2013 regulates the requirements for the applicant and accredited party when carrying out activities in a certain area of accreditation.

The accreditation criteria for legal bodies or entrepreneurs performing work and (or) providing services for ensuring the uniformity of assessment measurements include [3]:
1) general criteria for accreditation, which are a set of requirements that all applicants and accredited parties must satisfy [3];

2) additional criteria for accreditation that provide for special requirements of the quality management system in certain areas of work performance and (or) provision of services to ensure the uniformity of measurements [3].

Assessment and identification of risk consists in identifying adverse events and assessing their likelihood of occurrence, as well as assessing possible consequences. One way to identify and evaluate risks is the event tree. On the basis of this technique, a method for monitoring the risk for the opetation of a verification laboratory was developed [5].

In accordance with Order No. 326 of the Ministry of Economic Development of the Russian Federation (May 30, 2014), the following accreditation criteria are established, which can be grouped into the following blocks:

1. An integral generalized criterion characterizing the overall risk in the refusal or discontinuation of a verification laboratory accreditation.

2. General criteria required for accrediting a varification laboratory are those corresponding to paragraphs 42-49, 55 [3].

3. Particular criteria with a digital index are No. 44.1-44.2, 49.1-49.21, 55.1-55.8 [3] /

4. Particular criteria with an alphabetic index [3].

To assess the adequacy of the risk monitoring methodology for a verification laboratory [4], it is necessary to assess the likelihood of the risk of failure to fulfill the accreditation criteria in several ways and compare the results obtained.

The assessment of the probability of the risk of non-fulfillment of the accreditation criteria was carried out by expert methods by three groups of experts, using the following ways:
1. The method of direct determination of risk probability, based on the method of direct ranking (MDR).

2. The method of paired comparisons [5] (MPC).

3. The method of analytical hierarchy [6] (MAH).

4. Experimental results

In accordance with the selected methods for assessing the likelihood of non-fulfillment the accreditation criteria risks, an expert survey was conducted on the basis of which the probability trees were compiled in accordance with the results obtained using the direct risk probability method (figure 1), the paired comparisons method (figure 2) , and the analytical hierarchy method (figure 3).

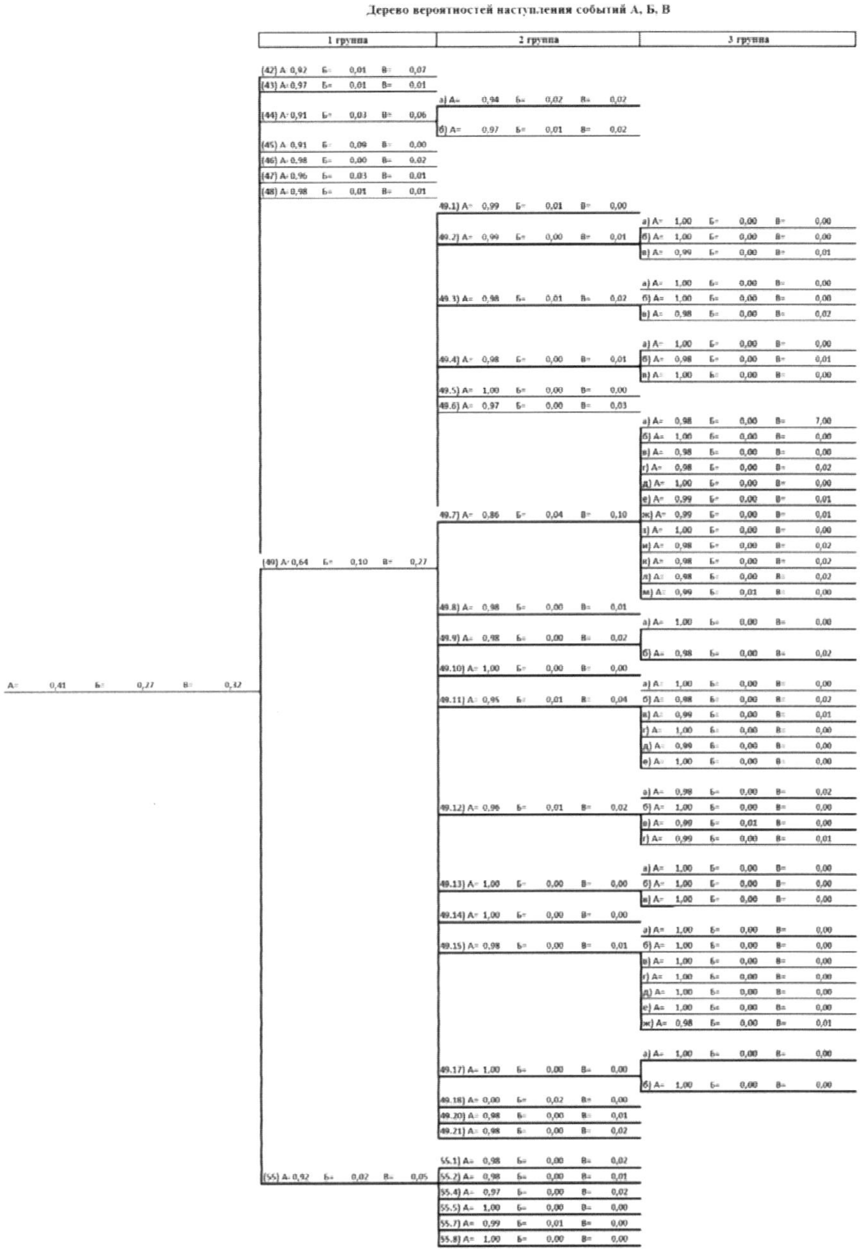

Figure 1. Probability ree of events occurrence by the method of direct ranking of risk probability.

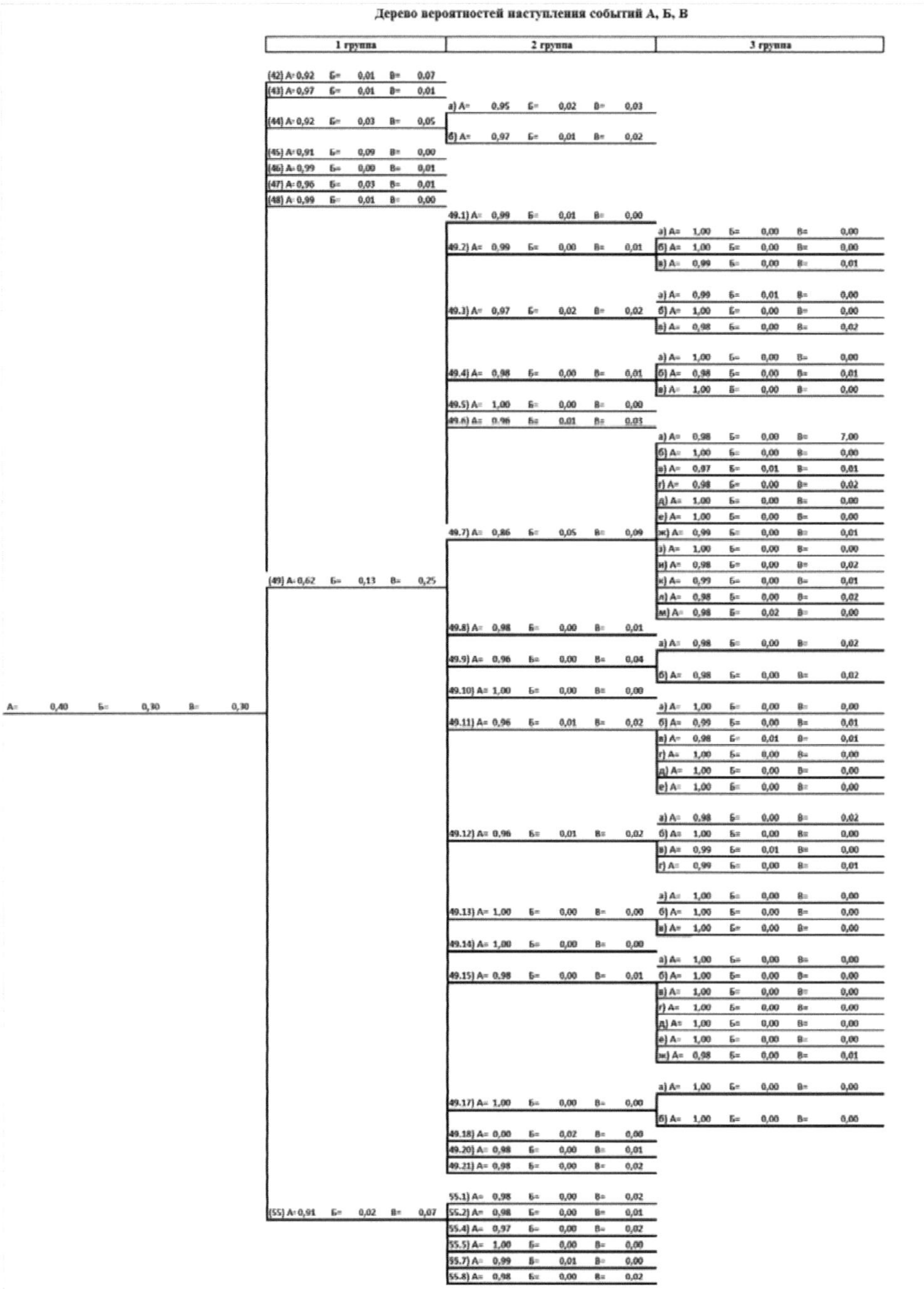

Figure 2. Probability ree of events occurrence by the method of of paired comparison.

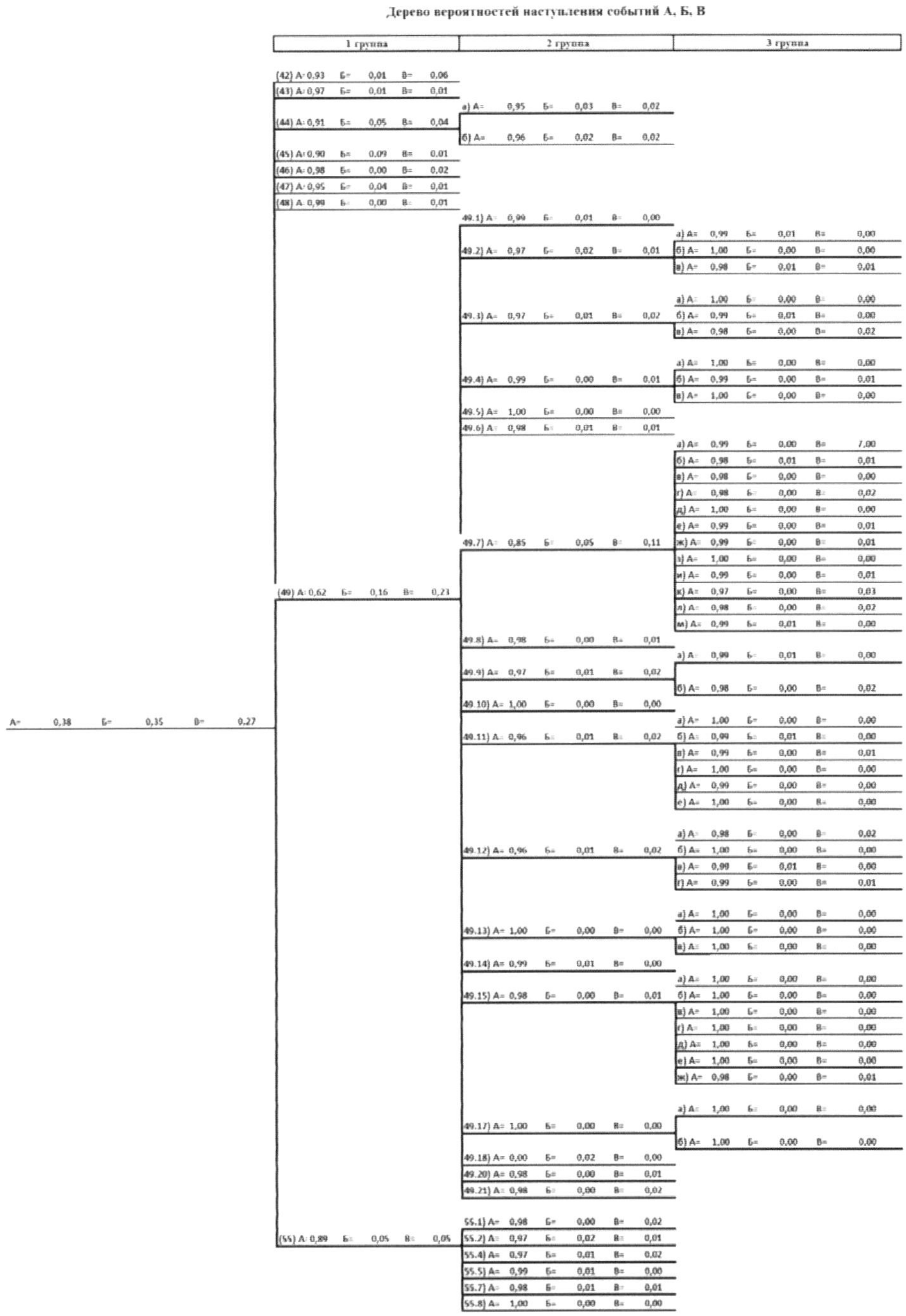

Figure 3. Probability ree of events occurrence by the method of analytical hierarchy.

In accordance with the presented results of the expert survey and their subsequent processing, the following estimates of the likelihood of adverse consequences in the form of refusal of accreditation in the declared area of a verification laboratory accreditation are presented in tables 1-3.

Table 1. Assessment of risk probability by method of direct definition

	Events		
Criteria groups	**A**	**B**	**C**
42	0.92	0.01	0.07
43	0.97	0.01	0.01
44	0.91	0.03	0.06
45	0.91	0.09	0.00
46	0.98	0.00	0.02
47	0.96	0.03	0.01
48	0.98	0.01	0.01
49	0.64	0.10	0.27
55	0.92	0.02	0.05
Total	**0.41**	**0.27**	**0.32**

Table 2. Assessment of risk probability by the method of pair comparison

	Events		
Criteria groups	**A**	**B**	**C**
42	0.92	0.01	0.07
43	0.97	0.01	0.01
44	0.92	0.03	0.05
45	0.91	0.09	0.00
46	0.99	0.00	0.01
47	0.96	0.03	0.01
48	0.99	0.01	0.00
49	0.62	0.13	0.25
55	0.91	0.02	0.07
Total	**0.40**	**0.30**	**0.30**

Table 3. Assessment of risk probability by the analytic hierarchy method

	Events		
Criteria groups	**A**	**B**	**C**
42	0.93	0.01	0.06
43	0.97	0.01	0.01
44	0.91	0.05	0.04
45	0.90	0.09	0.01
46	0.98	0.00	0.02
47	0.95	0.04	0.01
48	0.99	0.00	0.01
49	0.62	0.16	0.23
55	0.89	0.05	0.05
Total	**0.38**	**0.35**	**0.27**

The assessment of the obtained risk probability results adequacy is carried out on the basis of testing the hypothesis on the dispertion homogeneity of events probabilities obtained by different methods using the Cochran's Q Test:

IOP Conf. Series: Journal of Physics: Conf. Series **998** (2018) 012015 doi:10.1088/1742-6596/998/1/012015

$$G_p = \frac{\max\limits_i(S\{p\})}{\sum\limits_i S\{p\}}, \tag{1}$$

where $S\{p\}$ is the dispertion of the risks probability assessment of not being accredited by the criterion p.

If the calculated value of the Cochran's Q Test is less than the one given in a table, then the hypothesis of dispertion homogeneity is accepted and it can be considered that the probabilities of the risks of not being accredited are adequate.
The results the dispersions $S\{p\}$ calculation are given in table. 4.

Table 4. Disperse evaluation of events probabilities

Criteria groups	**Methods**			$S\{p\}$
	MDR	**MPC**	**MAH**	
Event			**A**	
42	0.92	0.92	0.93	0.000033
43	0.97	0.97	0.97	0
44	0.91	0.92	0.91	0.000033
45	0.91	0.91	0.90	0.000033
46	0.98	0.99	0.98	0.000033
47	0.96	0.96	0.95	0.000033
48	0.98	0.99	0.99	0.000033
49	0.64	0.62	0.62	0.000133
55	0.92	0.91	0.89	0.000233
Event			**B**	
42	0.01	0.01	0.01	0
43	0.01	0.01	0.01	0
44	0.03	0.03	0.05	0.000133
45	0.09	0.09	0.09	0
46	0.00	0.00	0.00	0
47	0.03	0.03	0.04	0.000033
48	0.01	0.01	0.00	0.000033
49	0.10	0.13	0.16	0.0009
55	0.02	0.02	0.05	0.0003
Event			**C**	
42	0.07	0.07	0.06	0.000033
43	0.01	0.01	0.01	0
44	0.06	0.05	0.04	0.0001
45	0.00	0.00	0.01	0.000033
46	0.02	0.01	0.02	0.000033
47	0.01	0.01	0.01	0
48	0.01	0.00	0.01	0.000033
49	0.27	0.25	0.23	0.0004
55	0.05	0.07	0.05	0.000133

The hypothesis of the dispertion homogeneity in assessment of risk probabilities is determined by the formula 1 and is equal to $G_p = 0.32$, comparing with the value given in a teble $G_p = 0.51$, it can be concluded that the dispertion is homogeneous, and therefore the adequacy of the obtained risk probability estimates does not correspond to the verification laboratory criteria for accreditation.

5. Discussion of the results

When identifying and assessing the risks of non-fulfilling verification laboratory accreditation criteria in the declared area of accreditation, one can use the following methodology:

1. Identification of risks based on the requirements for verification laboratory accreditation.
2. Assessment of risk probability by expert methods, in particular MDR, MPC, MAH
3. Checking the adequacy of the resulting risk assessment.
4. Determination of the expediency of the verification laboratory accreditation [4].

6. Conclusion

Thus, the suggested methodology for assessing the adequacy of the methodology for monitoring risks in the verification laboratory allows to determine the adequacy of assessing the probability of risk to not filfill the accreditation criteria in the declared area of a verification laboratory accreditation.

References

[1] Russian State Standard GOST R ISO 9001-2015 *Quality management systems. Requirements* 2015-11-01 (Moscow: Standartinform)

[2] Russian State Standard GOST R ISO/MJeK 31010 2011. Risk management. Methods of risk assessment. 2011-12-01 (Moscow: Standartinform)

[3] Order RF No 326 of Ministry of Economic Development of the Russian Federation. publ. 30.05.2014

[4] Grinevich V A, Chumachenko I M, Popov A A, Ivanov R N 2016 *Actual problems of electronic instrument engineering* 13th international scientifictechnical conference **1** p 175-177

[5] Eureka W E, Ryan N E (ed.) 1995 *Quality up, Cost down. A Manager's Guide to Taguchi Methods and QFD* (Dearborn, Michigan: ASI Press)

[6] Saati T Decision making 1993 (Moscow: Radio i svjaz')

Approaches to decrease the level of parasitic noise over vibroacoustic channel in terms of configuring information security tools

A V Ivanov[1], I L Reva[1] and A A Babin[1]

[1] Novosibirsk State Technical University, 20, K Marksa av., Novosibirsk, 630073, Russia

E-mail: andrej.ivanov@corp.nstu.ru

Abstract. The article deals with influence of various ways to place vibration transmitters on efficiency of rooms safety for negotiations. Standing for remote vibration listening of window glass, electro-optical channel, the most typical technical channel of information leakage, was investigated. The modern system "Sonata-AB" of 4B model is used as an active protection tool. Factors influencing on security tools configuration efficiency have been determined. The results allow utilizer to reduce masking interference level as well as parasitic noise with keeping properties of room safety.

1. Introduction

With a room being prepared for private negotiations, certain levels of speech intelligibility outside the room must be provided. With insufficient of sound/vibration isolation of enclosing structures, masking interference (i.e. applying active protection means – noise generators) is the most commonly used method. Applying active protection means is granted to be the simplest way to solve the problem, consisting of purchase and configuring costs only. However the disadvantage is that active protection means create «parasitic noise» during their operation.

Electro-optical channel is considered to be the most often protected channel of information leakage (remote vibration listening of window glass). Being created on the glass surface by vibration transmitters, vibration noise reemits inside the room in the form of acoustic noise which is regarded to be parasitic. Low-quality configuring of active protection means can result in essential level of parasitic noise, preventing from comfortable negotiations and being incompatible with Russian Sanitary Norms [1].

2. Research Objective

Firstly we will define factors influencing on active protection tool operation efficiency in terms of vibration transmitters SV-4B, the part of «Sonata-AV», model 4B. Interference spectrum and parasitic noise level are affected by:

- interference spectrum configuration (active protection tool is equipped with equaliser);
- ways to fit transmitters on the surface;
- the number of transmitters being used.

3. Theoretical Treatment

Noise level at the workplace is subject to Russian Sanitary Norms SN 2.2.4/2.1.8.562-96 «Industrial vibration, vibration of residential and public buildings» [1].

The Sanitary Norms establish the classification of noises, allowed noise levels and rate parameters. The Norms are valid for residential and public buildings, as well as for the area of residential blocks. The Norms are mandatory to follow.

The noise workplace characteristics is the equivalent (in terms of energy) sound level in dBA. [1] provides with allowed sound levels as well as with allowed sound pressure levels in octave bands, equivalent and maximal sound levels of noise penetrating in rooms of residential and public buildings and noise at the area of residential blocks (table 1).

Table 1. Allowed levels of sound pressure, sounds levels, and equivalent sound levels for most typical types of labour activity and workplaces.

Sounds pressure level, [dB], in octave bands with the centre frequencies, [Hz]									**Sound levels and equivalent sound levels, [dBA]**
31.5	63	125	250	500	1000	2000	4000	8000	
86	71	61	54	49	45	42	40	38	50

The levels are given for the following types of labour activity and workplaces:

- creative work, management work with increased demands, research activity, design and engineering, software development, instruction and training, medical practice;
- workplaces located in management offices, planning and design office, designers offices, programming stuff offices, computer engines room, in laboratories for theoretical work and data processing, reception of the patients in medical clinics.

Measurements of noise levels in areas of residential blocks, in residential and public buildings are conducted according to Russian methodological guidelines MUK 4.3.2194—07 «Noise monitoring within the residential development area, in residential and public buildings and premises» [2].

4. Experiments and Results

The noise level meter ZET 110, microphone BC501, and accelerometer BC111 are used for parasitic noise measuring. The complex is absolutely appropriate for noise measurements at residential and public buildings according to the Sanitary Norms [1] requirements.

4.1. Influence of interference spectrum

White and pink noise are used as the interference in the experiment. The vibration transmitter and the test point are placed as shown in Fig.1.

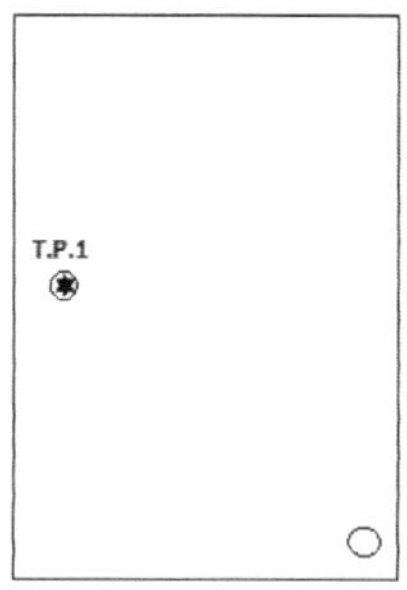

Figure 1. Placement of the vibration transmitter and the test point.

Values of interference levels in octave bands, word intelligibility and parasitic noise levels have been measured. The results are presented in tables 2 and 3.

The word intelligibility of 20% level has been achieved. As it is known, conversation subject is difficult to identify if the word intelligibility level is lower than 20-30% [3].

Background noise level (with turned off active protection means) in the experiment room was 38.2 dBA.

Table 2. Measured results for white noise levels in the test point.

f, [Hz]	125	250	500	1000	2000	4000	8000
Lnoise, [dB]	68.4	75.5	80.1	84.4	87.7	89.3	72.6

Word Intelligibility W=0.22. Parasitic noise level Leq =53.4 dBA.

Table 3. Measured results for pink noise levels in the test point.

f, [Hz]	125	250	500	1000	2000	4000	8000
Lnoise, [dB]	68.6	86.2	86.9	86.6	86.6	85.4	68.7

Word Intelligibility W= 0.21. Parasitic noise level Leq=51.3 dBA.

Parasitic noise level is found to be 2.1 dB lower with using pink noise in comparison with using white noise. Based on the experiment results, we can make a conclusion that applying of different interference types makes decrease of parasitic noise level by several dB possible. More essential results can be obtained by using formant-like interference, its effectiveness is confirmed by [4].

4.2 Influence of ways to fit the vibration transmitters on the surface

Two most common ways to fit the vibration transmitters on the glass surface are considered:
- with use of glue;
- with use of special-purpose fixing device «window frame» (Fig.2).

Obviously, the way to fit the transmitters influences on emitted parasitic noise level as well as on final interference spectrum in test point along with its level. We have corrected the interference level to obtain the same word intelligibility level as achieved for the first way of fitting, since the aim is to make word intelligibility value meet the Norms.

Figure 2. Fixing device «window frame».

Measured results are presented in tables 4 and 5.

Table 4. Measured results for noise levels in the test point using glue fixation.

f, [Hz]	250	500	1000	2000	4000	8000
Lnoise, [dB]	37.9	36.0	41.9	44.3	53.1	61.0

Word intelligibility W=0.20, parasitic noise level Leq = 51.5 dBA.

Table 5. Measured results of noise levels in the test point using the fixing device fixation.

f, [Hz]	250	500	1000	2000	4000	8000
Lnoise, [dB]	41.5	40.0	47.8	47.7	57.7	57.1

Word intelligibility W=0.21, parasitic noise level Leq = 56.2 dBA

As a result, change of ways to fit the vibration transmitter has allowed us to decrease parasitic noise level almost by 5 dB, and the same word intelligibility level has been achieved.

4.3. Influence of the number of the transmitters being used

Two experiments with one and two transmitters being used are conducted. Placement of vibration transmitters and the test point is shown in Fig.3. Measured results are presented in tables 6 and 7.

Obviously, in case with two transmitters their interference levels should be decreased, but the word intelligibility value should be the same.

Figure 3. Placement of vibration transmitters and the test point

Table 6. Measured results of noise levels in the test point with use of one transmitter.

f, [Hz]	250	500	1000	2000	4000	8000
Lnoise, [dB]	37.9	36.0	41.9	44.3	53.1	61.0

Word intelligibility W=022, parasitic noise level Leq = 51.5 dBA.

Table 7. Measured results of noise levels in the test point with use of two transmitters.

f, [Hz]	250	500	1000	2000	4000	8000
Lnoise, [dB]	37.4	35.9	40.4	43.1	50.5	65.0

Word intelligibility W=021, parasitic noise level Leq = 45.2 dBA.

Parasitic noise level is proved to be decreased by more than 6 dB by using of higher amount of the vibration transmitters.

4.4. Results Analysis

Summarizing all the obtained results, we can make a conclusion that all considered factors have influence on parasitic noise level, therefore, can be used for its decrease. The results are presented in the bar chart (Fig. 4).

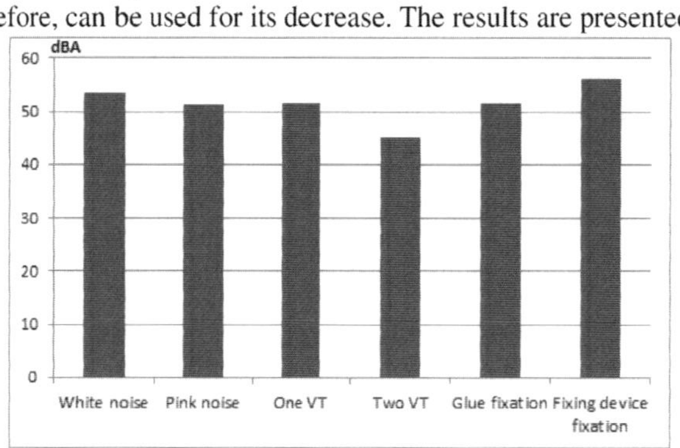

Figure 4. Comparison of the empirical results.

5. Conclusion

During the tests, influence of different factors on the parasitic noise levels was quantitatively estimated: interference spectrum – 2 dB (The results would be more essential with use of formant-like interference);
way to fix – 5 dB; the number of the transmitters being used – 6 dB;

Taking the obtained results into account allows essential decrease of parasitic noise level in setting active protection means with making it meet the Norms.

IOP Conf. Series: Journal of Physics: Conf. Series **998** (2018) 012016 doi:10.1088/1742-6596/998/1/012016

References

[1] State Committee on Sanitary and Epidemiology Surveillance 1996 The Sanitary Norms SN 2.2.442.1.8.562–96. Industrial vibration, vibration of residential and public buildings

[2] Federal Hygiene and Epidemiology Centre of Federal Service for Surveillance on Consumer Rights Protection and Human Wellbeing 2007 Methodological guidelines MUK 4.3.2194–07 Noise monitoring within the residential development area, in residential and public buildings and premises

[3] Horev A A and Makarov J K 2001 *Quest. of inf. sec.* **1** (Moscow: Scientific and technical centre of military-industrial complex "Compas") p 21–28

[4] Ivanov A V, Reva I L and Trushin V A 2011 *Sc. Bull. of Novosibirsk St. Techn. Univ.* **45** (Novosibirsk: Printing house of Novosibirsk State Technical University) p 151-154

Development of the information model for consumer assessment of key quality indicators by goods labelling

S Koshkina[1, a*] **and L Ostrinskaya** [2, b]

[1]Omsk State Technical University, 11, Mira av., Omsk 644050, Russia
[2]Siberian State Automobile and highway University, 5, Mira Pr. Omsk 644080, Russian

[a] E-mail: ksn_omsk@inbox.ru, [b] E-mail: l_i_romanova@mail.ru

Abstract. An information model for "key" quality indicators of goods has been developed. This model is based on the assessment of f standardization existing state and the product labeling quality. According to the authors' opinion, the proposed "key" indicators are the most significant for purchasing decision making. Customers will be able to use this model through their mobile technical devices. The developed model allows to decompose existing processes in data flows and to reveal the levels of possible architectural solutions. In-depth analysis of the presented information model decomposition levels will allow determining the stages of its improvement and to reveal additional indicators of the goods quality that are of interest to customers in the further research. Examining the architectural solutions for the customer's information environment functioning when integrating existing databases will allow us to determine the boundaries of the model flexibility and customizability.

Keywords: information model, labeling, quality of goods, specialized environments of data processing.

1. Introduction

National systems of standardization and goods control quality are not harmonized. The systems of certification and labeling in various countries are different. The existing imperfection of information environments, the lack of integration between them, the incompleteness of data bases on the composition and quality of goods require the creation of unified information environment that allows controlling the quality of goods and their compliance to standards by labeling. There are no integrated information environments for the support of the customer's purchasing decision. There is a conflict of interests of consumers and producers regarding their interest in observing of all the quality standards for goods.

Labeling is the main carrier of the current information about the product. It performs an information function, transferring information about the product characteristics, its quality composition and shelf life to the potential buyer. Combining the efforts of various scientific knowledge branches representatives will allow managing the quality of goods at various levels of impact: at mega-, meta- and micro levels.

The study of the subject area revealed an inadequate degree of investigation of the issues analyzed in the scientific literature. There are some works devoted to individual and several specific aspects of product quality marking standardization. But a harmonious system of scientific views on the problem under investigation has not been created yet. There is no data processing environment which will provide controlling the quality of goods and monitoring their compliance with standards through labeling on the territory of individual states and at the international level. There is no integrated information environment for it. This problem is serious and requires immediate action, because the quality control products are becoming a state goal now. The important task of each state is to ensure the safety of its citizens. Labeling should provide the customer with the necessary information through the use of applied information and advising the environment, which will be able to provide the customers with a solution in the form of advice. It will be possible using the information model, describing the process of forming information about consumer goods properties.

2. Statement of the problem

In the subject area of the study are included the existing information systems which are including the information about labeling and quality control of products. The subject of research is the information space of the customers in the decision-making process of the purchase of goods. On the basis of the interdisciplinary approach to the improvement of the information function of marking is being developed the Information model, describing the process of forming information about consumer goods properties. The using of this information will enable the integration of databases on the composition and quality of goods. It will help to make a purchasing decision which is based on the use of modern information technologies by consumers and the mobile technical devices available to them.

3. Theory

Various aspects of the need to improve the information function of marking have been studied by some researchers. We pay special attention to the labeling of consumer goods and food products. Perception of quality is investigated in a number of works. The information function of marking contributes to the evaluation of the quality of consumer goods [1]. Now it is increased the attention to eco-labels [2]. Many consumers give their preference to environmental and organic products. It is important to obtain reliable information about the quality of these products. The importance of eco-labeling in the decision to purchase is devoted to the publication of the results of the scientific research of Grankvist G. and Anders B. [3].

In the article of Tihonova O. and Reznichenko I. the possibility of assessing the quality of food products labeling through the development of consumer criteria for evaluation based on the systematization and classification of specific requirements, imposed to it both by regulations and by consumers in general. The assessment of consumer properties labelling on the proposed evaluation criteria will more fully realize all of its functions, provide high quality and meet all the consumers information needs [4].

An information carrier is a production and trademarks of the goods. The labeling contains the following elements: text, image and information. The commodity-sign symbolism in the modern food industry uses in its development a large amount of visual information as an important tool for marketers in the competition for the potential consumer [5].

There is the point of view on the color in labeling role in making a purchasing decision. Meng Shen, Lijia Shi and Zhifeng Gao examined the impact of label colors (red or blue) on consumers' cognitive effort and willingness-to-pay (WTP). This study found that red labels led to longer response time in decision-making than blue labels did. The research results indicate that color influences significantly on the WTP assessment concerning the certain attribute on food labels [6]. Therefore, we have to recognize the subjectivity of the decision to buy goods. It is necessary to improve the information function of labelling, because it is difficult for the buyer to make the right decision about the purchase.

As a result, we can say that the integration of databases on the composition and quality of goods for decision-making has been poorly studied. The ability to verify the quality of goods by labelling based on the use of modern information technologies by consumers and the mobile technical devices available to them have not been studied in scientific literature. We emphasize that insufficient attention has been paid to the modeling of the process of regulating information about the consumer properties of the product by its labeling. This problem is urgent for each country and the world as a whole. The extent of available data which is intended for use is very impressive. The available information contain the key characteristics of the goods and the factors of these characteristics providing. Their processing is possible only through the using of modern information technologies such as big data, cloud computing, expert systems, support systems and decision-making, monitoring and others. The ability of modern producers and consumers to immediately apply the skills and use digital environment will help them to apply all the collected, processed and stored information for the control of the goods quality.

4. Results and Discussion

The analysis of the subject area has identified the currently existing types of information systems for the support of standardization, labelling and quality control of products. Information model, describing the process of forming information about consumer goods properties was developed according to the IDEF0 methodology.

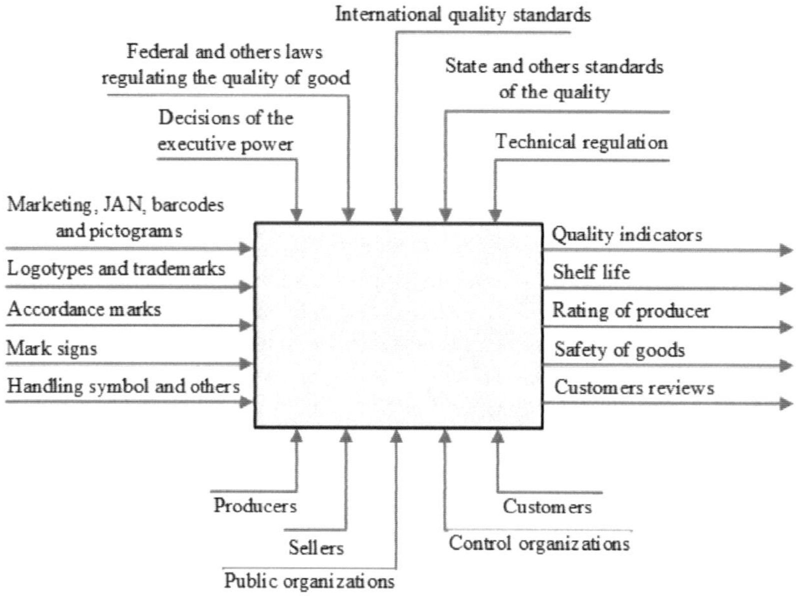

Figure 1. Information model, describing the process of forming information about consumer goods properties.

We offer an essential definition for the description of the developed information model. This model is shown in figure 1. This definition will describe key quality indicators of goods. The abbreviation KQI means "The goods key quality indicators". It means the indicators which characterize the basic consumer properties of the product and those which are most important for the customer decision to purchase.

In the developed Information model there are several KQI:

1) quality indicators declared by the goods manufacturer;
2) shelf life which is confirmed by the seller;
3) rating of the producer which is obtained on the basis of information received from controlling and public organizations;
4) product safety based on the goods composition. It is characterized by the useful and dangerous properties of the constituent ingredients;
5) customer reviews based on the specialized social network data and information.

For the buyer, the final assessment should be formed based on the evaluation of the goods KQI. In this case, one should take into account that the customer is not interested in large data flows, especially when he is in a trading organization (in store or in office etc.). It will be important for him to get a minimum of useful information through a single request format. This format should contain very concise information. Its purpose is to help customer in quickly making decision. His decision can be "to buy" or "not to buy".

The information and advisory environment should be implemented taking into account the customer's mobility and his possibly of changing geo-location. This environment should be oriented towards the use mobile devices of customers. The customer should receive a minimum useful amount of information for making his decision. For this purpose it is necessary to distinguish a number of key parameters. Key parameters should be essential. On the basis of these selected parameters the universal algorithm for the customer "Advice" information environment development will be developed. This " Advice " will be used for making a decision on purchasing the desired product.

For the formation of the " Advice ", a matrix with the values of the goods key quality indicators can be processed in the algorithm. This algorithm mathematical model is to develop an opinion in the form of a possible " Advice ". This advice can have one from the three following variants:

- This product is recommended for purchase;
- This product is not recommended for purchase;
- This product has doubtful consumer qualities.

The recommended matrix format is shown in figure 2, where X is the value of a particular quality.

$$\textbf{KQI of the goods} = \begin{pmatrix} X & X & X \\ X & X & X \\ X & X & X \\ X & X & X \\ X & X & X \end{pmatrix} \begin{matrix} \textbf{Quality indicators} \\ \textbf{Shelf Life} \\ \textbf{Manufacturer rating} \\ \textbf{Safety of goods} \\ \textbf{Customer reviews} \end{matrix}$$

Figure 2. The elements of the key quality indicators of goods matrix.

These key quality indicators of goods can have one of the value described below:

0 – Its level is low. Its qualitative properties have a questionable nature or they were lost.
1 – Its level is neutral. Its quality properties are close to the standards.
2 – Its level is high. Consumer properties of this product satisfy the KQI.

First and foremost, the minimum required amount of data at the output of the information system is determined. Then the question arises, what will be the basis of their receipt. It is also important to determine what available information resources should become sources of data at the input.

At this stage, these questions are not sufficiently studied yet. The authors are going to analyze the subject domain of the input data streams in the nearest future. Currently, the integration of databases is studied. Realization of the data flows architecture is carried out at the stage of existing data processing decomposition.

This presented information model has a lot of input values. All these values are regulated legally and by the regulatory framework existing in every country.

Business process modeling starts with the information model development. This model describes the process of obtaining optimal and useful information about consumer properties of goods. The users of this process are customers, producers, sellers, control organizations and public organizations.

5. Conclusion

The inconsistencies of national standardization and certification systems and of information environment imperfection were revealed. As the analysis shows, there is no integration between them. The databases of goods ingredients and quality are incomplete. The authors developed an information model, describing the process of forming information about consumer goods properties by its labeling. This model allows us to decompose the existing processes in data streams. On this basis the architectural solutions for different levels of information processing can be constructed.

There is a need to continue the research in the chosen direction. In the future, a more in-depth analysis of the decomposition levels of the presented information model will be possible. Various quality indicators of goods can be added into this model later. These added indicators will be relevant for customers.

The development of architectural solutions for the information environment functioning will help to define the boundaries of flexibility and customizability of the model. It is important to create an improved information environment that will ensure the goods quality control. This control is based on the using of modern information technologies and mobile technical devices by customers.

References

[1] Kenyon G N and Sen K C 2015 The Perception of Quality (Springer-Verlag London)
[2] Kümpel Norgaard M.and Brunso K 2009 *Food Quality and Preference* **20** (**8**) p 597-606
[3] Grankvist G and Biel A 2007 Predictors of purchase of eco-labelled food products: A panel study. *Food Quality and Preference* **18** (**4**) p 701-708
[4] Tihonova O and Reznichenko I 2013 *Research Journal of International Studies* **8-3** (**15**) p 73-77
[5] Popova I 2015 *Bulletin of NGIEI* **7** (**50**) p 66-79
[6] Meng Shen, Lijia Shi and Zhifeng Gao 2018 *Food Quality and Preference* **64** p 47-55

Considering the effect of the logging cable stretching under the influence of its own mass in the well on the accuracy of geophysical parameters determining

A A Kozhushko[1], A N Pahotin[1] , V N Mal'tsev[1], L V Bojarnikova[1] and E P Stepanova

[1]Omsk State Technical University, 11, Mira av., Omsk, 644050, Russia

E-mail: omgtu_148@mail.ru

Abstract. The technology of carrying out various geophysical studies of oil wells using a carrying well-logging cable for the delivery of geophysical instruments and equipment is considered in the article. The relevance of the topic results from the need to evaluate the effect of well-logging cable stretching in the well under the influence of its own mass and mass of geophysical instruments and equipment on the accuracy and adequacy geophysical studies. Calculation formulas for determining the well-logging cable stretching under the influence of the geophysical tools and equipment mass without taking into account and taking into account the influence of the carrying well-logging cable mass are also presented. For three types of carrying well-logging cables, calculations were made of their stretching in the oil well under the mass of geophysical instruments and equipment and its own mass, depending on the depth of the investigated well. The analysis of the obtained results made it possible to numerically evaluate the extension of the carrying well-logging cable depending on the depth of the investigated well and by correcting the obtained results it allows to provide the required depth accuracy and reliability of the interpreted results of the geophysical studies.

Keywords - results accuracy, results correction, geophysical tools and equipment, geophysical oil well studies, logging truck, carrying well-logging cable, mechanical characteristics of carrying logging cable, extension of well-logging cable, hydrocarbons production

1. Introduction

At present, hydrocarbons are the main source of energy and raw materials for the production of a variety of products. Despite the development of alternative energy sources, hydrocarbon fuels remain the most widespread and demanded source of energy in the world due to high energy intensity and ease of transportation.

Almost all hydrocarbons in the world are produced by the well method. Bore wells are used for oil and gas production. During the drilling, arrangement and operation of wells, a variety of geophysical instruments and equipment are used. The tasks solved in oil and gas wells by means of geophysical studies in the wells are divided into geological, related to the study of the composition and characteristics of rocks in the sections of wells, technical and technological. The last two groups of tasks include surveying the technical state of open and cased wells, locating the equipment used for hydrocarbon production, and studying the composition of fluids entering the well. The content of the tasks of these groups is significantly overlapped. The same data (for example, information on the spatial position and profile of the open hole trunk or on the composition of the fluids coming from the perforation interval) are used to immediately correct the drilling and production technologies or fix them for later use (for geomodeling, reservoir development analysis and others) [1-4].

One of the most important among geophysical operations is the determination of the depth of the tool or apparatus location in the well, i.e. determination of the depth of the objects under study.

Various geophysical tools and equipment lowering and lifting is carried out with the help of a lift, a block-balance and a cable. Trigger equipment placed permanently or on the chassis of a car is called a hoist. For carrying out geophysical works, hoists with winches of different sizes and designs are used, depending on the type and length of the cable [4-6]. The general structure of a self-propelled logging truck is shown in Fig. 1.

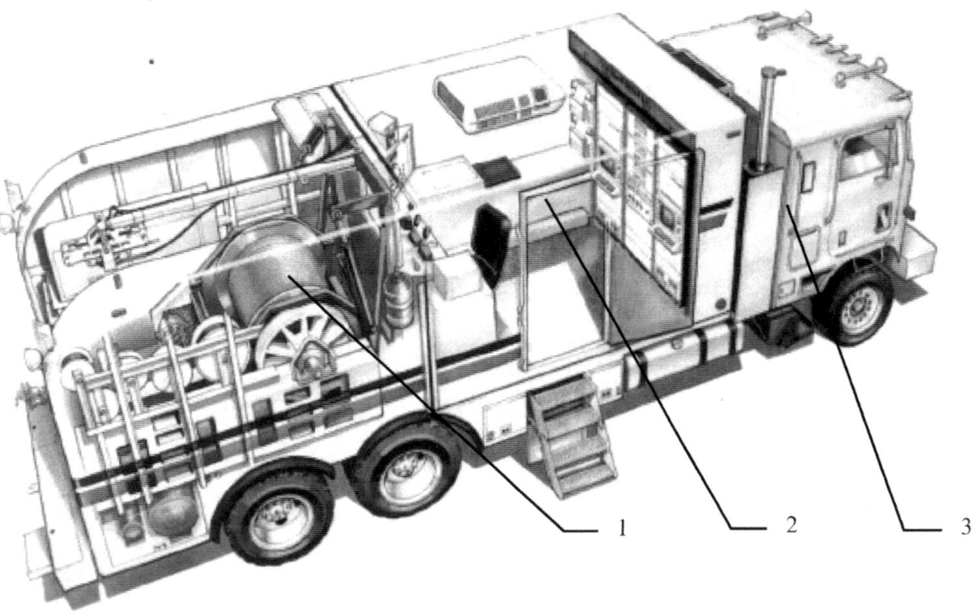

Figure 1. Self-propelled logging truck

1. a log hoist with a cable; 2. compartment with geophysical equipment; 3. automobile chassis.

The main means of delivering geophysical instruments and equipment to the well in geophysical studies and shooting-blasting operations is a carrying well-logging cable. A typical design of a carrying well-logging is shown in Figure 2. The well-logging cable consists of current-carrying survey conductors designed to supply power to downhole tools and converters, control devices, and the transmission of electrical signals from these devices (converters) to ground-based recording and processing equipment. Over the insulation of current-carrying survey conductors, armor of two steel wire layers is applied. The lower and upper layers have different direction and thickness, which reduces the unwinding of the cable under the influence of the axial load. The lower and upper layers carry a mechanical load and protect the survey conductors from damage when the cable is lowered into the borehole and wound on the winch.

IOP Conf. Series: Journal of Physics: Conf. Series **998** (2018) 012018 doi:10.1088/1742-6596/998/1/012018

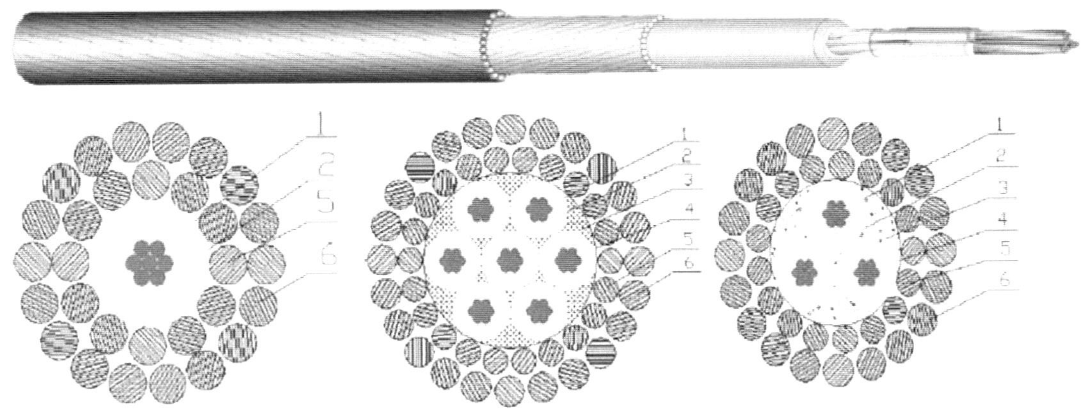

Figure 2. Typical design of a geophysical cable
1. conductor core; 2. the first layer; 3. filler; 4. winding; 5. insulation; 6. the second layer.

The general scheme of carrying out geophysical studies of wells using cable technologies (during wells construction and operation) is as follows:

1. Downhole tools with physical field sensors and electronic components designed to amplify the sensors signals and transmit them to the earth surface via a logging cable, are connected to the well-logging cable wound on the hoist winch drum using standard cable end.

2. The tools are lowered into the well by the well-logging cable through the securely fixed lower roller of the balance block so that the plane of its wheel is perpendicular to the axis of the winch drum and points to its center, and the cable passed through the upper roller of the balance unit is freely lowered into well head. The axis of the lower pulley block balance is connected to the mechanical and electronic depth sensors; on the pulley frame, near the well-logging cable, a depth marker sensor is placed. The depth marker sensor and the electronic depth sensor are connected by cables to the equipment of the logging station.

The logging cable is, in turn, connected to the recording equipment of the ground geophysical laboratory, which, together with the log lift, is a part of the field information measuring complex called the logging station.

3. During the lowering of downhole tools into the oil well, the following activities are performed:

- control of their operability, the lowering smoothness and absence of stops;
- continuous depth control with the help of mechanical and electronic sensors - depth gauges.

4. The measured parameters are recorded when lifting downhole tools (to avoid errors in depth determination due to the effect of "floating" of the cable and downhole tools in oil wells with a high density of washing liquid). This rule exceptions are geophysical studies of wells by using thermometry, which, in a complex of geophysical studies, are performed first (to prevent disturbance of the temperature field in the well and to avoid temperature errors due to the mixing of the washing liquid during tripping operations).

The purpose of this paper is to assess the effect of stretching of a well-logging cable under the influence of its own mass and mass of geophysical instruments and equipment on the accuracy and quality of performing geophysical surveys of wells.

2. Statement of the problem

To record the geophysical survey curves as a function of the well depth and control of tripping operations, it is necessary to continuously measure the length of the downhole cable or the well depth and transfer the movement of the cable or drilling tool to the registrars (loggers) and control devices. Such measurements are carried out in two ways: by means of a measuring roller and by reading magnetic marks.

The accuracy of the measuring roller, related to the slip of the cable along the measuring roller and the discrepancy between the estimated roller diameter and the actual cable diameter, reaches ± 5 ÷ 10 m in deep bores. Therefore, the measurement results are used only for visual control of tripping operations.

To determine the depth at which the downhole tool is located in the well, the logging cable is supplied with magnetic or real marks at regular intervals (usually 10 meters). Magnetic marks are the magnetized parts of the cable, while real ones, small bandages made of the insulating tape, are superimposed on the cable.

For a more reliable anchoring to the depths, the marks, multiples of 100 meters, are made different from the ordinary ones. The of magnetic marks reading is carried out with the help of the of depth magnetic marks sensor, which forms the electrical signal "depth - mark" at the moment of passage of the magnetized part of the cable past it. The real marks are visually controlled, while at the time of their passage between the rollers of the cable laying machine a mark is placed on the log chart. The error in measuring depths with magnetic marks does not exceed ± 1.5 m in wells with a depth of 2 km and ± 2.5 m in wells up to 4 km deep [3, 4, 6].

When running operations, the well logging cable is subjected to a complex of various factors. A part of the cable in the well is pulled out under the action of its own mass, the mass of geophysical tools and equipment, and the temperature, which increases by about 1° C for every 30 m. When lifting, geophysical tools and equipment pulling and sticking, friction of the cable against the well walls and high viscosity of the environment (liquid) foster the cable stretching. For these reasons, the cable stretching equals to several meters in wells with a depth of 2 ÷ 5 km and can reach tens of meters in ultra-deep wells [4, 6].

The cable marking is made taking into account its uneven stretching under the influence of tension toprovide a sufficiently clear logging of the well-logging records for the depth of depth determination for various operations in the well. Tensions that need to be maintained in accordance with the interval of the cable to be measured are determined from the results of the tension measurements in the wells of the area where the cable will be used, or according to calculated data. The problem of improving the technical and software-methodological support for research and processing of geological and geophysical survey data carried out in horizontal wells is still relevant today. Therefore, the task of improving the calculation methods for determining the logging cable stretching in a well is up-to-date.

3. Theory

It was discussed earlier, that the logging cable is a complex structure of various materials. As a simplifying assumption, we will consider a well-logging cable as a cable made of a homogeneous material. Consider the problem of determining the stretching of a logging cable under the influence of a mass of geophysical tools and equipment suspended on it. According to Hooke's law, the elastic force in the material of the logging cable will be:

$$F = \frac{E \cdot S}{l_0} \cdot \Delta l , \qquad (1)$$

where F is the elastic force in the material of the well-logging cable; Δl is the stretching of the well-logging cable; l_0 is the initial length of the well-logging cable; E is the modulus of elasticity of the well-logging cable; S is the the cross-sectional area of the well-logging cable.

At equilibrium, the elastic force is equal to the mass of the load suspended to the well-logging cable:

$$\frac{E \cdot S}{l_0} \cdot \Delta l = m \cdot g , \qquad (2)$$

where m is the mass of suspended geophysical equipment and tools; g is the gravity acceleration.

IOP Conf. Series: Journal of Physics: Conf. Series **998** (2018) 012018 doi:10.1088/1742-6596/998/1/012018

Then, the stretching of the well-logging cable will be described by the following formula:

$$\Delta l = \frac{m \cdot g \cdot l_0}{E \cdot S}. \tag{3}$$

We determine the stretching of the logging cable in the well under the influence of its own mass and the action of mass of geophysical devices and equipment suspended on it. According to the Hooke's law, the voltage in the material of the well-logging cable is described by the following expression:

$$\sigma(x) = E \cdot \varepsilon(x) = E \cdot \frac{dl}{dx}, \tag{4}$$

where $\varepsilon(x) = \frac{dl}{dx}$ -is the elastic deformation of the logging cable at depth x.

From formula (4), we find:

$$\varepsilon(x) = \frac{du}{dx} = \frac{\sigma(x)}{E}, \tag{5}$$

where u is the displacement of the logging cable point during deformation.

Let us define the voltage in the well-logging cable at some depth x:

$$\sigma(x) = \frac{m \cdot g + \rho \cdot g \cdot S(l_0 - x)}{S}, \tag{6}$$

where ρ is the average density of the well logging cable material.

Substituting (6) into (5), express du as:

$$du = \frac{\sigma(x)dx}{E} = \frac{(m \cdot g + \rho \cdot g \cdot S(l_0 - x))dx}{E \cdot S}. \tag{7}$$

Integrating (7) and performing some transformations, we obtain:

$$\int_0^{\Delta l} du = \int_0^{l_0} \frac{(m \cdot g + \rho \cdot g \cdot S(l_0 - x))dx}{E \cdot S} \Rightarrow \Delta l = \frac{m \cdot g \cdot l_0}{E \cdot S} + \frac{m \cdot g \cdot l_0^2}{2 \cdot E}. \tag{8}$$

Thus, the stretching of the logging cable in the well under the action of its own mass is determined by the addend according to the formula (8)

4. Experimental results

Figures 3 - 5 show the results of calculating the stretching of the well-logging cable in a vertical well filled with a liquid. In Figure 1, the curve is obtained by using the formula (3) and the assumption that the entire mass of the well-logging by the action and the mass of the geophysical instruments and equipment suspended on it are concentrated at its free end. Figure 2 shows a curve obtained by using formula 8. In calculations, the mass of geophysical instruments and equipment suspended on a well-logging cable was assumed equal to 100 kg. The mechanical characteristics of the well-logging cables under consideration are given in Table 1.

Table 1. Mechanical characteristics of armored cables

Cable brand	Breaking strength min, KN	Stretch ratio, m/km/N	Cable outer diameter, mm	Mass of 1 km in air, kg/km	Mass of 1 km in water, kg/km
KG1x0.75-55-260	55	0.3	8.7	329.1	259.5
KGl3x0.2-30-150	30	0.5	6.25	160.6	141.7
KG3x0.75-70-150	70	0.2	10.85	462.2	344.9

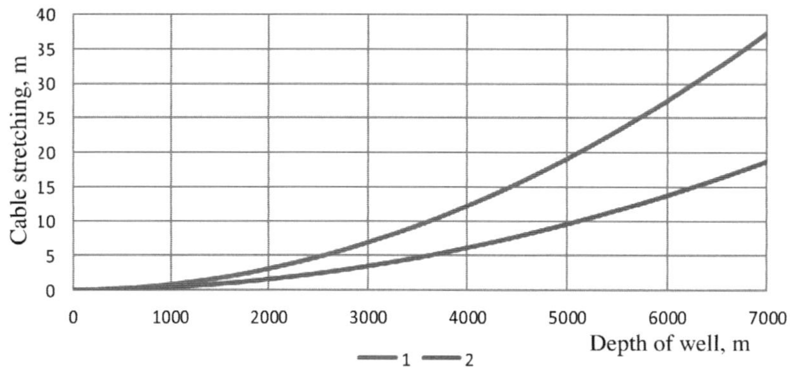

Figure 3. Stretching of well-logging cable brand brand KG1x0.75-55-260.

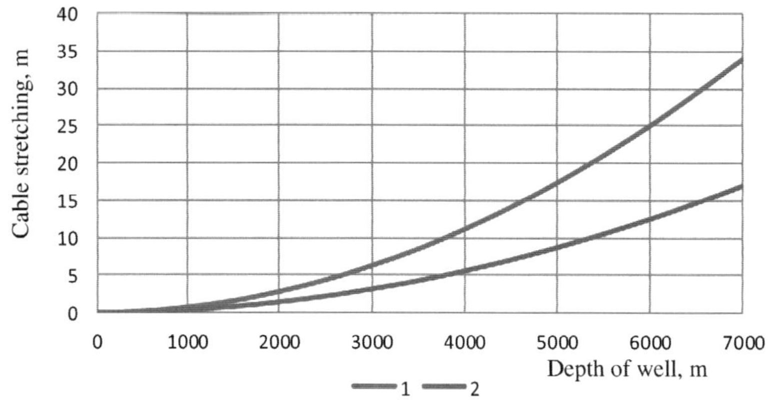

Figure 4. Stretching of well-logging cable brand KGl3x0.2-30-150.

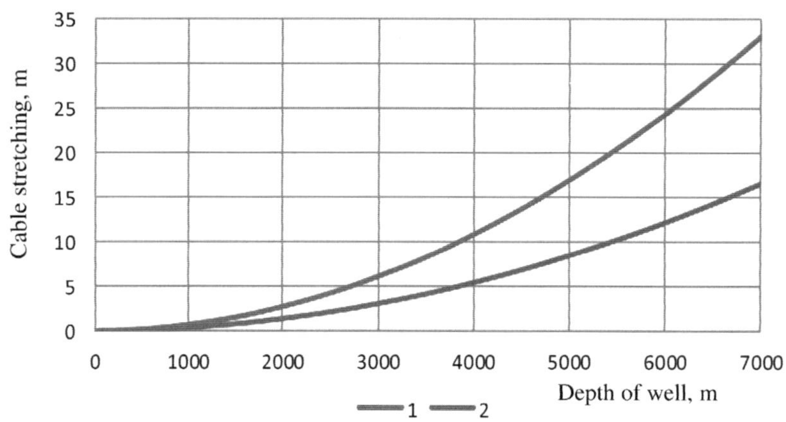

Figure 5. Stretching of well-logging cable brand KG3x0.75-70-150.

5. Results and discussion

As it can be seen from the graphs given in Fig. 3 - 5 in the cases under consideration to a well depth of 1000 meters, the calculations according to formulas (3) and (8) have comparable results. At well depths of more than 1000 meters, the use of formula (3) and assumptions about the application of the cable own mass to its free end, result in the almost doubled value for the logging cable stretching, which can lead to a very significant error in the geophysical research results interpretation. When the mass of the cable exceeds the mass of the geophysical tools and equipment attached to it, the application of formula (3) to correct the results of geophysical studies is incorrect. In such cases, the intrinsic mass of the well-logging cable makes the main contribution to its stretching and the use of formula (8) is more correct.

6. Conclusion

In modern economic conditions, it is impossible to organize effective exploration and exploitation of hydrocarbon fields without intensive application of geophysical studies. The leading role in the performance of a variety of geophysical studies is played by the accuracy of determining the position of the tools and equipment used in the investigated well. Increasing the accuracy of determining the position of geophysical tools and equipment in the well during geophysical surveys allows to significantly reduce the time spent on exploration, to increase the reliability of interpreted geophysical research results and to improve the operational characteristics of the developed hydrocarbon field.

The analysis of the obtained results allowed us to numerically estimate the stretching of the carrying logging cable mass depending on the depth of the investigated well. The proposed technique for determining the stretching of the carrying logging cable in the well under the influence of its own mass and the mass of the geophysical tools and equipment suspended on it makes it possible to correct the obtained results and to provide the required depth accuracy and reliability of the interpreted results of geophysical studies.

References

[1] Kearey Philip, Brooks Michael, and Hill Ian 2002 An Introduction to Geophysical Exploration, 3rd ed., Blackwell Science Ltd

[2] Milsom John, Field geophysics. 3rd ed. p. cm. (The geological field guide series)

[3] Kuznecova M A, Cherkovskij N L, and Bahtijarov G A 2014 Method for the determination of corrections of absolute marks in inclined wells and water-oil contact of a multi-layer deposit Geology, geophysics and development of oil and gas fields **11** p 49-51(in Russian)

[4] Gninev I V., Korolev A B, and Sitnikov V N 2016 Increase of inclinometric measurements accuracy in subvertical wells: Karotazhnik journal **12 (270)** p 98-113

[5] Korjakin A G, Larin Ju T, Mesenzhnik Ja Z 2015 Some actual problems of submersible cables. Part II: Photon-Express **4 (124)** p 32-34

[6] Konjuhov V M, Krasnov S V, Konjuhov I V 2016 Computer modeling of implosion processes in the oil reservoir system - well - mobile implosion chamber in: Automation, telemechanization and communication in the oil industry **5** p 32-37

Measurement of machine parts dimensions positional deviation with regard to their geometric accuracy

D B Martemyanov[1], V V Pshenichnikova[1], V A Penner[1], and A E Zemtsov[1]

[1]Omsk State Technical University, 11, Mira av., Omsk, 644050, Russia

E-mail: den-79-03@yandex.ru

Abstract. Real surfaces of the parts, obtained with the help of technological processes, are always characterized by deviations from a nominal (regular) form. When analyzing a nominal cylindrical surface or a prismatic component element, the interrelation between current dimensions in various sections and a surface form, as well as surfaces position, can be found.

Key words. tolerance, positional deviation, form deviation, geometric model, qualified tolerance, element dimensions.

1. Introduction

Qualitative indicators, reliability and durability of hydraulic units depend to a large extent on the correctness of the normalization, the accuracy of manufacturing and the accuracy of each geometric element, each part of the unit arrangement tolerances control.

It is very difficult to ensure the given tolerances of channel dimensions and their positional deviations in the machine parts, because making accurate measurements of the through holes and their numerous positional deviations actual dimensions with respect to the set of bases in each mating part is necessary.

As practice shows, the measurement of holes and ports in the parts of hydraulic units is very laborious. Besides, it is not always possible to detect deviations in the arrangement of all the holes in the parts of hydraulic units with an error not exceeding the permissible value (30% of the position tolerance). The values of the hydraulic units channels axes positional deviations influence the area of the flow sections for the part-to-part passage of liquid, which in turn affects the intensity of increasing or decreasing the liquid pressure along its flow through a spool valve or a sleeve. In accordance with this, the characteristics of the sleeve ports opening by the spool valve ports influence the hydraulic drive accuracy and sensitivity. The tolerances of the channel axes position are among the most important indicators of the parts quality by geometric accuracy [1].

2. Statement Of The Problem

The analysis of the form deviations of standard surfaces presupposes the following statements:

1. As the form deviations are automatically limited by the pre-set fields of the dimensions tolerances, they should be intentionally normalized only in cases when they have to be made more rigorous compared with values which were pre-set during dimensions tolerancing.

2. The form tolerances system should include tolerances for the most frequent standard cases. Above all, form tolerances of nominally flat surfaces and the surfaces of a rotary body type should be normalized [3].

3. The element complex dimensions include the dimensions of the adjoined element, the deviations of the element surface form and the deviations of the considered element position towards the bases.

As the measurement of the position deviations are carried out by the subtraction, element size and form deviations do not influence the accuracy of the position qualified tolerances measurement results.

Position deviations are geometric values, so the two-dimensionality principle should also be applied to them. The application of the two-dimensionality principle makes it possible to implement a systematic approach to the classification of position deviations, to the account of the elements service purpose, to establishing the optimal composition of deviations, to shortening the extent of the dimensional chains of elements and, as a consequence, to increasing the geometric accuracy of the elements arrangement with respect to the parts' bases [4].

4. In order that the two boundary values of the elemental dimensions do not exceed the limits of the tolerance field, they should be understood as the dimensions of two regular elements nominally located with respect to the set bases and covering the real element surfaces, touching them: one - outside the material of the part (determines the size of the element material maximum), the second - from the element material (determines the size of the element material minimum). These definitions determine the dimensions of the element dimension field, i.e. the zone in the part area, which occupies the real surfaces of the element due to the deviations in form and position with respect to the coordinate system of the elements set [2].

Thus, the reason for the two-dimensionality of elemental dimensions are the form deviations of the elements surfaces and the elements position deviation in the reference frame of the elemental dimensions – in the coordinate system materialized by the set of elements bases. If one of the set elements shows its maximum informativeness, it will not have any position deviations with respect to the bases of other elements, since it determines the position of the reference frame coordinate axis or coordinate plane itself. In this case, the reasons for the two-dimensionality of its dimensions are only the form deviations of the element surfaces.

3. Methods

The basis for analyzing and measuring position deviations of the machine part is a geometric model that is a graphic representation of all the part components together with the deviations of the working surfaces in the generalized coordinate system of the part as a whole (Figure 1). The reference system of the deviations in the position of the normalized spool valve elements is the generalized OXYZ coordinate system, which is materialized by a set of three bases: by the Z4 axis of the cylindrical base A4 double guide, which deprives the object of 4 degrees of freedom measurement, by the end datum base B1, which deprives the part of the translation along the Z4 axis, and by the datum base B1 symmetry plane of the prismatic spool, which deprives the part of the rotation about the Z4 axis.

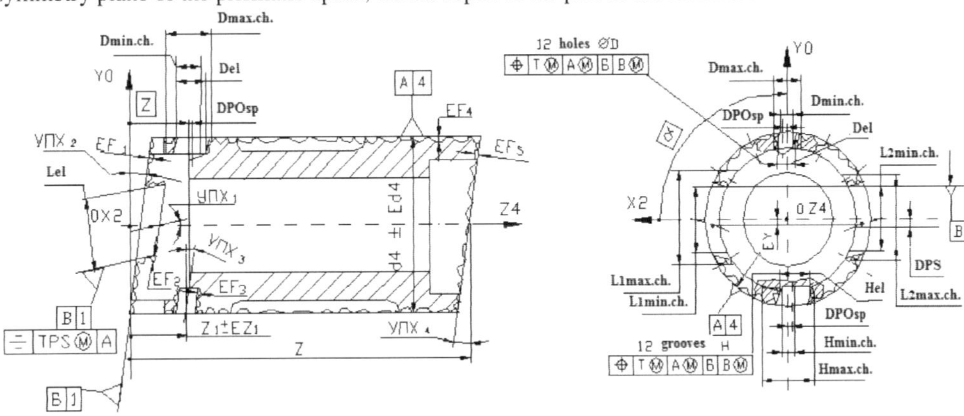

Figure 1. The spool valve geometric model.

In the Cartesian coordinate system, the dependent tolerance of the base B1 symmetry tolerance to the base A4 is normalized, while in the polar coordinate system, the dependent positional tolerances for the twelve cylindrical holes Ø D and for twelve prismatic slots of width H fulfilling the service purpose of the hydraulic channels are given, since in the first case the base B1 has spent only one

IOP Conf. Series: Journal of Physics: Conf. Series **998**(2018) 012019 doi:10.1088/1742-6596/998/1/01201

degree of freedom (from the maximum possible 2) on the materialization of the Cartesian frame of reference, while the channels, being the execution surfaces, - none [4].

4. Implementation

Full dependent position deviation of the channel axis (DPOhole or DPOport) is determined as the half-difference of the maximum and the minimum complex channel dimensions, while their half-sum is equal to the element dimensions (Del or Hel). Geometric models of the parts (sleeve and spool) channels interaction prove, that minimum Dmin.ch. and maximum Dmax.ch. channel complex dimensions include adjoined element dimensions Del and the double value of the position deviation DPO of the considered element with respect to the bases (Fig. 2):

Dmin.ch. = Del – 2DPO, Dmax.ch. = Del + 2DPO.

Qualified tolerance of the position depends not only on the normalized minimum position deviation value, but also on the deviations of the actual dimensions of the base and analyzed elements from the maximum material sizes, or maximum material limits:

Tqual = Tmin + Tadd,

where Tmin is the tolerance, specified in the draft; Tadd is the additional variable tolerance constituent, which value depends on the actual dimensions of the analyzed elements:

Tadd = Ti + Tt,

where: Ti is the intrinsic tolerance of the channel size; Tt is the base tolerance [5].

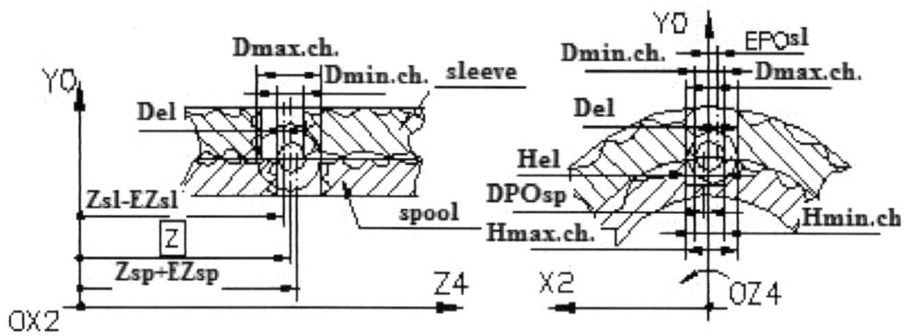

Figure 2. Geometric models of sleeve and spool channels interaction.

Additional constituent of the qualified tolerance can be found only after measuring the actual dimensions Dmin.ch. and Dmax.ch. of the basic and normalized elements (figure 3). The tolerance field of the coordinated channel complex dimensions widens into the two full dependent position tolerances in terms of diameter, which spikes the machine parts processability.

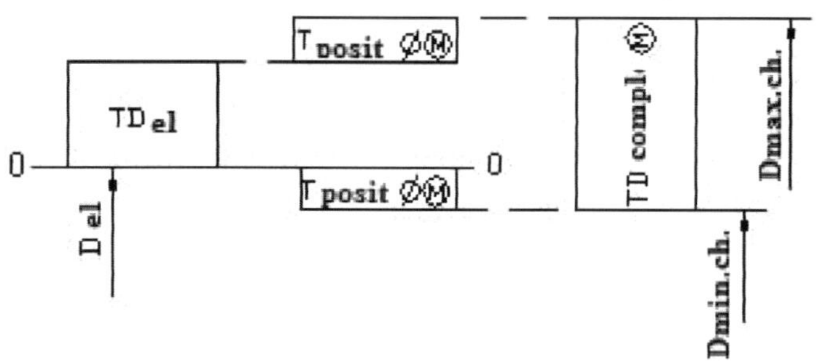

Figure 3. Position of the dependent dimensions and position tolerances fields.

5. Conclusion

The principle of the proposed method for measuring position deviations of the channel axes is to measure the half-difference of the maximum and minimum complex dimensions of the element in question, that is, the dimensions of two regular elements nominally located with respect to the parts' bases and the tangential real surfaces of the elements outside and from the material of the parts. Since measurements of position deviations are made using the difference method, the intrinsic deviations of the form and dimensions of the element do not influence the accuracy of the arrangement dependent tolerances measurements results. When measuring, the minimum Dmin.ch., Hmin.ch. and the maximum Dmax.ch., Hmax.ch. values of the complex channel dimensions, in fact, only the minimum complex dimensions (Dmax.ch., Hmin.ch.) will determine the area of the channel cross-sections. In order to increase the accuracy and performance of the channel position dependent deviations measurements it is proposed, instead of six differentiated measurements (three position deviations and three influencing dimensions of normalized base elements), to determine two values of one normalized element complex dimensions, taking into account its position with respect to the set of base elements (figure 1).

In conclusion, when conducting the technological processes metrological examination, special attention should be paid to the choice of measuring techniques (MT) of the units geometric values. The MT must provide control in accordance with the requirements to the permissible measurement errors.

The proposed method of reliable complex measurements of the pass channels effective dimensions will increase the reliability and durability of the hydraulic units' details. Knowing the actual position of the channels deviations, it is possible to ensure the adjustment of the machine program, which will allow obtaining a higher precision of manufacturing hydraulic components, their quality and reliability in operation.

References

[1] Henke R P, Summerhays K D, Baldwin J M, Cassou R M and Brown C W 1999 *Precision Engineering* **23 (4)** p 273-292

[2] Summerhays K D, Henke R P, Baldwin J M, Cassou R M, and Brown C W 2002. *Precision Engineering* vol **26 (1)** p 105-121

[3] Qingzhen Bi, Nuodi Huang, Chao Sun, Yuhan Wang, Limin Zhu, and Han Ding 2015 *Journal of Machine Tools and Manufacture* **89** p 182-191

[4] Glukhov V I and Martemyanov D B 2013 *The world of measurement* **7 (149)** p 7-13

[5] Russian State Standard 28187-89 Basic norms of interchangeability. Deviations of the form and arrangement of surfaces. *General requirements for measurement methods* p 19

Supervised Multi-Authority Scheme with Blind Signature for IoT with Attribute Based Encryption

O V Nissenbaum , K Y Ponomarov and A A Zaharov

Department of Information Security, Institute of Mathematics and Computer Science, University of Tyumen, Perekopskaya st., 15a, Tyumen, 625003, Russia

E-mail: o.v.nissenbaum@utmn.ru, k.y.ponomaryov@utmn.ru, azaharov@utmn.ru

Abstract. This article proposes a three-side cryptographic scheme for verifying device attributes with a Supervisor and a Certification Authority (CA) for attribute-based encryption. Two options are suggested: using a message authentication code and using a digital signature. The first version is suitable for networks with one CA, and the second one for networks with several CAs, including dynamic systems. Also, the addition of this scheme with a blind signature is proposed to preserve the confidentiality of the device attributes from the CA. The introduction gives a definition and a brief historical overview of attribute-based encryption (ABE), addresses the use of ABE in the Internet of Things.

1. Introduction

Attribute-based Encryption is a public key cryptographic scheme for realizing scalable and fine grained access control systems, where flexible rights can be assigned to individual users. This conception originates from the idea of Identity-based Encryption (IBE) that was introduced by A. Shamir [1] in order to ease public-key encryption and certificate management. IBE does not have necessity to authenticate user's certificates, as user's public key in fact their identities, such as e-mail or passport data. In this case sender can create a ciphertext under the receiver's identity without any prior information. Afterward, A. Sahai and B. Waters [2] introduced the concept of Attribute-based Encryption (ABE) by replacing the identity with an attribute set. It turned out that ABE is quite efficient for realizing fine grained access control systems.

There are two approaches for creating ABE schemes namely Key-Policy (KP-ABE) and Ciphertext-Policy (CP-ABE). They differ by position of access policy: by the key or by the ciphertext. In a KP-ABE ciphertext is associated with attribute set, and each user private key is associated with an access structure that defines ciphertexts which could be decrypted by this key. In contrast, in CP-ABE each user's key is associated with attribute set, and ciphertext determines access policy over attributes. More specifically, encrypted message is associated with an access structure, and user is able to decrypt this message if and only if his key attributes satisfy this access structure. The main difference between these approaches relies on who determines the access policy: sender or receiver, ciphertext or user key.

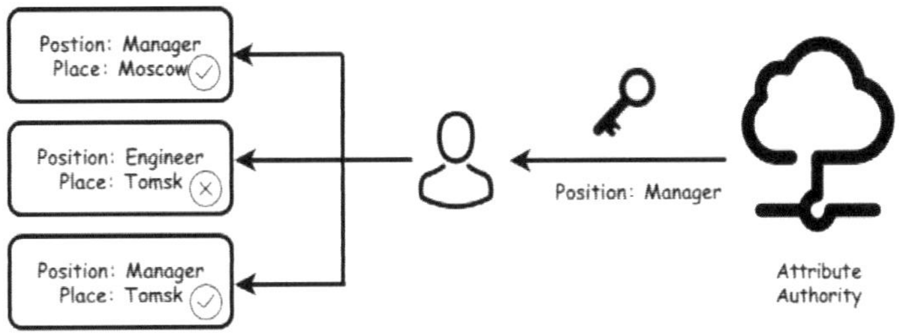

Figure 1. Key-Policy (KP-ABE) Scheme

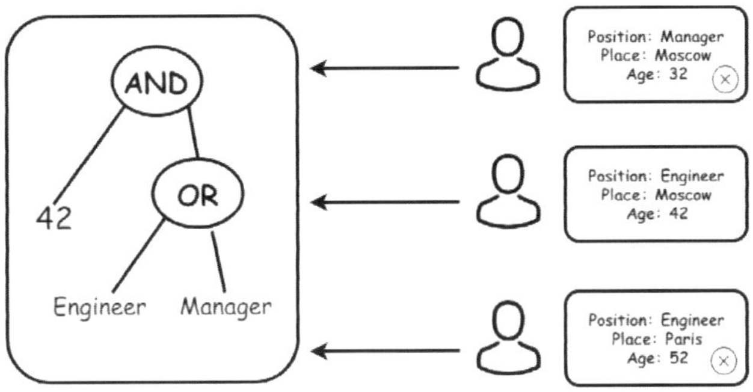

Figure 2. Ciphertext-Policy (CP-ABE) Scheme

There is the specific entity named Attribute Authority (AA) in ABE, that is responsible for key management and generating private keys for users by theirs attributes. In many situations single authority cannot be responsible for the whole attributes universe. So we need to share its responsibilities between different authorities, each of them has their own attributes subset, generate and distribute private keys on its basis. Such ABE schemes are called Multi-authority ABE (MA-ABE), one of them was proposed by M. Chase in [3]. Usually MA-ABE contains main Attribute Authority that distributes system keys and acts like single point of failure. Decentralized multi-authority Attribute-based encryption system was proposed by A. Lewko and B. Walters in [4]. In such systems any party can become an authority and there is no any requirement for any coordination, only create public key and issue private keys to users that reflect new attributes set.

Attribute-based encryption schemes also classifying by its access structure: threshold [2,5], monotonic [6] or AND-gate [7]. At a threshold model users are able to decrypt only and only if they have at least t attributes among a certain universe of attributes, for some threshold t chosen by sender. Monotonic structures developed the idea of threshold parameter. They are using logical conditions to define access rules through access trees. Each non-leaf node of the three obtains threshold gate value, and if it's equal to the number of children of this node, then node is realizing «AND» condition. When threshold value equals one, the threshold gate is «OR». Each leaf node describes certain attribute value. AND-gate access structure describes by attributes list, and secret key looks like array with specific values: positive – user has attribute, negative – user hasn't attribute, wildcard – this attribute does not matter for message decryption.

Next, we describe the main problems that make it difficult to use ABE in practice. At many realizations question about revoking attributes is still opening, because private key does not expire. Publication [8] provides the solution to attributes revoking problem, authors introduced concept of semi-trusted authority (STA) and proxy re-encryption servers for CP-ABE. STA should store the data and on revocation event generate several proxy re-encryption keys. Proxy servers will update secret keys for all users but the one to be revoked.

It's necessary to pay attention on the problem of key cloning. User can break access policy, just gives away their private key to another user. This problem makes sense in all known ABE schemes. The solution was proposed in work [9], the idea is that secret keys should be personalized and directly linked to a user's personal information. So, users will think about sharing the keys because it will reveal their personal data. To achieve it authors introduced new entities: trusted key generator (TKG) and token server (TS). TGS generate private keys and store and contains some user-specific information to create token information. TS create tokens for every decryption message event in the system, this tokens directly connected with personal information. Similar problem is possibility for users to break private key on parts and transfer them to third parties [10].

At CP-ABE schemes for that receiver will be able to successfully decrypt the message, it's necessary to transmit access structure with ciphertext. It can be dangerous for user's privacy, if attributes are confidential [11]. Adversary can intercept the message, get knowledge about using attributes. Based on the user's behavior adversary is able to discover message decrypt result. By this way adversary have chances to know users attributes.

ABE schemes are quite demanded in the Internet of Things (IoT) as they allow building decentralized dynamic systems that combine many sensors and servers [12,13]. ABE has advantages for IoT environment [14]:

- fine-grained access control based on context and device attributes;
- scales independent from the number of authorized users;
- does not require key sharing or key management algorithms between the participating parties (data owner does not need to identify the destination client).

The main aspect of using ABE in IoT is performance on resource-constrained devices: execution time, memory and energy consumption. In [14] these metrics were calculated for widely used board computers: Intel Galileo, Intel Edison and Rasberry Pi. Authors evaluated CP-ABE encryption and decryption operations with monotonic access trees. For example, encryption with 15 attributes and 112 bits security level (number of bits in AES symmetric key for plaintext encryption) shows average execution time of 9.68 sec. Authors of [15] calculated the above metrics with data and network overhead additionally for Android-based smartphone devices. They evaluated CP-ABE and KP-ABE with access trees too. Performance analysis noticed that classic ABE schemes have unacceptable metrics on smartphone devices.

Work [16] focused on CP-ABE lightweight security protocol for buttery-limited devices, such as laptops and smartphones. CP-ABE schemes require long decryption keys, ciphertexts, so this publication presented RSA-based ABE-model. It contains constant size secret keys and constant size ciphertexts with AND gate access structure. Also this work provides comparative summary for many ABE schemes with description of their security models, access structures and size of keys and ciphertexts. Publication [17] provides comprehensive overview of ABE implementations and shows their libraries, mathematical foundations, programming languages, descriptions. Authors noticed the lack of real-world implementations but the variety of mathematical and theoretical models.

Many researchers have been inspired by ABE and started to use it in their projects. For example, in [18] WI-FAB was proposed - Attribute-based WLAN access control mechanism without pre-shared keys. The idea is that WPA2 secret encrypt by CP-ABE and divided into several chunks. Each of them broadcasted into the networks through WLAN beacons. On the users who can rebuild the information included in beacons and retrieve secret, are authorized to connect the network.

2. Supervised Attribute Authority Scheme

Despite the fact that ABE is useful in IoT schemes, there is the problems with ABE when changing the device attributes. Consider next example: an electronic bracelet is issued to the every patient. The attributes of which are the patient's ID, his placement (ward, bed), his doctor, information for the access to his/her medical history etc. The patient can be transferred to another ward, the bracelet can be given to another patient or transferred to another ward, etc. Some attributes should be able to change only the attending doctor, and others - the senior nurse or medical registrar. Therefore we need so-called *Multi-Authority Scheme*, that should be simple, fast and realized by standard cryptographic primitives.

To solve this problem we propose an attribute verification scheme with a supervisor. To verify the ABE-attributes, we enter two nodes: a Supervisor (SV) and a Certification Authority (CA). Supervisor is a person or a center with the authority to change the attributes of devices (D). There can be several

SV in the system, that corresponds with typical situation in IoT; each SV may have different authority rights. For example, it is possible to set limits on the ranges of devices, types, and attribute values that the Supervisor is able to modify. The Supervisor is registered with a Certification Authority, which stores information about the rights of this Supervisor towards Device attributes. Upon registration, the SV and CA generate a common secret key for the MAC (Message Authentication Code).

A set of Device attributes are stored and transmitted as an *attr* string. The process of changing device attributes occurs in the proposed protocol using digital signature algorithms (S), the authenticator (MAC), and the following notations are used:

- $attr_{old}$, $attr_{new}$ – old and new device attributes, respectively;
- MAC(x) – SV authenticator of the message x;
- S(x) – CA digital signature of the message x;
- T – time stamp;
- * indicates optional fields in the transaction.

The protocol consists of the following steps (figure 3).

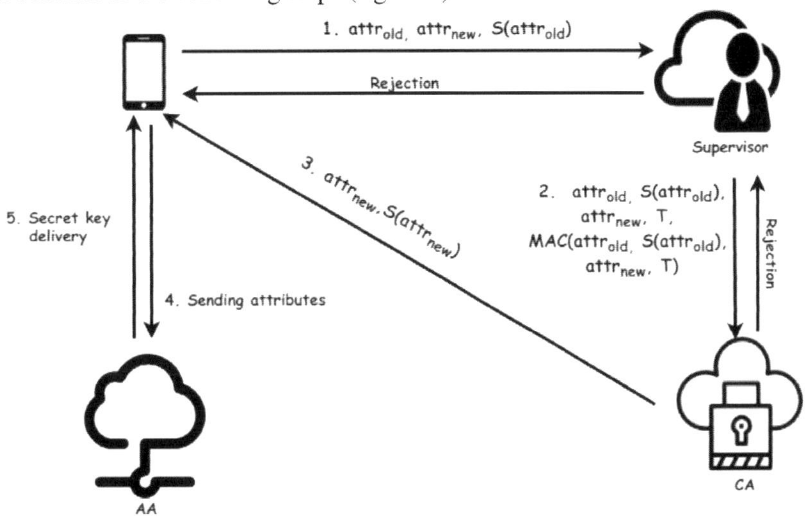

Figure 3. Supervised Attribute Authority Protocol

1. The context in which the device is located is changed - it is necessary to get a new secret key to read messages intended for new attributes. The Device sends to the Supervisor a signed old and (if the change is initiated by the Device itself) new attribute values:

$$D \rightarrow attr_{old}, S(attr_{old}), attr_{new}{}^{*} \rightarrow SV.$$

2. The Supervisor verifies the signature of old attributes to make sure they are correct. If not, it refuse to continue a transaction. If signature is correct and Supervisor is agree with changes in attributes and Supervisor is allowed to make these changes then S adds a time stamp and calculates authenticator of the whole resulting message. Then S sends to CA this message and authenticator:

$$SV \rightarrow attr_{old}, S(attr_{old}), attr_{new}, T, MAC(attr_{old}, S(attr_{old}), attr_{new}, T) \rightarrow CA.$$

3. Each Supervisor is limited by his rights to modify or confirm attribute changes. If the Supervisor has exceeded his authority, or the authenticator and (or) time stamp is incorrect, CA responds with a refusal. Otherwise, it signs a set of new attributes and passes them and a signature of $attr_{new}$ to the Device:

$$CA \rightarrow attr_{new}, S(attr_{new}) \rightarrow D.$$

In case the Device needs to change several attributes belonging to different supervisors, steps 1-3 are repeated sequentially for each of the supervisors.

4. Further, according to KP-ABE or CP-ABE paradigm, the Device presents attributes and digital signature of CA to the trusted center AA:

$$\mathrm{D} \rightarrow attr_{new}, \mathrm{S}(attr_{new}) \rightarrow \mathrm{AA}.$$

5. Finally, AA verifies the signature and, if it's correct, begins a pass of the secret key to Device using some appropriate protocol.

Note that scheme above can be implemented in decentralized environment with several CAs using cross-certificates to verify each other's signatures. Also various trusted centers can respond for different sets of attributes, so the construction of MA-ABE (Multi-Authority ABE) schemes is in progress.

It should be noted that MAC is evaluated and verified using a secret key, which must be unique for each pair (SV, CA), so in the case of a distributed system with multiple SVs and CAs, it is advisable to replace it with the digital signature of the supervisor S_{SV}. Then steps 1-3 of the protocol above will look like this:

1. $\mathrm{D} \rightarrow attr_{old}, \mathrm{S}(attr_{old}), attr_{new}{}^* \rightarrow \mathrm{SV}$;
2. $\mathrm{SV} \rightarrow attr_{old}, \mathrm{S}(attr_{old}), attr_{new}, \mathrm{T}, \mathrm{S}_{SV}(attr_{old}, \mathrm{S}(attr_{old}), attr_{new}, \mathrm{T}) \rightarrow \mathrm{CA}$;
3. $\mathrm{CA} \rightarrow attr_{new}, \mathrm{S}(attr_{new}) \rightarrow \mathrm{D}$.

The digital signature, although calculated slower than MAC and has a larger size, is a public-key algorithm and therefore only one open-secret key is sufficient for each SV.

Note that in proposed scheme it's possible to merge CA and AA, as well as the presence of several SV and CA.

In this scheme, as in most ABE schemes, the set of attributes becomes known to third parties: the Certification Authority, even if encrypted communication channels are used. To hide the set of attributes from CA, we propose to use a blind-based approach [19-21].

3. Blind Signature in Supervised Attribute Authority Scheme

Now we will describe the interaction between the Supervisor and the CA in the framework of the new approach, that is, we complete the scheme given above. Now we assume that attributes are confidential and should not be known to anyone other than the Device, AA and Supervisor, hence all communication channels are considered protected. The purpose of the proposed modification of the protocol is to hide the attribute values from the CA. We assume that there is K attributes in *attr* string, the supervisor has the right to confirm the changes in k of them, each of which has n_1, n_2, ..., n_k possible values, respectively.

1. The supervisor creates l identical sets of attributes, each set contains d different attribute values ($d \leq n_1 \cdot n_2 \cdot ... \cdot n_k$), and $attr_{new}$ among them. All the changed attributes in the set should be available to the supervisor within the scope of his rights. In each set, Supervisor mixes the attributes randomly and encrypts them according to a blind signature scheme (for example, using a blinding factor if Chaum scheme is used [19]). The encryption keys are random, one-time and are different for each set.

2. By second step of the protocol above, instead of the $attr_{new}$ the Supervisor sends all sets of encrypted attributes.

3. CA randomly selects the set to sign and asks secret keys of others.

4. SV sends to CA all the secret keys except one and the position of $attr_{new}$ in the encrypted set.

5. After receiving the keys from the SV and decrypting the l-1 set, the CA makes sure that all the sets contain the same attribute values that are within the SV rights. Probability of correctness of encrypted set is $(l\text{-}1)/l$.

If all the parameters are correct, CA signs blindly the specified value in the selected encrypted set and sends a signature to SV.

6. SV decrypts the received blind signature and in this way receives a CAs digital signature of attributes $attr_{new}$. Afterwards instead of step 3 of the protocol above, transmits $attr_{new}$ and signature to the device.

Thus, although CA is convinced with a probability $(l-1)/l$ that Supervisor has not exceeded his authority, he does not recognize these new attributes of the device. From his point of view, there are d possible values of attributes.
Note that in the above approach there should be a lot of attributes, they must be independent of each other and have a large set of valid values. This option is often not always achievable in practice. Also not only one attribute should be of particular importance for an attacker, since it is easy to guess, and a certain aggregate with the condition of independence of attributes from each other.

We just described a scheme in brief. Let describe a process in detail.

3.1. Blinding of Attributes (Steps 1-2 of Blind Signature Protocol)

Let $attr=[attr_1, attr_2, ..., attr_d]$ is a set of random attributes, containing true attribute $attr_{new}$, where $attr_m=attr_{new}$, and $m\in\{1,2,...,d\}$ selected randomly too. Every array element in $attr$ is possible modification of $attr_{old}$ that Supervisor allowed to perform.

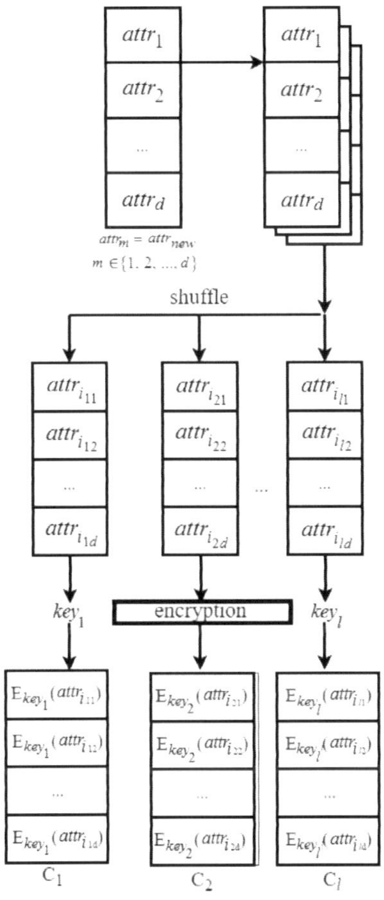

Figure 4. Blinding of attributes in blind signature (steps 1-2).

After creation of $attr$ set Supervisor forms l random permutations $(i_{11}, i_{12}, ..., i_{1d})$, $(i_{21}, i_{22}, ..., i_{2d})$, ..., $(i_{l1}, i_{l2}, ..., i_{ld})$ and makes l shuffled arrays

$$P_1=(attr_{i11}, attr_{i12}, ..., attr_{i1d}),$$

$$P_2=(attr_{i21}, attr_{i22}, ..., attr_{i2d}),$$

$$...$$

$$P_l=(attr_{il1}, attr_{il2}, ..., attr_{ild})$$

through implementation them to attribute array.
Afterwards Supervisor creates l random secret keys key_1, key_2, ..., key_l and encrypts every single P_j (j=1,..., l) by corresponding key key_j using a symmetric-key encryption algorithm E (that commutates with digital signature S according the blind signature principle: S(E(x))=E(S(x)) $\forall x$). Note that every array element of P_j is encrypted separately:

$$C_1=E_{key1}(P_1)=(E_{key1}(attr_{i11}), E_{key1}(attr_{i12}), \dots, E_{key1}(attr_{i1d})),$$

$$C_2=E_{key2}(P_2)=(E_{key2}(attr_{i21}), E_{key2}(attr_{i22}), \dots, E_{key2}(attr_{i2d})),$$

$$\dots$$

$$C_l=E_{keyl}(P_l)=(E_{keyl}(attr_{il1}), E_{keyl}(attr_{il2}), \dots, E_{keyl}(attr_{ild})).$$

Set C_1, C_2, ..., C_l is entity Supervisor creates on Step 1 and sends to CA on Step 2 of protocol above.

3.2. Blind Selection of Signed Element (Steps 3-5 of Blind Signature Protocol)
CA receives a set C_1, C_2, ..., C_l and randomly selects index $x \in \{1, 2, ..., l\}$ that indicates array CA going to sign. CA sends x to Supervisor who respond wit set of secret keys key_1, ... , key_{x-1}, key_{x+1}, ..., key_l (excluding key_x) and index of true attributes in selected array y: $i_{xy}=m$.

CA decrypts every single C_j (j=1,...,x-1,x+1,...,l) with corresponding key_j, and every array element of C_j decrypted separately. CA make sure that every decrypted array is some permutation of each other and every array element is legal modification of $attr_{old}$. If not, CA refuses to continue this protocol and sends to Supervisor message about error.

3.3. Blinding Signing and Unmask (Steps 5-6 of Blind Signature Protocol)
If all decrypted arrays are correct, CA picks true encrypted attributes $E_{keyx}(attr_{new})$ = $E_{keyx}(attr_{ixy})$ = $C_x[y]$ and digitally signed it. Then CA sends a resulted digital signature $S(E_{keyx}(attr_{new}))$ to Supervisor.

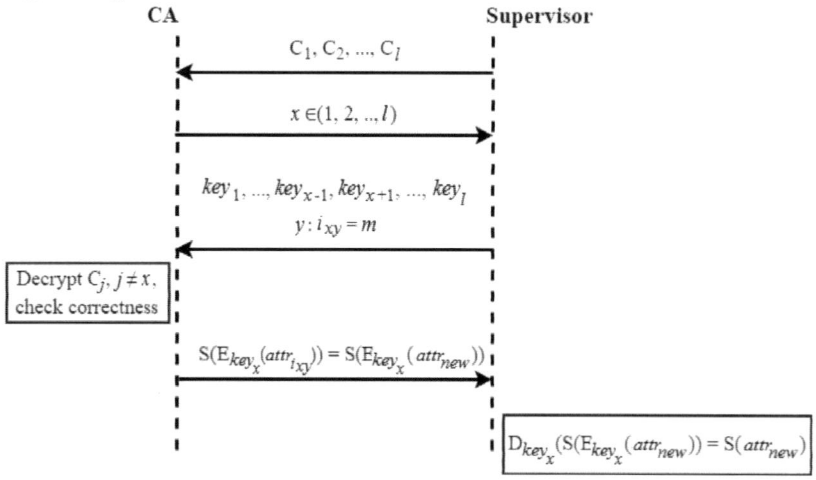

Figure 5. Blind selection of masked element, signing and unmask the digest (steps 3-6).

Finally Supervisor decrypts received value using decryption algorithm D with secret key key_x: $D_{keyx}(S(E_{keyx}(attr_{new})) = S(attr_{new})$. Afterwards Supervisor verifies resulting signature with open key of CA to acknowledges signature is correct. If yes, Supervisor pass attributes and signature to Device.

3.4. Security
A proposed Blind Signature protocol has vulnerabilities typical for all Blind signature schemes and provide to Supervisor some possibilities for overwhelming his boundary.
First of possible attack is follows. If Supervisor has a possibility of multiple tries in case of discrepancy on Step 5 (if some of array is different with other and/or has attributes out of Supervisor's rights), it may purposely sent exceeding attribute inside one custom array in hope it will not be decrypted. To avoid this vulnerability there must be logging and limited tries for each supervisor.

Second vulnerability is relevant if Chaum scheme if used. Since RSA signature in Chaum scheme is used, Supervisor can attempt to find several legal attributes $attr_{new}$*, $attr_{new}$** so that their product modulo RSA will provide $attr_{new}$ that transcends Supervisor's authority. Then after Supervisor receives CA's signatures of $attr_{new}$* and $attr_{new}$** it can calculate CA's signature of $attr_{new}$ by multiplying them modulo RSA.

We cannot avoid this vulnerability because in Chaum blind signature scheme the signer signs the message directly. By contrast, in an usual signature scheme the signer would typically use a cryptographic hash function applied to the message, instead of signing the message itself. But according blind signatures, implementation of hash-function would produce an incorrect value when unblinded. Probability of this attack is definitely small if attribute notation is redundant and therefore it's difficult to find appropriate false attribute values. Anyway, we do not recommend use of RSA signature. Much better are ElGamal-based schemes like [20,21].

4. Conclusion

A supervised attribute-authority protocol with optional blind signatures for verification of device attributes for use in the Internet of Things is proposed. A blind signature scheme for attributes is intended to hide new attribute values from the Center Authority which may be useful if the device attributes contain confidential information, such as personal or medical data. Brief security analysis of proposed protocols is done.

5. References

[1] Shamir A 1984 *Proc. Of CRYPTO 84. LNSC* **196** 47–53 Springer
[2] Sahai A and Walters B 2005 *Eurocrypt* **3494** 457–73 Springer
[3] Chase M 2007 *Proc. of the 4th Theory of Cryptography Conference* 515–34 Springer, Berlin, Heidelberg
[4] Lewko A and Walters B 2011 *Proc. of the 30th Annual international conference on Theory and application of cryptographic techniques: advances in cryptology* 568–88. Springer-Verlag
[5] Herranz J, Laguillaumie F and Rafols C 2010 *Public Key Cryptography - PKC* 19–34. Springer
[6] Goyal V, Pandey O, Sahai A and Waters B 2006 *Proc. of the 13th ACM conference on Computer and Communications Security* 89–98 ACM
[7] Cheung L and Newport C 2007 *Proc. ACM Conference on Computer and Communications Security (CCS)* 456–65 ACM
[8] Shusheng Y, Wang C and Ren K 2010 *Proc. Of the 5th ACM Symposium on Information, Computer and Communication Security* 261-70 ACM
[9] Jiang S, Safavi-Naini R and Shahandashti S.F. 2012 *International Journal of Applied Cryptography* **2(3)** 250–70
[10] Jiang Y, Susilo W, Mu Y ang Guo F 2017 *Future Generation Computer Systems.* Elsevier
[11] Hishide T, Yoneyama K and Ohta K 2008 Attribute-Based Encryption with Partially Hidden Encryptor-Specified Access Structures *ACNS* 111–29 Springer, Berlin, Heidelberg
[12] Internet of Things Global Standards Initiative Available at: http://www.itu.int/en/ITU-T/gsi/iot/Pages/default.aspx
[13] Lee J, Oh S and Wook Jang J 2015 *The 10th International Conference on Future Networks and Communications, Procedia Computer Science* **56** 271–75. Elsevier
[14] Ambrosin M, Anzanpour A, Conti M, Dargahi T, Moosavi S R, Rahmani A M and Liljeberg P 2016 *IEEE Micro* **36(6)** 25–35
[15] Wang X, Zhang J, Schooler E M and Ion M 2014 *IEEE International Conference on Communications*
[16] Odelu V, Das A K, Khan M K, Choo K R and Jo M 2017 *IEEE Access* **5** 3273– 83

[17] Zickau S, Thatmann D, Butyrtschik A, Denisow I and Kupper A 2016 *19th International ICIN Conference*
[18] Pisa C, Caponi A, Dargahi T, Bianchi G and Blefari-Melazzi N 2016 *HotPOST16* 31–36, Paderborn, Germany
[19] Chaum D 1983 *Advances in cryptology* 199–203 Springer US
[20] Camenisch J, Piveteau J M and Stadler M 1995 *Advances in Cryptology - EUROCRYPT'94* 428–32 Springer Berlin/Heidelberg
[21] Chakraborty K and Mehta J 2012 *IJ Network Security* **14(6)** 316–19

Geometric model of pseudo-distance measurement in satellite location systems

K L Panchuk, A A Lyashkov and E V Lyubchinov

Omsk State Technical University, 11, Mira av., Omsk, 644050, Russia

E-mail: panchuk_kl@mail.ru

Abstract. The existing mathematical model of pseudo-distance measurement in satellite location systems does not provide a precise solution of the problem, but rather an approximate one. The existence of such inaccuracy, as well as bias in measurement of distance from satellite to receiver, results in inaccuracy level of several meters. Thereupon, relevance of refinement of the current mathematical model becomes obvious. The solution of the system of quadratic equations used in the current mathematical model is based on linearization. The objective of the paper is refinement of current mathematical model and derivation of analytical solution of the system of equations on its basis. In order to attain the objective, geometric analysis is performed; geometric interpretation of the equations is given. As a result, an equivalent system of equations, which allows analytical solution, is derived. An example of analytical solution implementation is presented. Application of analytical solution algorithm to the problem of pseudo-distance measurement in satellite location systems allows to improve the accuracy such measurements.

1. Introduction

The distance between satellite and receiver is the main sought parameter of satellite coordinate metrology. Due to substantial bias in distance measurement, the acquired values of distance are called pseudo-distances. A pseudo-distance is a value of distance between satellite and receiver at the moment of transmission and reception of radio signal with use of pseudorandom code packages generated on satellite and receiver [1]. The former are generated using satellite clock and the latter – using receiver clock.

Satellite and receiver clock readings diverge, which causes inaccuracy in distance measurement. Additional inaccuracies are caused by delays on the path of the radio signal in the atmosphere (ionosphere and troposphere).

With consideration of these singularities in pseudo-distance measurement theory [1] the quadratic equation for pseudo-distance measurement is acquired. The equation includes four unknown values comprising the coordinates (x_r, y_r, z_r) of receiver location and receiver clock adjustment δt_r alongside the coefficients. In order to define these unknown values at least four satellites are surveyed and a system of four quadratic equations is formed, where each equation corresponds to the distance between the receiver and a satellite.

In order to solve the system of equations, linearization of comprising quadratic equations is performed, which decreases accuracy of the solution. As a result, coordinate metrology has inaccuracy level of several meters [1], and the method itself has mostly navigational application.

2. Problem definition

The objective of this paper is to eliminate linearization of quadratic equations of the system on the basis of geometric interpretation of the equations and the system, which leads to simplification of the existing mathematical model of pseudo-distance measurement and increasing the accuracy of measurements.

3. Theory

1. Pseudo-distance measurement mathematical model simplification

Working equation for pseudo-distance measurement is of the following form [1]:

$$R = (t_r - t_s)\cdot c = \rho + c\cdot(\delta t_r - \delta t_s) + \delta R, \quad (1)$$

where t_r and t_s represent clock readings on receiver and satellite registered at the moment of pseudo-distance measurement; δt_r and δt_s represent clock reading drift relative to standard clock on receiver and satellite at the moment of pseudo-distance measurement; c represents electromagnetic wave propagation velocity in vacuum; ρ represents geometric distance between satellite and receiver at the moment of pseudo-distance measurement; $\delta R = c\cdot\delta t_{atm}$ represents adjustment to the measured pseudo-length caused by the influence of the atmosphere.

Additionally:

1) The value of δt_s is determined with the aid of surveillance stations for each individual satellite and receiver and is contained within navigational packages sent to the customer (receiver). In the absolute method of receiver location the value of δt_s is considered known.

2) The value of δt_{atm} is preliminary calculated on the basis of modeling of delays on the path of the radio signal in ionosphere and troposphere.

3) The value of δt_r is not preliminary determined and is among the unknown parameters.

4) $\rho = \sqrt{(x_s - x_r)^2 + (y_s - y_r)^2 + (z_s - z_r)^2}, \quad (2)$

where (x_s, y_s, z_s) represent the known coordinates of the satellite.

Consequently, the equation (1) includes four unknown values comprising the coordinates (x_r, y_r, z_r) of receiver location and receiver clock adjustment δt_r. In order to define these unknown values at least four satellites are surveyed and a system of four quadratic equations (1) is solved.

As follows from the equation (1), taking into the account four satellites:

$$R_{1i} = R_i + c\cdot\delta t_{si} - c\cdot\delta t_{atm} = \rho + c\cdot\delta t_r, \qquad i = 1,2,3,4.$$

From the latter equation the equation $\rho = R_{1i} - c\cdot\delta t_r$ is derived, from which follows the system of equations

$$(x_r - x_{si})^2 + (y_r - y_{si})^2 + (z_r - z_{si})^2 - (c\cdot\delta t_r - R_{1i})^2 = 0, \qquad i = 1,2,3,4.$$

For further convenience of algebraic transformations we substitute designation:

$$x_r \to x_1, y_r \to x_2, z_r \to x_3, x_{si} \to a_{i1}, y_{si} \to a_{i2}, z_{si} \to a_{i3}, R_{1i} \to a_{i4}, c\cdot\delta t_r \to x_4 .$$

The latter system of equations after designation substitution is as follows:

$$F_i(x_1, x_2, x_3, x_4) = (x_1 - a_{i1})^2 + (x_2 - a_{i2})^2 + (x_3 - a_{i3})^2 - (x_4 - a_{i4})^2 = 0, \quad i = 1,2,3,4. \quad (3)$$

Consider the geometric sense of the system of quadratic equations (3). Assuming that (a_{i1}, a_{i2}, a_{i3}) are the coordinates of a certain known sphere S_i, while $r_i = a_{i4}$ is its radius, there is no difficulty in ascertaining that the equation (3) describes a set (∞^3) of spheres centered at (x_1, x_2, x_3) with radiuses $r_4 = x_4$, tangent to the known sphere S_i.

Consequently, the system of equations (3) describes a finite amount of spheres tangent to four known spheres $S_i(a_{i1}, a_{i2}, a_{i3}, r_i = a_{i4})$, $i = 1,2,3,4$.

Thus, the solution of the system of equations (3) determines a finite number of quadruples of numbers $(x_1, x_2 x_3, x_4)$, that determine the coordinates of the receiver and the value of the corresponding adjustment δt_r.

Let us implement the following algebraic transformations of the system of equations (3):

1. We subtract the second equation of the system from the first:

$$F_1 - F_2 = 2[(a_{21} - a_{11})\cdot x_1 + (a_{22} - a_{12})\cdot x_2 + (a_{23} - a_{13})\cdot x_3 - (a_{24} - a_{14})\cdot x_4] + (a_{11}^2 + a_{12}^2 + a_{13}^2 - a_{14}^2) - (a_{21}^2 + a_{22}^2 + a_{23}^2 - a_{24}^2) = 0.$$

Further designations are introduced:

$$2(a_{21} - a_{11}) = A_{11};\ 2(a_{22} - a_{12}) = A_{12};\ 2(a_{23} - a_{13}) = A_{13};\ -2(a_{24} - a_{14}) = A_{14};$$
$$({a_{11}}^2 + {a_{12}}^2 + {a_{13}}^2 - {a_{14}}^2) - ({a_{21}}^2 + {a_{22}}^2 + {a_{23}}^2 - {a_{24}}^2) = A_{15}.$$

Consequently, we receive:

$$F_1 - F_2 = A_{11}x_1 + A_{12}x_2 + A_{13}x_3 + A_{14}x_4 + A_{15} = 0.$$

2. We subtract the third equation of the system from the first:

$$F_1 - F_3 = A_{21}x_1 + A_{22}x_2 + A_{23}x_3 + A_{24}x_4 + A_{25} = 0,$$

where $A_{21} = 2(a_{31} - a_{11})$; $A_{22} = 2(a_{32} - a_{12})$; $A_{23} = 2(a_{33} - a_{13})$; $A_{24} = -2(a_{34} - a_{14})$; $A_{25} = ({a_{11}}^2 + {a_{12}}^2 + {a_{13}}^2 - {a_{14}}^2) - ({a_{31}}^2 + {a_{32}}^2 + {a_{33}}^2 - {a_{34}}^2)$.

3. Likewise, we subtract the fourth equation of the system from the first:

$$F_1 - F_4 = A_{31}x_1 + A_{32}x_2 + A_{33}x_3 + A_{34}x_4 + A_{35} = 0,$$

where $A_{31} = 2(a_{41} - a_{11})$; $A_{32} = 2(a_{42} - a_{12})$; $A_{33} = 2(a_{43} - a_{13})$; $A_{34} = -2(a_{44} - a_{14})$; $A_{35} = ({a_{11}}^2 + {a_{12}}^2 + {a_{13}}^2 - {a_{14}}^2) - ({a_{41}}^2 + {a_{42}}^2 + {a_{43}}^2 - {a_{44}}^2)$.

The following system of equations is the result of the foregoing transformations:

$$\begin{aligned} F_1 &= 0, \\ F_1 - F_2 &= 0, \\ F_1 - F_3 &= 0, \\ F_1 - F_4 &= 0. \end{aligned} \tag{4}$$

The first equation of the latter system is quadratic, the other three are linear. Thus the initial system of four quadratic equations (3), which is currently used as a mathematical model of pseudo-distance measurement, is replaced by an equivalent system of equations (4), which includes only one quadratic equation. The simplification of the initial system of equations (3) is obvious.

Let us perform the geometric interpretation of the new system of equations (4), which represents the simplified mathematical model of pseudo-distance measurement. The first equation of the system $F_1 = 0$ describes a set (∞^3) of points of hyperquadric of space R^4, which represents a hypercone of revolution with body angle at the vertex equal to 90° and a spherical base$(a_{11}, a_{12}, a_{13}, r_1 = |a_{14}|)$. Let us call this cone an α-hypercone K_α^3. The other three equations of the system (4) describe hyperplanes P_i^3, $i = 1,2,3$ of space R^4.

Let us analyze the solution of the system (4) on the basis of its geometrical representation. Obviously, the equation of intersection $K_\alpha^3 \cap P_i^3 = K_i^2$, $i = 1,2,3$ takes place. This equation defines two-dimensional quadrics K_1^2, K_2^2, K_3^2. Consider the following pairwise intersections: $K_1^2 \cap K_2^2 = k_{12}^4 = (k_{12}^2 + k_{12(\infty)}^2) \subset P_1^3 \cap P_2^3 = P_{12}^2$; $K_2^2 \cap K_3^2 = k_{23}^4 = (k_{23}^2 + k_{23(\infty)}^2) \subset P_2^3 \cap P_3^3 = P_{23}^2$. Where $k_{12(\infty)}^2 = k_{23(\infty)}^2 = k_\infty^2$ is ideal conic. As follows from the equations of intersection, k_{12} and k_{23} are curves of second order situated correspondingly in planes P_{12}^2 and P_{23}^2, while the quadrics K_i^2 themselves represent α-cones with the same feature of intersection.

Thus the geometric solution of the system of equations (4) is a pair of points belonging to the line of intersection of the planes P_{12}^2 and P_{23}^2 and to the intersecting conics k_{12} and k_{23} simultaneously.

4. Results of the experiment

The computing experiment initial data is taken up as follows:

$$\begin{aligned} a_{11} &= 2;\ a_{12} = 0;\ a_{13} = 2;\ a_{14} = 3; \\ a_{21} &= 5;\ a_{22} = 6;\ a_{23} = 5;\ a_{24} = 4; \\ a_{31} &= -3;\ a_{32} = 3;\ a_{33} = 5;\ a_{34} = 2; \\ a_{41} &= 12;\ a_{42} = 0;\ a_{43} = 4;\ a_{44} = 5. \end{aligned}$$

Based on the system of equations (4), the following equations of the projections of curves k_{12} and k_{23} on coordinate plane $(x_1 x_2)$ are obtained:

$$k_{12}:\quad (x_1 - 2)^2 + x_2^2 + (\frac{1}{3}x_1 - \frac{3}{2}x_2 + \frac{29}{4})^2 - (4x_1 + \frac{3}{2}x_2 - \frac{43}{4})^2 = 0,$$

$$k_{23}:\quad (x_1 - 2)^2 + x_2^2 + (x_1 - 3x_2 - \frac{5}{4})^2 - (6x_1 + 3x_2 - \frac{145}{4})^2 = 0.$$

Rendering of the curves is presented on fig. 1.

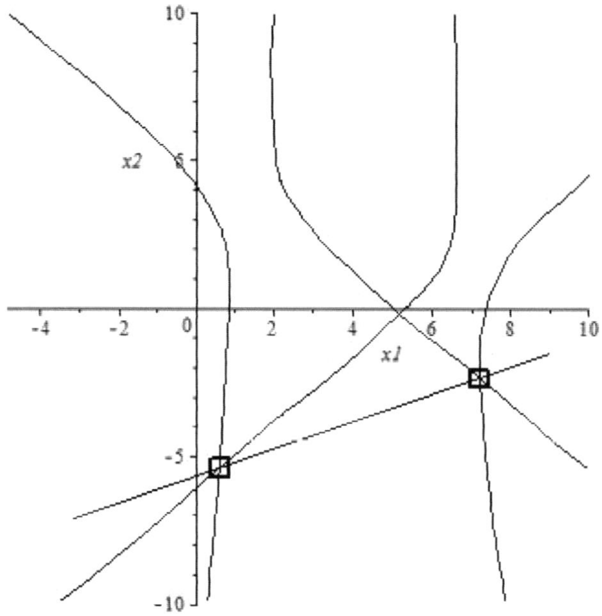

Figure 1. Projections of the conics and projections of their intersection points on the plane (x_1x_2)

The points of intersection of the projections of second order curves k_{12} and k_{23} allow us to pass on to the coordinates of these points in space R^3, and further to rendering of two spheres centered in these points and tangent to the given four spheres (fig. 2, fig. 3).

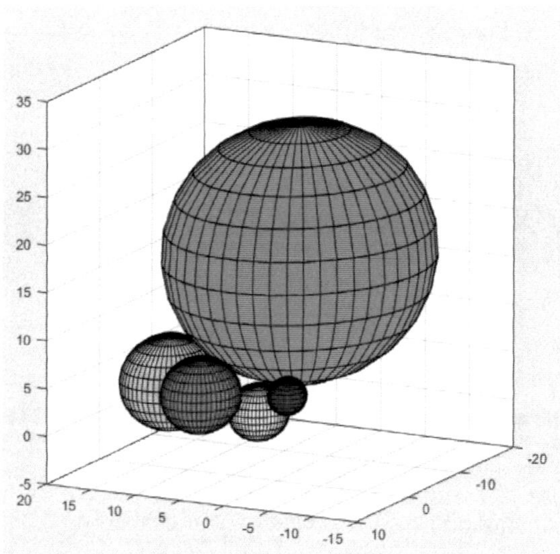

Figure 2. Geometric interpretation the first solution of the system of equations

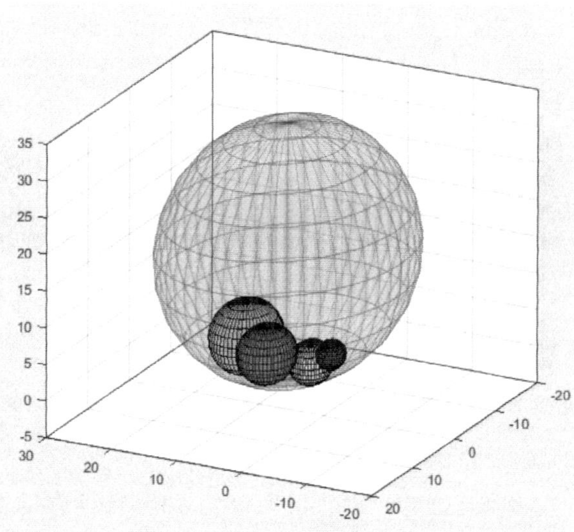

Figure 3. Geometric interpretation of the second solution of the system of equations

The existence of two spheres means that the system of equations (4) has two solutions, only one of which is considered valid based on the conditions of the problem of pseudo-distance measurement. The valid solution represents the coordinates (x_1, x_2, x_3) of the receiver and the value of the adjustment δt_r.

5. Consideration of the results

Computing experiment has confirmed the advantage of the suggested simplified mathematical model of pseudo-distance measurement over the existing mathematical model. The mathematical model suggested in this paper makes it possible to obtain the analytic solution of the system of algebraic equations as opposed to the existing model, the solution of which is based on linearization of the equations. This significantly improves accuracy of pseudo-distance measurement.

6. Conclusion

The results of the study evolve the existing theoretical research aimed on solving the known Fermat problem of finding the set of spheres tangent to the four given spheres [2,3,4]. The development consists in acquiring the analytical solution of the mathematical model of the problem by means of its significant simplification on the basis of geometric representation and interpretation of the equations of the existing model. Application of analytic solution algorithm to the problem of pseudo-distance measurement in satellite location systems allows to improve the accuracy of such measurements.

7. References

[1] Genike A A and Pobedynisky G G 2004 *Global satellite location systems and their appliance in geodesy*. (Moscow, Kartgeotsentr Publ)

[2] Stachel H 1996 Why shall also teachthe theory behind engineering graphics *TU-Wien, Institute for Geometry Technical Report* **35**

[3] Pottmann H and Wallner J 2001 *Computational Line Geometry* (Berlin Heidelberg: Springer Verlag)

[4] Peternell M 2000 Assessment Geometric properties of bisector surfaces *Graphical Models and Image Processing* **11** p 202-236

IOP Conf. Series: Journal of Physics: Conf. Series **998** (2018) 012022 doi:10.1088/1742-6596/998/1/012022

Pyrometer with tracking balancing

D B Ponomarev[1], V A Zakharenko[2] and A G Shkaev[3]

[1] Omsk State Technical University, 11, Mira av., Omsk, 644050, Russia

E-mail: dimi34@gmail.com

Abstract. Currently, one of the main metrological noncontact temperature measurement challenges is the emissivity uncertainty. This paper describes a pyrometer with emissivity effect diminishing through the use of a measuring scheme with tracking balancing in which the radiation receiver is a null-indicator. In this paper the results of the prototype pyrometer absolute error study in surfaces temperature measurement of aluminum and nickel samples are presented. There is absolute error calculated values comparison considering the emissivity table values with errors on the results of experimental measurements by the proposed method. The practical implementation of the proposed technical solution has allowed two times to reduce the error due to the emissivity uncertainty.

1. Introduction

When using pyrometric devices one of major metrological problems is the control object emissivity uncertainty, due to the fact that this coefficient depends on measured temperature, the test object surface texture and the range of wavelengths in which radiation pyrometric measurement is checked. For example, the integral value of emissivity at the same thermodynamic temperature of the brick lining is 0.8...0.9, and aluminum is 0.05...0.2 [1], and by the steel surface temperature range 730...1080°C at a wavelength of 0.9 μm, it changes more than 50% [2]. Consequently, the thermometer indications without exact emissivity specification in incorrect metering values of this coefficient can vary significantly. Along with this the instrumental error of such a pyrometer does not exceed one percent.The emissivity values, most often, are given in special tables or graphs, with accuracy a few percents. Calculations of the emissivity values specifying errors effect on the magnitude of measured temperature error show that at a deviation of emissivity predetermined value from the true 1%, for example, at the wavelength of 7-14 μm in the range of measured temperatures from 250 to 500 °C can lead to absolute error of temperature from 0.5 % to 2.5 %.

2. Problem statement

There are various ways to reduce the emissivity influence of the control object on the pyrometer readings so in [3] a pyrometer with automatic emissivity correction is proposed, and in [4] a new emissivity measurement device for the infrared spectrum is proposed. In [5] classification ways of the emissivity influence considering, used in pyrometry is proposed. The greatest application have methods that require prior information on emissivity of the control object [6], which is not always convenient technologically. Principle of spectral relations is currently the most widely used instrumental method in pyrometric devices [7]. However, it is not possible to take into account the temperature dependence of the emissivity as a wavelength function.In this article the pyrometer, in which emissivity influence decrease is implemented by the method proposed in [8], is suggested. In this method the process of pre-information on the control object emissivity is actually automated. The device implements the method of measured surface radiation temperature comparison with the sample surface temperature, which is identical to controlled surface. The temperature of controlled surface is measured by the contact thermometer.

3. Theory

Functional diagram of the pyrometer shown in Figure 1.

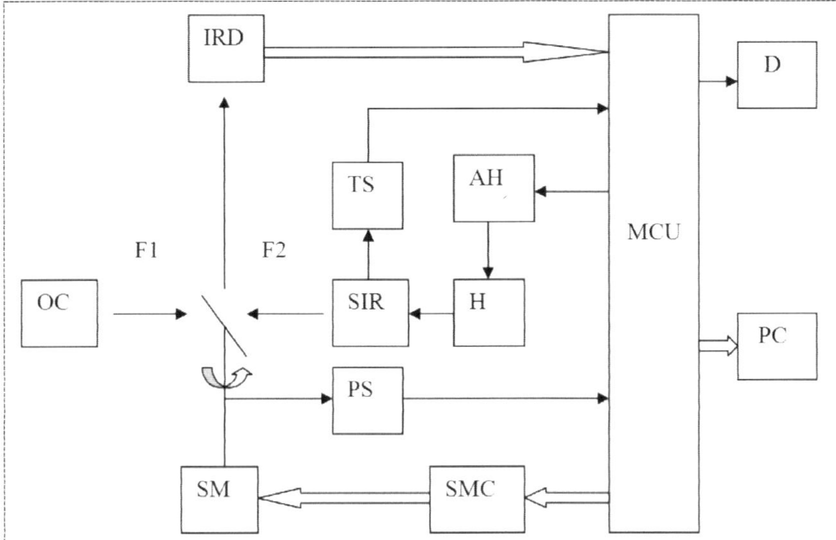

Figure 1. A functional diagram of a pyrometer

OC – object of control; SIR – the source of infrared radiation with a surface identical to the object of control; SM – stepper motor; SMC – stepper motor controller; PS - position sensor; H – heater; AH – actuator heater; TS – a contact temperature sensor; IRD – infrared detector; MCU – microcontroller; D – display; PC – personal computer.

The thermometer works as follows. Before the measurements beginning a sample of material with the surface identical to the surface of the test object is established into the source of infrared radiation SIR integrated in device. On the infrared detector IRD via a modulator made in the form of the rotating mirror alternately receives the flows F1 from the object of control and F2 from the internal radiation source. The modulator rotates by the stepper motor SM through a stepper motor controller SMC control signal from the microcontroller. The position of the modulator is controlled by an optical position sensor PS. Flows F1 and F2 measured by the light receiver are compared with the help of microcontroller and their difference signal is generated by negative feedback control action via the actuator to the heater of the SIR.

In the approximation of the Stefan-Boltzmann flows F1 and F2 can be described by the dependence

$$\Phi = \sigma \cdot \varepsilon \cdot T^4 \cdot A \,, \tag{1}$$

where: T is the temperature controlled radiating surface, K; σ – Stefan-Boltzmann constant, (5,6697±0,0029)·10-12 W/(m^2·K^4); ε is the emissivity of heated controlled surface; A is the geometrical parameter visiwave square heated surface, m^2.

Because the emissivity of the control object ε_{OC} and material sample placed in the source of infrared radiation ε_{SIR} , are the same

$$T_{OC} = \sqrt[4]{\frac{\varepsilon_{SIR}}{\varepsilon_{OC}}} \cdot T_{SIR} = T_{SIR} \,. \tag{2}$$

Thus the equality of the control object surface temperature T_{OC} and the surface temperature of the sample material placed in the source of infrared radiation T_{SIR} is ensured.

At achieving flows equality F1 and F2 the true temperature of the source of infrared radiation is measured by the contact temperature sensor *DT* microcontroller and displayed on the temperature indicator *I* either a personal computer *PC* as the temperature of the test object.

When measuring the temperature of real objects, a common way of accounting for the control object emissivity value is the use of reference or pre-defined values. The values of the emissivity given in reference literature and documentation on poromerics the transducers have an uncertainty value $\varepsilon=\pm\Delta\varepsilon$. Calculations of the temperature measurement absolute error from the emission coefficient parametric error $\Delta\varepsilon$ produced by the formula [5]:

$$|\Delta T| = \frac{1}{n}\left[1-(\frac{T_O}{T})^n\right]\cdot\left|\frac{\Delta\varepsilon}{\varepsilon}\right|\cdot T \quad (3)$$

where: T – true surface temperature of OK, K; T_O = 27 K – reflected background temperature; $\Delta\varepsilon$ - parametric error of the task of emissivity; ε is the integral value of emissivity; n = 4.83 – coefficient determined by the spectral interval $\Delta\lambda$ = 7-14 microns.

4. The results of the experiments

In work researches of the pyrometer absolute error in measuring the temperature of aluminum and nickel samples with temperature T_{OC} =100°C±0.1°C were made. In this connection the absolute error of temperature measurement by the pyrometer, operating on the proposed method amounted to aluminium 3.5°C. When determining the estimated temperature of the aluminum sample by the formula (3) using the emissivity value equal to 0.2, taken from the reference tabulated data [1], with the introduction coefficient parametric error of the of the $\Delta\varepsilon$=0,05 the estimated absolute error in the surface temperature of 100°C was 7.5°C. And at temperatures of 200°C and 500°C this error was 15°C and 124°C, respectively. Similarly, when determining the nickel sample estimated temperature with the emissivity value 0.07 with the introduction coefficient parametric error $\Delta\varepsilon$=0.02 the estimated absolute error in the surface temperature of 100°C was 12°C, and at temperatures of 200°C and 500°C this error was 12°C and 97°C, respectively.

As an infrared detector in the pyrometer thermoelectric receiver MLX90614ACC with built-in ADC was applied. The spectral sensitivity of the receiver in accordance with the passport data of the manufacturer lies in the range 7...14 microns. As contact temperature sensor is used the the p-n junction of transistors KT814A, calibrated in a thermostat within ±0.1 C. Temperature measurement instrumental error of DT is amounted to 0.2 °C

5. Conclusions

The implementation of the proposed technical solution in the pyrometer has allowed, as shown by the results of the experiment, more than twice to reduce the error due to the blackness coefficient uncertainty. In addition, the infrared detector is a null-indicator in the pyrometer according to the functional diagram with tracking balancing, is shown in Fig. 1. The infrared detector is a measuring tool for balancing two streams: from the test object and from the built-in pyrometer the source of infrared radiation. Therefore, the scheme of such device is not sensitive to temperature and time instability of infrared detector parameters and the enhancer circuit.

References

[1] Kriksunov L Z 1978 *A Guide to the basics of infrared technology* (Moscow: Owls. Radio)

[2] Dolganin Yu N, Zavyalov V M, Kozlov Yu, Mikhailov A A, Mikhailov Al 1997 *Measurement techniques* The spectral ratio pyrometer for the true temperature of carbon steels **2** 23-25

[3] Dobesch A and Poliak J 2013 *Radioengineering* IR thermometer with automatic emissivity correction **22** 1301-1306

[4] Feng Zhanga, Kun Yub, Kaihua Zhangb, Yanlei Liua, Kaipin Xub and Yufang Liu 2015 *Infrared Physics & Technology* An emissivity measurement apparatus for near infrared spectrum **73** 275-280.

[5] Vavilov V P 2013 *Infrared thermography and thermal control* (Moscow: Publishing house

"Spectrum")

[6] Vavilov V 2001 *The basics of transient IR thermographic NDT. Short Course Notes* (Bellingham: SPIE The International Society for Optical Engineering)

[7] Delgrande N, Clark G A, Durbin P F et al. 1991 *SPIE Surveillance Technologies* Buried object remote detection technology for law enforcement **1479** 335-352

[8] RF patent №2456557, MKI G 01 J 5/06. *The method of temperature measurement* / V. A. Zakharenko, Yu. Klikushin, S. A. Orlov, A. G. Shkaev, applicant and paten holder of the State educational institution of higher professional education "Omsk state technical University". No. 2011109417/28; Appl. 11.03.2011; publ. 20.07.2012, bull. **20**

IOP Conf. Series: Journal of Physics: Conf. Series **998** (2018) 012023 doi:10.1088/1742-6596/998/1/012023

Errors in Air Permeability Rationing as Key Sources of Construction Quality Risk Assessment

A A Popov[1], A A Nitievski[2] and R N Ivanov[3]

[1]FBE "Omsk FMC", Omsk, Russia
[2]LTC "IRBest", Latvia
[3]Omsk State Technical University, 11, Mira av., Omsk, 644050, Russia

E-mail: irnsoft@mail.ru

Abstract. The article deals with different approaches to the valuation parameters of air permeability n_{50} and q_{50}. Examples of erroneous conclusions about the state of the building are presented as well as the ways to obtain reliable results. There are obtained comparative data of the air permeability parameters on examples of buildings with different configuration and with different values of compactness factor.

1. Introduction

Estimation of microclimate parameters in the premises is one of the main topics of current researches in the field of construction, energy saving and hygiene. Particular attention is paid to the parameters of air supply to residential and public buildings. The relevance of these studies is explained with two factors. Firstly, one is working to reduce the ventilation rate standards. [1,2]. Secondly, the role of qualitative air composition rises.

The special significance of the qualitative air exchange characteristics proves itself in modern buildings as a person spends most of his life in this room. Air environment of modern buildings has a complex composition and it depends on the air pollution level and intensity of internal contamination sources. These sources are mainly in the decomposition products of polymeric materials, which are part of finishing materials, as well as objects of the domestic situation. Furthermore, the concentration of carbon monoxide gas is constantly changing. Prolonged exposure of these factors, particularly in combination with other (temperature and humidity, the electric and magnetic field, radioactive background, etc.) can have an extremely negative impact on human health, his performance and general health.

2. Formulation of the problem

Reducing ventilation rate normative values in the design is aimed at reduction of heat losses due to convective heat exchange component. Having set (according with [1, 3, 4]) calculated value of air exchange equal to 3 m^3/h with 1 m^2 the amount of thermal energy required for heating the incoming air can be easily calculated. The mode provision is a significant problem, since modern construction standards include reducing of air exchange rate to save energy. By reducing the "safety margin" of living space, importance of fidelity design conditions increases.

Currently there are two methods for determining the minimum required ventilation, sufficient to provide an acceptable indoor air quality.

1. A method based on specific standards of ventilation,

In this method the amount of air coming from the environment is determined depending on the purpose of the room and its mode of operation. This method is used to calculate the parameters of ventilation in the premises not intended to change their destination, volume and nature of entering the room contaminants during the operation period.

2. A method based on the calculation of allowable concentrations of pollutant substances,

In this method, the desired amount of air is determined depending on the size and nature of the contaminants in the room. This method is used to calculate ventilation parameters in the premises intended to change their destination during operation, and where intense sources of pollutants may be present or may appear.

Both methods are based on established standards. However, in fact builders often do not follow the requirements of the project in favor of the existing materials or due to insufficient qualification.

Due to the fact that air parameters can not be controlled with organoleptic method, special measuring technology is required to quantify the ventilation rate and to calculate the cost of thermal energy on the convective heat loss.

Field test method for determining air permeability parameters of buildings has been defined in GOST 31167-2009 [5]. The essence of this standard method is that air in the test room is pumped or sucked off out, and after establishment of steady air flow through the fan at a fixed pressure differential between the tested premises and the outside environment, air flow through the fan is measured and equated it to the air flow, filtered through fences, limiting the test room. Generalized characteristics of air permeability of the test space enclosures are calculated using the measurement results.

3. Rationing of air permeability parameters

When rationing air permeability parameters usually two parameters are determined:

1. Air changes - a value, indicating how many times within 1 hour indoor air is completely replaced to a new one:

$$n_{50} = \frac{Q}{V_{pr.}} [h^{-1}]; \quad (1)$$

where Q is measured flux [m^3/h];
$V_{pr.}$ is the volume of the measured premises [m3].

2. Air permeability:

$$q_{50} = \frac{Q}{S_{encl}} [\frac{m^3}{m^2 \cdot ч}]; \quad (2)$$

where Q is measured flux [m^3/h];
S_{encl} is area of enclosing structure [m^2]

On the territory of the Russian Federation, the following regulations on the air permeability valuation parameters: CR 50.13330.2012 (updated SNiP 23-02-2003), GOST 31167-2009.

Valid values of air exchange multiplicity n_{50} according to CR 50.13330.2012 are presented in Table 1.

Table 1

№	purpose of the building	Value n_{50},h^{-1}
1.	For all types of buildings	≤ 4.0
2.	Buildings with mechanical ventilation	≤ 2.0
3.	passive house	≤ 0.6

According to GOST 31167-2009 the following classes of air permeability of building envelopes are determined:

IOP Conf. Series: Journal of Physics: Conf. Series **998** (2018) 012023 doi:10.1088/1742-6596/998/1/012023

Table 2

Multiplicity of air at Dp = 50 Pa (n_{50}, h^{-1})	**The class name**
$n_{50} < 1$	Very low
$1 < n_{50} < 2$	Low
$2 < n_{50} < 4$	Normal
$4 < n_{50} < 6$	Moderate
$6 < n_{50} < 10$	High
$10 < n_{50}$	Very high

Multiplicity of air exchange rate with pressure difference n_{50}. is used for estimating the air permeability parameters both as in CR 50.13330.20121 and as in GOST 31167-2009. Using this setting can increase the risk of commissioning buildings with low energy efficiency and possible defects in the quality of construction.

These risks can be reduced by normalization air permeability parameter q_{50}, used as a main component for the objective evaluation of building envelopes air permeability.

This conclusion can be drawn as air permeability q_{50} (1m^3 air, passing through 1 m^2 surface of the enclosing structure for 1 hour at pressure difference 50 Pa) describes the properties of the enclosing structure against heat transfer by convection exchange. Thus, the use of multiplicity of air exchange coefficient n_{50} for estimating the air permeability is not correct and inherently does not allow evaluating objectively the properties of the building structure for the entire wide range of buildings this standard to be used.

The main reason for not using n_{50} is the building factor compactness. Coefficient of the building compactness is a ratio of the total outer surface area of the building envelope to entrapped heated volume of the building (the ratio of the enclosing structure area to the internal volume of the building). This ratio varies widely depending on the linear dimensions of the building.

However, it is possible to calculate the rated value $q_{50н}$ using the following formula:

$$q_{50н} = \frac{\Delta p \cdot T}{353 \cdot R_u^{mp}},$$

where Δp is the difference in air pressure on the outer and inner surfaces of the building envelope, Pa, Па

T is the ambient temperature during testing.

R_u^{mp} is normalized total resistance of air permeability to building envelope determined on CR 50.13330.20121

There are considered examples of measurement parameters n_{50} and q_{50} for buildings with different values of compactness coefficient.

Example 1

There are three typical apartment blocks (1 section) with a total volume 27 304 m^3 and building envelope area 7072 m^2. Compactness coefficient is 0.26.

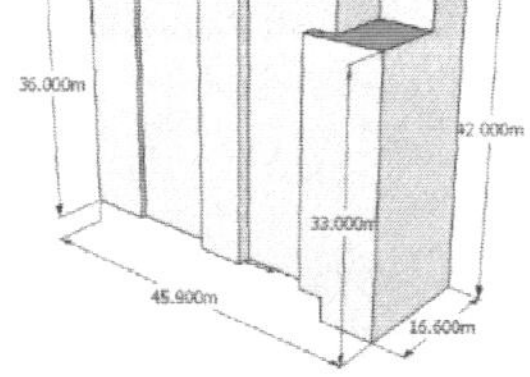

Figure 1. Model of a typical apartment block.

IOP Conf. Series: Journal of Physics: Conf. Series **998** (2018) 012023 doi:10.1088/1742-6596/998/1/012023

Table 3.

	n_{50}	q_{50}	$q_{50н}$
Block №1	0.75	2.85	2.47
Block №2	0.61	2.35	
Block №3	0.85	3.35	

Example 2.

There are four typical logistic centers with total volume 1 150 000 m^3 and building envelope area 207 215 m^2. Compactness coefficient is 0.18.

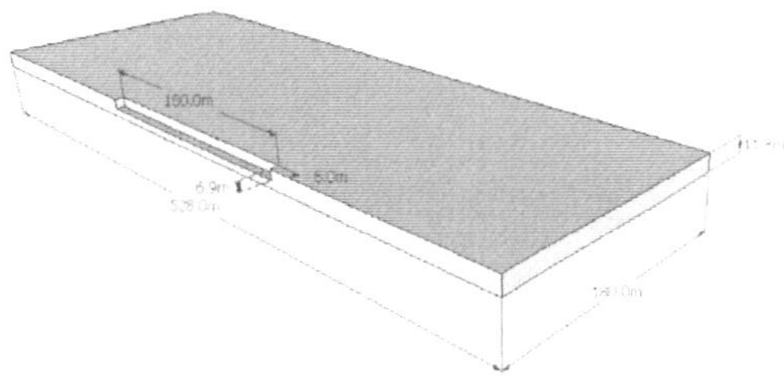

Figure 2. Model of a logistic center

Table 4.

	n_{50}	q_{50}	$q_{50н}$
Building №1	0.23	1.24	0.81
Building №2	0.25	1.37	
Building №3	0.20	1.10	
Building №4	0.14	0.78	

Example 3

There are four typical grocery supermarkets with a total volume 13570 m^3 and building envelope area 5792 m^2

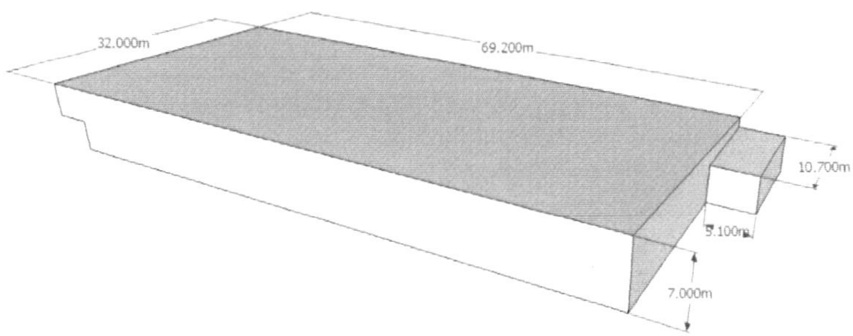

Figure 3. Model of a grocery supermarket.

Table 5.

	n_{50}	q_{50}	$q_{50н}$
Building №1	0.39	0.91	0.74
Building №2	0.29	0.68	
Building №3	0.36	0.84	
Building №4	0.49	1.25	

4. Discussion of the results

Analyzing the experience of countries where the measurement of air permeability (including required) have been practiced for many years and have accumulated enough experience to evaluate the air permeability the parameter q_{50} is used.. These are the United States, Canada, UK, Ireland, France (q_4), Denmark (w_{50}). Conversely, countries that have taken as a basis for regulations on the multiplicity of air exchange n_{50}, have problems with measurement of medium and large buildings, as according to the regulations required to obtain a sufficiently high power setting Blower Door. It leads to the impossibility of carrying out a test, or an increase in its value, but the actual measurement results show significantly better results with modern constructions than required by national regulations. Underreporting standards are indicated and it does not stimulate an increase in the quality of design and construction.

5. Conclusion

These examples demonstrate the disparity of existing national standards to the real state of the air permeability of design and modern technologies in construction. In each group of buildings, only one actually corresponds with the standards. Furthermore, with the increase in linear dimensions the compactness factor tends to decrease and, accordingly, the apparent severity of standards in reality leads to a significant weakening of the requirements for air permeability. Also large buildings generate large heat losses in absolute values. Accordingly, they require greater control and an objective factor for evaluating the properties of heat-shielding constructions.

In conclusion, it should be noted that the use of n_{50} increases the probability of errors in calculations, since to calculate internal volume the net floor area calculation is required (in accordance with EN13829). When changing the actual internal volume of the building in the process of completion of construction or operation (re-plan of internal structures) respectively and calculating volume will change for measuring and it will lead to different results for the same measured flow.

References

[1] STR NP "AVOK" 2.1-2008 *Residential and public buildings. Norms of air exchange* 5-8
[2] ANSI/ASHRAE Standard 62.1-2007
[3] SNiP 41-01-2003 Heating, ventilation and air conditioning
[4] ASHRAE 62-1999 Ventilation for Acceptable Indoor Air Quality
[5] GOST 31167-2009 2011 Buildings and constructions. Methods for determination of air permeability of walling in natural conditions.-(Moscow) *Publishing house standards* 20
[6] Matrosov Yu A and Butovski I N 1991 *Res.build.* **8** 22
[7] Popov A A, Nitievski A A, Vavilov V P and Ivanov R N 2017 *Meas.World* **3**

Statistical Methods in Assembly Quality Management of Multi-Element Products on Automatic Rotor Lines

V V Pries [1], **N E Proskuriakov** [1]

[1] Tula State University, 92, Lenin av., Tula, 300012, Russia

E-mail: vippne@mail.ru

Abstract. To control the assembly quality of multi-element mass-produced products on automatic rotor lines, control methods with operational feedback are required. However, due to possible failures in the operation of the devices and systems of automatic rotor line, there is always a real probability of getting defective (incomplete) products into the output process stream. Therefore, a continuous sampling control of the products completeness, based on the use of statistical methods, remains an important element in managing the quality of assembly of multi-element mass products on automatic rotor lines. The feature of continuous sampling control of the multi-element products completeness in the assembly process is its breaking sort, which excludes the possibility of returning component parts after sampling control to the process stream and leads to a decrease in the actual productivity of the assembly equipment. Therefore, the use of statistical procedures for continuous sampling control of the multi-element products completeness when assembled on automatic rotor lines requires the use of such sampling plans that ensure a minimum size of control samples. Comparison of the values of the limit of the average output defect level for the continuous sampling plan (CSP) and for the automated continuous sampling plan (ACSP) shows the possibility of providing lower limit values for the average output defects level using the ACSP-1. Also, the average sample size when using the ACSP-1 plan is less than when using the CSP-1 plan. Thus, the application of statistical methods in the assembly quality management of multi-element products on automatic rotor lines, involving the use of proposed plans and methods for continuous selective control, will allow to automating sampling control procedures and the required level of quality of assembled products while minimizing sample size.

Key words. Assembly, automatic rotor line, statistical methods, continuous sampling.

1. Introduction

Modern assembly production of mass-produced products is characterized by the continuity of the technological process, high degree of automation and high productivity of technological assembly equipment. Automated rotor lines are used to assemble multi-element mass-produced products in various branches of machine and instrument engineering, both in our country and abroad [1 – 3]. The productivity of automatic rotor lines, depending on the type and size of the assembled multi-element product, can be from 100 to 1000 pieces/min.

The quality of multi-element products assembly is ensured by high requirements to the component parts quality, as well as to the reliability of the main technological systems in the automatic rotor line. From the standpoint of reliability, automatic rotor lines belong to complex reconstructed systems, the features of which are a multi-level structure and the ability to function with different levels of actual performance. Technological rotors form a multi-channel part of the line, which has the property of increased survivability, characterized by the fact that if a functional device or tool fails in one of the rotors, the line retains the ability to function, but with less productivity.

The reliability of the main technological systems of the automatic rotor line is ensured by constructive methods, for example, by reserving the automatic loading of component parts, and organizational methods, in particular, by choosing the optimal maintenance strategy for the multi-channel part of the automatic rotor line [4].

To control the assembly quality of multi-element mass-produced products on automatic rotor lines, control methods with operational feedback are required. The quality control of automated assembling of multi-element products is carried out by using control devices (rotors) and an information-control system in the automatic rotor line structure that provide complete control over the availability of component elements in the process stream, control of automatic loading systems (refusal to load the subsequent element to the assembly in the absence of the previous element), removal of incomplete products from the process stream, etc.

However, due to possible failures in the operation of the devices and systems of automatic rotor line, there is always a real probability of getting defective (incomplete) products into the output process stream. Therefore, a continuous sampling control of the products completeness, based on the use of statistical methods, remains an important element in managing the quality of assembly of multi-element mass products on automatic rotor lines [5 – 9].

The feature of continuous sampling control of the multi-element products completeness in the assembly process is its breaking sort, which excludes the possibility of returning component parts after sampling control to the process stream and leads to a decrease in the actual productivity of the assembly equipment. Therefore, the use of statistical procedures for continuous sampling control of the multi-element products completeness when assembled on automatic rotor lines requires the use of such sampling plans that ensure a minimum size of control samples.

2. Theoretical part

In general, the continuous sampling procedure is an alternation of periods of continuous and spot checks. In a one-stage inspection, the transition from continuous monitoring to selective control is carried out under the condition that n products are successfully received in a stream. Selective control is carried out with a frequency f and at the first appearance of a defective (incomplete) product go to a continuous control. In multi-stage control, several different frequencies are used.

For the first time, Dodge and Romig [10] proposed the models of continuous sampling plan (CSP). Therefore, when using the one-stage CSP-1 plan, the control begins with a continuous inspection of the products, beginning with the first product being manufactured. Such verification is carried out until the control device has passed a contract of i good products. After this, the sampling control begins with a frequency f, continuing until a defective product is found. Immediately a continuous inspection is resumed, which continues until the demand for passing through the control device in a row of i suitable products is fulfilled.

The alternation of selective control procedures and continuous monitoring of the appearance of incomplete products in the process stream leads to the fact that the control devices of the automatic rotor line must work with variable capacity.

In order to eliminate this drawback, models for the automated continuous sampling plan (ACSP) were developed [11]. These plans allow the control to be carried out rhythmically, which greatly simplifies the design of the control devices themselves and their integration into automatic rotor lines. For example, in the one-stage ACSP-1 plan, each f^{-1} collected product is monitored. At the same time, the current sequence of i products, produced by the automatic rotor line, is in the store. When an incomplete product appears among the selectively controlled products, the volume of the storage is taken from the process stream to be scrapped, and the line is adjusted.

IOP Conf. Series: Journal of Physics: Conf. Series **998** (2018) 012024 doi:10.1088/1742-6596/998/1/012024

Since the products in the automatic rotor line are in the process of continuous transportation, when use of the plan ACSP-1 in the volume of the store can also be included the product, subjected to automatic control, but inside the group of i accumulated products. In this case the volume of the storage can be calculated according to the expression: $[(1-f)\,i-1]$.

In the case when suitable ones, do not replace the defective (incomplete) products, which is typical for the automated assembly of multi-element mass-produced products, the expression for the average output defects level will take the form:

- for one-stage plan CSP-1 $$q_{av}=\frac{(1-f)(1-q)^{i-1}}{(1-f)(1-q)^{i-1}+f}q; \qquad (1)$$

- for one-stage plan ACSP-1 $$q_{av}=\frac{1-(1+iq)f}{1-(1+iq)fq}q, \qquad (2)$$

where q - the probability of detection of defective (incomplete) product.

The expression for the average sample size will take the form:

- for one-stage plan CSP-1 $$n_{sam}=\frac{f}{f+(1-f)(1-q)^{i}}\times 100\ \%; \qquad (3)$$

- for one-stage plan ACSP-1 $$n_{sam}=\frac{(1+iq)(1-q)}{1-(1+iq)fq}f\times 100\ \%. \qquad (4)$$

3. Discussion of results

We perform a comparative analysis of the average level of output defects and the average sample size for the above plans. Analysis of the dependencies (1) and (2) shows the possibility of providing lower limit values for the average output defects level using the ACSP-1 (Figure 1).

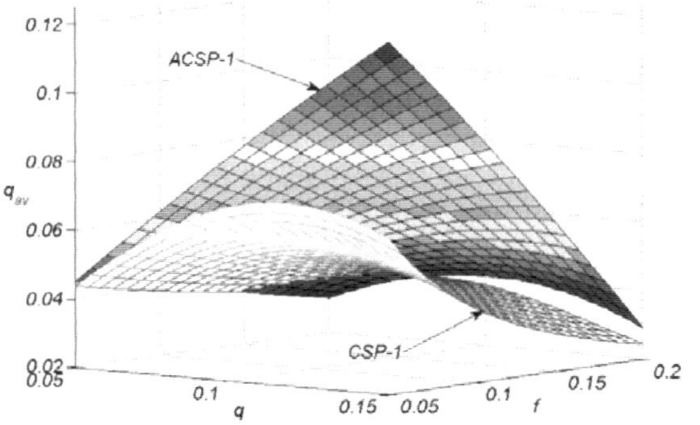

Figure 1. Comparison of the values of the limit of the average output of defect level when using the CSP-1 plan and the ACSP-1 plan (for i =20).

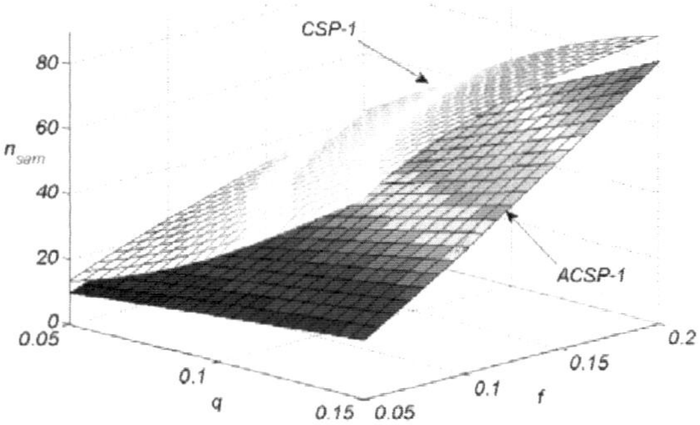

Figure 2. Comparison of the average sample size when using the CSP-1 plan and the ACSP-1 plan (for i =20).

Analysis of the dependencies (3) and (4) shows, that the average sample size in the ACSP-1 plan is smaller than in the CSP-1 plan (Figure 2).

Thus, the simulation results showed that the application of the ACSP-1 plan improves the quality control for both parameters – q_{av} and n_{sam}.

4. Applications of research results

The application of continuous sampling control plans for the management of the automated assembly quality of multi-element mass products on automatic rotor lines made it possible to propose a new method for the implementation of automated statistical control. The aim of the method is to ensure the limit of the average output of defective results. The essence of the proposed method is that the collected products are taken from the process stream for monitoring at a certain frequency and accumulates uncontrolled products in the storage. The storage is located behind the control device. If the monitoring device detects one or several defective products, the products contained in the storage are removed from the stream and must to discard. The volume of the storage exceeds the number of products manufactured during the period between the sampling control, and selective control is performed by several similar or different control devices with separate storages or shared storage.
In Figure 3 shows a scheme for implementing the proposed method for continuous sampling assembly quality control of multi-element products on an automatic rotor line.

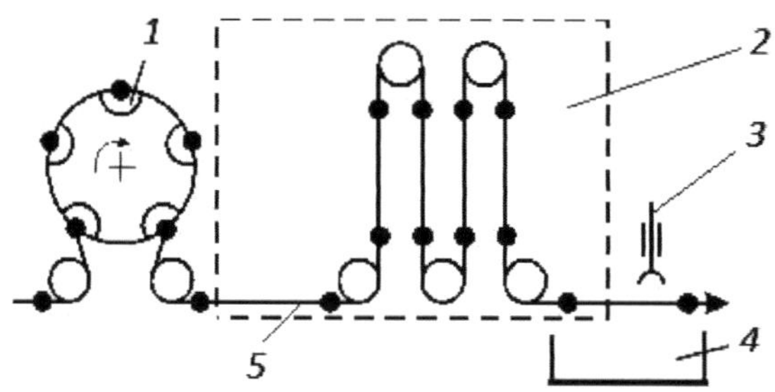

Figure 3. Scheme for implementing the proposed method for continuous sampling quality assembly control on an automatic rotor line: 1 – control rotor; 2 – storage; 3 – device for removing products; 4 – storage and sorting device; 5 – conveyor.

The assembled products are loaded from the conveyor 5 to the appropriate control devices located in the control rotor 1 with frequency f. After the sampling in the control devices, the unchecked products remain inside the storage 2, which may be a continuation of the conveyor 5, or a separate transport element of the automatic rotor line. If a defective product is found, the device for removing products 3, discards uncontrolled products in the storage and sorting device 4. The products in device 4 are scrapped with or without return of suitable products to the conveyor 5 in accordance with the control technology.
The proposed solutions will allow us to proceed to the creation of automated quality control systems that provide a comprehensive solution to the problem of improving the assembly quality of multi-element products in mass production based on the application of statistical methods and procedures, automated technical means for managing the technological stream and computerized information management systems [12].

5. Conclusion

Thus, the application of statistical methods in the assembly quality management of multi-element products on automatic rotor lines, involving the use of proposed plans and methods for continuous selective control, will allow to automating sampling control procedures and the required level of quality of assembled products while minimizing sample size.
The proposed method for the statistical assembly quality control of multi-element products allows for controlled products limit the level of defects to a certain value inherent in the customer requirements.

6. The source of funding

The results of the research project are published with the financial support of Tula State University within the framework of the scientific project № 2017-39. Scientific Director V.V. Preis.

References

[1] Product Catalogue. OJSC Design Bureau of automated lines named after L.N. Koshkin, available at http://kbal.ru/page-165.html
[2] Telerobot S.P.A. High Speed Machines division, available at http://www.telerobot.it/sezioni~phpidSez,18.html
[3] High speed assembly machines, Norwalt Design, Inc. Randolph, available at http://www.norwalt.com/
[4] Preis V V 2003 Reliability of automatic rotor-conveyor lines for the assembly of multi-element products *Assembly in engineering, instrumentation.* **10** p 17–22
[5] Schilling E G, Neubauer D V 2009 *Acceptance Sampling in Quality Control* 2nd ed. (CRC Press, Boka Raton).
[6] Montgomery D C 2009.*Introduction to Statistical Quality Control* 6th edition (New York: John Wiley & Sons)
[7] Hastie T, Tibshirani R and Friedman J 2009 *The Elements of Statistical Learning: Data Mining, Inference and Prediction* 2nd ed. (Springer Science+Business Media, LLC)
[8] Neubauer D V and Luko S N 2011 *Statistical standards and ASTM. Part 2, Quality Engineering.* **23** (1) p 100–104
[9] Gomboeva S G, Hamhanova D N, Shishkin I F, Vasendina E A, Plotnikova I V and Efremova O N 2016 Sampling control planning on the base of operating characteristics and with a grapho-analytical method *Key Engineering Materials* **685** p 463–466
[10] Dodge H F and Romig H G 1959 *Sampling Inspection Tables, Single and Double Sampling* 2nd ed. (John Wiley & Sons, New York)
[11] Gorelov A S, Preis V V and Savvina E A 2007 Automated Statistical Monitoring of Manufacturing Products *Russian Engineering Research* **27** No.11 p 791–795
[12] Gorelov A S, Preis V V and Soskov V B 2008 Design Principles for Integrated Automated Statistical Quality-Control Systems in Manufacturing *Russian Engineering Research* **28** No.3 p 251–254

IOP Conf. Series: Journal of Physics: Conf. Series 998 (2018) 012025 doi:10.1088/1742-6596/998/1/0

Features calibration of the dynamic force transducers

M Yu Prilepko, D. Sc. V G Lysenko

Russian Federation Institute for metrological service (Russian VNIIMS), 46, Ozernaya st., Moscow, 119361, Russia

E-mail: prilepko@vniims.ru

Abstract. The article discusses calibration methods of dynamic forces measuring instruments. The relevance of work is dictated by need to valid definition of the dynamic forces transducers metrological characteristics taking into account their intended application. The aim of this work is choice justification of calibration method, which provides the definition dynamic forces transducers metrological characteristics under simulation operating conditions for determining suitability for using in accordance with its purpose. The following tasks are solved: the mathematical model and the main measurements equation of calibration dynamic forces transducers by load weight, the main budget uncertainty components of calibration are defined. The new method of dynamic forces transducers calibration with use the reference converter "force-deformation" based on the calibrated elastic element and measurement of his deformation by a laser interferometer is offered. The mathematical model and the main measurements equation of the offered method is constructed. It is shown that use of calibration method based on measurements by the laser interferometer of calibrated elastic element deformations allows to exclude or to considerably reduce the uncertainty budget components inherent to method of load weight.

1. Introduction

Now the status of machines, mechanisms, their nodes and separate details is estimated by measurements of their vibration characteristics (vibration rate, vibration acceleration, vibration displacement). However, in some cases it is more expedient to measure directly the dynamic forces arising by operation of this or that mechanism [6]. Calibration methods of dynamic forces transducers known today do not allow to define the valid metrological characteristics researched transducers. It is explained by the distinctions of the metrological characteristics taking place in real operating conditions where their response frequency characteristics do not match that, defined in the lab. Distinctions are caused by the fact that in lab determination of dynamic forces transducers metrological characteristics produced by forces which different from those in real operating conditions. In lab make calibration impulse, quasistatic or by a load mass method that does not provide the broad picture of the dynamic response transducer under test (TUT). The conducted researches showed that the metrological characteristics of dynamic forces transducers determined by a range of their impulse response on different forms of exciting signals have more reliability in comparison with the characteristics determined by a method of the load mass where excitation is made at single frequency on the sinusoidal law of force change. The current situation restricts a possibility of creation full-fledged system of metrological support in the field of dynamic forces measurements.

2. Problem definition

Thus, the task of creation new calibration methods and the measurement techniques allowing to make determination of dynamic force transducers metrological characteristics in the conditions reproducing conditions of real operation on subject to measurements in case of excitation by harmonic complex waves, including, impulse in all the range of operating frequencies in case of broad static load variability is relevant. For achievement of a theoretical measurements accuracy limit, as reference quantities of comparing it is necessary to use physical constants, such as wavelength of laser radiation, matter elastic modulus of the reference transducer "force - deformation".

3. Theory

Methods of static calibration [1], [5] do not allow to define response dynamic characteristics calibration of dynamic force transducers with an acceptable accuracy in the range of operating frequencies therefore in this article are not considered. Now for calibration of dynamic forces transducers the main method is the method of the load mass providing idea of dynamic properties transducers under test. According to this method, for determination of metrological characteristics (a sensitivity at a basic frequency, frequency response) of dynamic force transducers, use the set loading mode by force changing under the sinusoidal law [3]. The transducer under test with the load mass of the known weight established on it is subjected to harmonic oscillations of the known frequency and acceleration, and force operating at the same time on transducer under test, determine in the estimated method by the second law of Newton. An acceleration of a sheaf "transducer under test + the load mass" in case of mechanical oscillations can be measured by means of the reference piezoelectric accelerometer, however, for achievement of higher accuracy, it is necessary to use methods and means of a laser interferometry.

The general diagram of a calibration method and its composition by the load mass is shown in Figure 1.

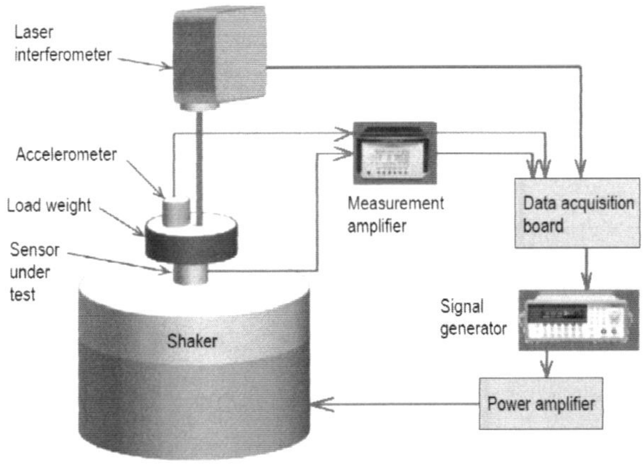

Figure 1. A general view of installation for calibration of dynamic force transducers by load weight method [4]

The mathematical model of calibration process by load mass can be presented in the following form:

$$S_f = \frac{q_f}{(m_i + m_l)\cdot[(x_b - x_t)]f''(t)} \tag{1}$$

S_f is the sensitivity of transducer under test (TUT);
q_f is the output signal TUT;
m_i is the internal mass of a sensitive element TUT;
m_l is the load weight;
x_b, x_t is the initial and final coordinates of a vibration displacement.

The main equation of measurements by this method can be presented in a general view as:

$$S_f = \frac{\sqrt{\sum_{i=1}^{n}(q)u_{sign}^2}}{\sqrt{\sum_{i=1}^{n}(m)u_{mass}^2 \cdot \sum_{i=1}^{n}(a)u_{acc}^2}} \tag{2}$$

$\sum_{i=1}^{n}(q)u_{sign}^2$ is the result of measurements for output signal (charge) TUT;

$\sum_{i=1}^{n}(m)u_{mass}^2$ is the result of measurements for weigh mass;
$\sum_{i=1}^{n}(a)u_{acc}^2$ is the result of measurements for acceleration.

Measurements uncertainty for acceleration consists of two groups: total standard uncertainty of measurements for acceleration by the laser interferometer and total standard uncertainty reproduction of acceleration by an electrodynamic shaker:

$$u_a = \sqrt{\sum_{i=1}^{n} u_{meas}^2 + \sum_{i=1}^{n} u_{rep}^2} \quad (3)$$

We will consider the structure of these groups in more detail:

$$u_a = \sqrt{\begin{matrix} u_{cos}^2 + u_{quad}^2 + u_f^2 + u_{noice}^2 + u_{phase}^2 + u_{Ab}^2 + u_{drift}^2 + u_{ref}^2 + \\ + u_{trans}^2 + u_{rock}^2 + u_{rel}^2 + u_{mag}^2 + u_{grad}^2 + u_{sign}^2 + u_{mass}^2 \end{matrix}} \quad (4)$$

u_{cos} is the Cosine uncertainty;
u_{quad} is the SDE uncertainty;
u_f the is uncertainty caused by instability of laser radiation frequency;
u_{noice} is the uncertainty caused by electronic noise of the laser interferometer;
u_{phase} is the uncertainty caused by phase noise of an interferometer output signal;
u_{ab} is the uncertainty caused of Abbe shift;
u_{drift} is the uncertainty of interferometer setup drift;
u_{ref} is the uncertainty of changes air refraction index;
u_{trans} is the uncertainty caused by cross motion of shaker;
u_{rock} is the uncertainty caused by unevenness of the shaker movement;
u_{rel} is the uncertainty caused by the relative movement of the interferometer;
u_{mag} is the uncertainty caused by influence of a shaker magnetic field on an TUT output signal;
u_{grad} is the uncertainty owing to an acceleration gradient along an axis of TUT;
u_{sign} is the uncertainty of TUT output signal measurements;
u_{mass} is the uncertainty of load weight measurements

On the basis of the analysis listed calibration uncertainty budget components of the of dynamic force transducers by load mass method it is possible to draw a conclusion that the main components of the calibration uncertainty budget are the specified method is consequence of acceleration reproduction and measurements of the transducer under test and the load mass. Therefore, for minimization of the most part of the specified uncertainty components, the failure from an inertial method of excitation of dynamic forces by means of the electrodynamic shaker and transition to their excitation by a inertialess principle would be logical. The inertialess principle of dynamic force excitation (the variable mechanical strength causing proportional deformations) identical to the fastened together reference transducer and transducer under test is the base of this method.

In the offered method the deformation direct measurement of the calibrated elastic element performing function of the reference transducing "force-deformation" by method of a laser interferometry is used. Inertialess dynamic force excitation by means of a piezoactuator use of the calibrated elastic element as the reference transducer "force-deformation" and measurement of this deformation by the laser interferometer allows to exclude most a part of the uncertainty budget components inherent in a method of the load mass. The functional diagram explaining the offered principle is figured in Figure 2.

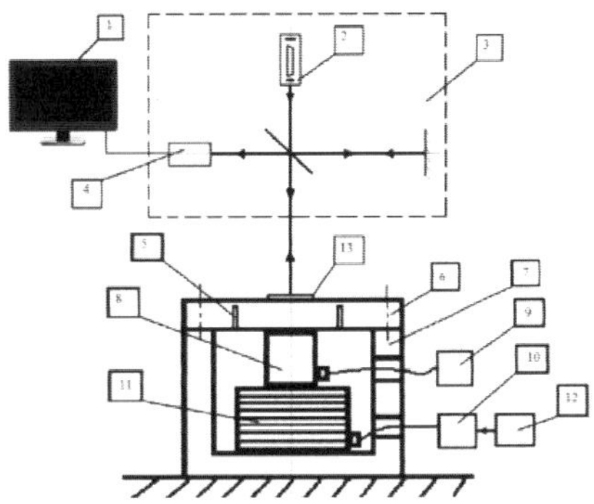

Figure 2. Calibration of dynamic forces transducers by the inertialess method, with use the laser interferometer [2]

1 – PC; 2 – coherent light source (laser); 3 – laser interferometer; 4 – photodetector and encoder of the laser interferometer; 5 – ring pro-point; 6 – calibrated elastic element; 7 – housing; 8 – dynamic force TUT; 9 – measuring amplifier; 10 – power amplifier; 11 – piezoactuator; 12 – signal generator; 13 – reflective coating.

The mathematical model of calibration process can be presented this method in the following form:

$$S_f = \frac{q_f}{s \cdot (E_{max} - E_{min}) \cdot d_{rev}} \quad (5)$$

S_f is the sensitivity of TUT;
q_f is the output signal (charge) of TUT;
s is the stifness of the reference converter "force-displacement";
E_{max}, E_{min} is the maximum and minimum strengths of the piezoactyuator electric field;
d_{rev} is the piezoelectric modulus.

The main equation of measurements can be presented in the deformation method in a general view as:

$$S_f = \frac{\sqrt{\sum_{i=1}^{n}(q)u_{sign}^2}}{\sqrt{\sum_{i=1}^{n}(k)u_{conv}^2 + \sum_{i=1}^{n}(s)u_{strain}^2}} \quad (6)$$

$\sum_{i=1}^{n}(q)u_{sign}^2$ is the measurements result of an TUT output signal (charge);
$\sum_{i=1}^{n}(k)u_{conv}^2$ is the transformation uncertainty of force to deformation coefficient;
$\sum_{i=1}^{n}(s)u_{strain}^2$ is the measurements result of elastic element deformation by laser interferometer.

Finally, it is possible to write down expression of total calibration standard uncertainty by the deformation method in the following look:

$$u_a = \sqrt{u_{cos}^2 + u_{quad}^2 + u_f^2 + u_{noice}^2 + u_{phase}^2 + u_{Ab}^2 + u_{drift}^2 + u_{ref}^2 + u_{hist}^2 + u_{creep}^2 + u_{sign}^2} \quad (7)$$

Apparently, in a formula (7) there are no uncertainty components inherent in a method of load weight caused by reproduction and vibration acceleration measurements: the uncertainty caused by a cross component of shaker motion u_{trans}, uncertainty, caused by unevenness of the shaker movement u_{rock}, the uncertainty caused by the relative movement of the interferometer u_{rel}, uncertainty, caused by influence of shaker magnetic field on an TUT output signal u_{mag}, uncertainty owing to an acceleration gradient along an axis of TUT u_{grad}. At the same time, there is measurements uncertainty of the reference converter "force-deformation" by the laser interferometer, measurements uncertainty of the TUT output signal and the uncertainty components of piezoactuator such as a hysteresis and drift. Uncertainty components inherent to a piezoactuator such as hysteresis, in control mode with a charge feedback it can be reduced to negligible size.

IOP Conf. Series: Journal of Physics: Conf. Series **998** (2018) 012025 doi:10.1088/1742-6596/998/1/012025

4. Results of experiments

The main components numerical values of the calibration dynamic force transducers uncertainty budget by load weight and by a deformation method listed in the section III are presented in comparative Table 1 in the form of the maximum values obtained as a result of an experiment and passport data of the laser interferometer.

Table 1.

Component of uncertainty	Method of load weight	Deformation method
a	b	c
Cosine uncertainty u_{cos}	0.017 %	
SDE uncertainty u_{quad} (passport data)	Periodic, 158 nм	
The uncertainty caused by instability of frequency of laser radiation u_f (passport data)	± 0.05 ppm/hour	
The uncertainty caused by electronic noise of the laser interferometer u_{noice} (passport data)	0.1 nм	
The uncertainty caused by phase noise of an interferometer output signal u_{phase}: uncertainty of an arctangent u_{arc}, time of a delay and jitter u_{del}, random uncertainty u_r (passport data)	1.4 %	
The uncertainty caused by Abbe shift u_{ab}	be absend for RLE10	
The uncertainty caused by working point drift of the interferometer due to conditions fluctuations of the environment u_{drift} (passport data)	100 nм/°C	
The uncertainty caused by fluctuations of air index refraction u_{refr} (passport data) temperature fluctuations atmospheric pressure fluctuations humidity fluctuations	 1 ppm/°C 0.3 ppm/мBar 0.1 ppm/%	
Total standard uncertainty of measurements by the laser interferometer (passport data)	1.2%	
Uncertainty caused cross component of shaker motion u_{trans}	0.5%	0%
Uncertainty caused by unevenness of shaker motion u_{rock}	0.27%	0%
The uncertainty caused by the relative movement of the interferometer u_{rel}	0.04%	0%
a	b	c
The uncertainty caused by influence of a shaker magnetic field on an output signal of the TUT u_{mag}.	0.06%	0%
The uncertainty caused by an acceleration gradient along an axis of the TUT excitations u_{grad}	0.01%	0%
Uncertainty of load weight measurements u_{mass}	0.01%	0%
Uncertainty of the output signal TUT measurements u_{sign}	1% (when using the instrument amplifier 2626 of Bruel & Kjaer, Denmark	
Total standard uncertainty	1.768%	1.2%

5. Discussion of the results

On the basis of the comparative analysis of uncertainty budget components for calibration method by load weight and a deformation method, it is possible to draw the following conclusion: along with the identical values of uncertainty components for the laser system [7] which is used in both methods and presented to items 1 – 8 of Table 1, the uncertainty components such as caused by a cross component of a shaker fluctuations u_{trans}, the uncertainty, caused by unevenness movement of shaker u_{rock}, the uncertainty caused by an acceleration gradient along an axis of TUT u_{grad}, the uncertainty caused by the relative movement of the interferometer u_{rel}, the uncertainty, caused by influence of shaker magnetic field on an TUT output signal u_{mag} , the measurement uncertainty of load mass u_{mass}, in the offered deformation method are reduced to negligible size or are absent thanks to the inertialess principle of dynamic forces excitation by means of a piezoactuator and to lack of TUT free oscillations and the most load weight.

6. Outputs and inference

In the present article a brief description of the scientific rationale and solution of urgent scientific and technical tasks of creating a new high-precision calibration method of dynamic forces transducers, the mathematical model is defined and listed the most significant components of the calibration uncertainty budget.

Comparative analysis of both dynamic forces transducers calibration methods by load weight and strain method determined the main components of the uncertainty budget for these methods.

It is determined that the main contribution to the uncertainty budget of a load mass method is made by the components directly or indirectly connected to reproduction and measurement of the movement characteristic (acceleration). The offered new calibration method for dynamic force transducers by inertialess force excitation principle by means of an piezoactuator with use the calibrated elastic element as the reference transducer "force-deformation" and a laser interferometer for measurements of this deformation allowed to reduce calibration uncertainty of on comparing to a load mass method, to expand the frequency range of calibration and made possibility calibration by forces which law of change is excellent from sinusoidal (the polyharmonic, a pulse sequence, etc.). The offered new calibration method for dynamic force transducers and the experimental gage installation created on its basis allow to make calibration of dynamic force transducers in the range of forces from 0,1 N to 300 N with uncertainty no more than 1.2% in the frequency range from 1 Hz to 10 kHz.

Theoretical reasons and the received experimental results allow to consider the offered calibration method for dynamic force transducers and the experimental gage installation as a prototype of special standard for unit of dynamic forces allowing to reproduce a unit of measure with the initial accuracy, with a binding to wavelength of the stabilized He – Ne laser and elastic modulus of the reference transducer "force-deformation".

References

[1] GOST R 8.640-2014 (Russian State Standard p 8.640-2014). State system of ensuring unity of measurements. The state testing scheme for force measuring instruments (Moscow: Standartinform) p 1-7

[2] Prilepko M Yu 2017 *Method of calibration of transformers of variable force.* Russian Federation patent No 2613583

[3] Christian Schlegel, Gabriela Kiekenap, Bernd Glöckner and Rolf Kumme 2011 *Dynamic Calibration of Force Transducers Using Sinusoidal Excitations* The 2nd Int. Conf. on Sensor Device Technologies and Applications (SENSORDEVICES 2011)

[4] Bartoli C, Beug M F, Bruns T, Elster C, Esward T, Klaus L, Knott A, Kobusch M, Saxholm S and Schlegel C 2012 Int. J. Metrol. Qual. Eng. **3**, p 127–135

[5] ISO 376: 2011 *Metallic Materials – Calibration of Force-Proving Instruments Used for the Verification of Uniaxial Testing Machines* (Int. Organization for Standardization: Geneve, Switzerland)

[6] Nieves Medina and Jesús de Vicente 2014 *Force Sensor Characterization Under Sinusoidal Excitations* Sensors, **14**, 18454–18473

[7] The Modal Shop, Inc. Standard Operating Procedure QS-P009.

Improving the Quality of Backup Process for Publishing Houses and Printing Houses

N E Proskuriakov [1], B S Yakovlev [1], V V Pries [1]

[1] Tula State University, 92, Lenin av., Tula, 300012, Russia

E-mail: vippne@mail.ru

Abstract. The analysis of main types for data threats, used by print media, and their influence on the vitality and security of information is made. The influence of the programs settings for preparing archive files, the types of file managers on the backup process is analysed. We proposed a simple and economical version of the practical implementation of the backup process consisting of 4 components: the command line interpreter, the 7z archiver, the Robocopy utility, and network storage. We recommend that the best option would be to create backup copies, consisting of three local copies of data and two network copies.

Key words. Backup, data protection, cloud computing, network storage.

1. Introduction

Digital technologies are actively used in all branches of human activity. In polygraphy, they are especially important at the stages of preparation of layouts, communication with customers, organization of printing on demand, etc. These technologies are well developed and proven effective. Without them it is difficult to imagine a modern cycle of production of books and other printed products.

During the work of the publishing house, the customer base is accumulating, their orders are executed, which are inexpedient to delete, because there is a possibility of their repeated order. In the course of working on projects, a creative search of ideas takes place, which very often turns into a collection of a clipboard that can nudge certain ideas. Collections of fonts, collages, backgrounds, sets of brushes and styles for various programs - is a very important part of the results of the work of the publishing house, their employees, to some extent, the history and portfolio of the organization. Also worth mentioning is the software, their versions, device drivers, setup color profiles, etc. It is also an integral part of the collection of technological developments.

In addition, you need to take into account that the software version is very important component of a successful re-release, re-release of products. For example, during publishing books in the 90s Lexicon, Photoshop 4.5 editors were used, and it is not necessary that modern versions of the same programs will be able to open older files. There may be conflicts of formats, settings, etc. Because of this, previously prepared projects may not be available for replication.

Moreover, the old software distributions cannot be installed on a modern operating system without additional and not reliable software. So Windows 98 or 95 need to install Photoshop 4.5, and this image of Windows should be available.

IOP Conf. Series: Journal of Physics: Conf. Series **998** (2018) 012026 doi:10.1088/1742-6596/998/1/012026

In addition, in the current realities the virus attacks and competitors can harm the publishing houses, by stealing, destroying, corrupting the collection, practically stopping their work. According to the data of [1], in the current realities the ways of infection have become more diverse due to the development of social networks, systems of communication between users, the affordability of creating websites, mailing messages. Since data exchange between authors and enterprises of printing provides for a large data flow, infection cannot be excluded due to the employee's inattention. This issue is particularly acute escalated now, when there were the flashes of cyber-attacks with virus-encryptors for Internet users. Many large private and public firms, services suffered.

Administrators of computer networks in the small and medium-sized organizations usually do data recovery after such attacks. However, the availability of special service for the information protection does not guarantee the success of data recovery. So, in the case of virus-encryption viruses, encryption takes place over a long enough password to be considered not to be opened. Besides, the password is changing in the course of infection; therefore, even in case of successful password guessing, it is suitable only for one specific file. Decoding in this case can last for months and years, and it will be too long for the company, and practically everyone, in such cases, consider then data is lost forever.

Under the influence of such circumstances, the creation of the archive and its protection somehow still rises before the majority of printing enterprises and publishing houses, which is an actual task in modern production.

2. Formulation of Problem

To date, the majority of various firms' and companies' experts believe that for more high quality and reliable preservation of information it is necessary to copy the files on the devices of different connections and locations. Most often, the right technique is to store the files on the logical drives (hard disk different from the original), on removable, external hard drives and network storages. In this case, the information is systematically distributing in three different data stores what there is a very positive effect on the reliability of its storage. It is also possible to disconnect the external hard drive and sync with the cloud service, which will greatly improve the stability of the archive to virus attacks and malfunctioning.

However, these statements are little supported by facts; the speed of the process has not been analyzed, which has become a very important factor recently.

In order to find out, it is necessary to understand which of the proposed or existing solutions will be most suitable for printing enterprises. Therefore, we have tested the three most popular software products and the rate of their work for copying files in two different directories of a personal computer (PC). The question was investigated - in what form backups should be created: as is or with the use of the archive; as multi-volume archive or not; what should the format of archive and its compression ratio be [1]. Beyond that, we also considered the question with network data repositories and their role in modern status of backup.

To identify the type of archive and the convenience of its use, we conducted a preliminary series of experiments, during which it was revealed that not all archivers could cope with the task assigned to them. Therefore, the archivers WinRAR and ZIP are not able to pack all files because of restrictions for file name length. Furthermore, the ZIP cannot be multi-volume archive and does not support large volume archives, such as our test directory of 31 GB 548 MB.

Consequently, we excluded archivers WinRAR and ZIP from further studies as unreliable because they cannot guarantee an error free backup with any degree of nesting objects.

As a result, we left with only one variant of the archiver - 7z. This is a basic archive of UNIX systems, including Linux. Its main advantage was the fact that it is able to back up any depth of nesting information. It also supports the creation of multi-volume archives and the command line, being practically a utility [2]. In addition, its compression ratio is equal to results of WinRAR [3].

Next, we need to figure out which method of backup would be most beneficial. For this, we need to determine the degree of compression and volume of the archive file.

The compression ratio directly bound to the types of files, which take part in process and sometimes may be the situation when the archive with the maximum compression will decrease insignificantly.
Therefore, it makes sense to choose the speed of backup process as the main benefit measure. The compression rate depends on the degree of compression of the file, for this reason, it is expedient to choose the no compression variant.
As for the division of the archive into volumes, this is not quite an explored area for making backup copies. Almost all well-known software solutions boil down to the fact that each project file is a separate ZIP archive. It is precisely unknown, how reasonably this provision is. Based on practice, we assume that this approach is wrong, because it does not reduce the number of files. Therefore, we believe that one of the main tasks of the backup operation is to create a backup reducing the number of files and to pack materials, which are heterogeneous in type, number and volume into blocks that strictly divide into a predetermined size. According to our observations, this clearly reduces the time of copying process. However, it is necessary to study in depth that we try to carry on.
In addition, the division of archives into volumes is a necessity in case of backup on the cloud services. The fact that almost all of the file storages limit the speed of file transfer because of the huge number of users, which constantly transfer data to the server. The communication channel cannot withstand such a stream, therefore the majority of servers limit the download speed to itself, while leaving the upload speed of files is almost the maximum. Moreover, this objective reality exists in the Internet. However, to verify this assertion it is necessary to conduct the study.
We investigated three types of software to synchronize with network storage: Google Drive, Yandex Disk and DropBox.
Synchronizing your directory with the network storage always comes at the expense of the software installed on the machine where the backup copy is. The task of software is to monitor the status of files in the controlled directory, and in case of their changes instantly to send files to the server. The server produces the creating of actual file versions, their storage and other service functions, such as cloud or network storage, but not the user software installed on PC.
Most of these programs have the standard settings: choice of directory for synchronizing, login and password for the connection. In addition, they can pause their work and embed into the system, reflecting in it as additional directories.
Network storages are comfortably to use and very well fit into the backup system. In such services there is version control, and even if something is infected on your PC and this will be reflected in network storages, then this backup can be rolled back using the version control feature.
Unfortunately, Yandex Disk does not support control of version. Moreover, it is not entirely clear how the history of file works. Therefore, we cannot guarantee the stability of backup in case of constantly changing files, used on Yandex Disk technologies. For example, we use Yandex Disk as additional hard disk drive.
In most cases, the network storages selected as one of the alternatives for the storage of backups, and usually the main reasons are:
1. The server can store a maximum of 5 versions of files that have been uploaded. This means that the owner of the archive can restore to an adequate version of the project at any time. Most often, this need arises from the malfunction of hardware, software, operating system, or when virus infected the OS.
2. Fulfill the rule: to store data on various media unrelated to the primary place of storage. I.e. you cannot store everything in one place; you need alternative data banks so that you can recover corrupted or lost data.
Based on this, we conducted a study of network storage for the best result. Various volumes of archives were created along the length of the archive and it was revealed which network storage was preferable and at what size of the archive volume.

IOP Conf. Series: Journal of Physics: Conf. Series **998** (2018) 012026 doi:10.1088/1742-6596/998/1/012026

3. Results of Experiments

We chose Google Drive, Yandex Disk and DropBox as network storages. We had to cut the archive into volumes of 100, 300 and 500 MB. In this case, we only considered the full volume size of the archive, because only they had a stable size, so we excluded from the studies the last volume of the archive. When splitting archive (volume of 31 GB 548 MB) into volumes at 100 MB, number of them in fact will be 300 pieces, 300 MB - 100 and 500 MB - 60.

The results of the calculations we give in Table 1 and in Figure 1. According to Figure 1, we can determine the speed of loading a test directory for each of the network storages. Based on the analysis of Figure 1 we can conclude that each software for file network storage creates different conditions for the processing of files depending on their volume. So both Google Drive and Yandex Disk increase the speed of receiving files for file sizes from 100 to 300 MB , but nearby to 500 MB - decrease. Most likely this is due to attempts by companies to make file downloads more stable and reduce the number of failures. DropBox uses a different strategy, simply increasing the speed of downloading files depending on their size.

Table 1. Estimated download time (min) of the whole archive on the network storage depending on size of volume (The total volume – 31.55 GB).

Splitting into volumes [MB]	Network storage		
	Google Drive	Yandex Disk	DropBox
100	111	219	231
300	80	165	188
500	82.2	174	154.8

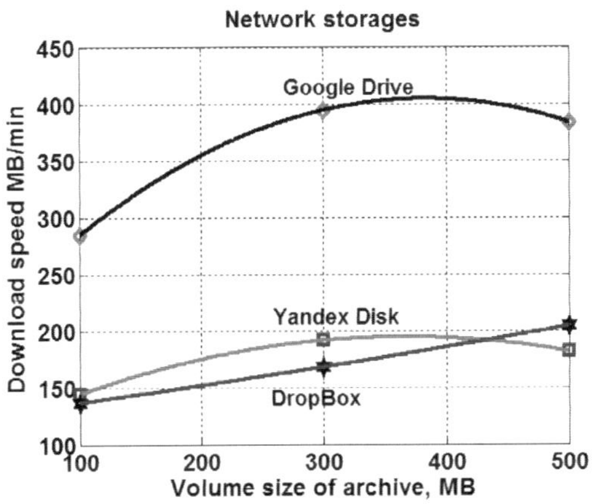

Figure 1. The influence of volume size on the download speed of whole archive in network storage (Total files volume – 31. 55 GB).

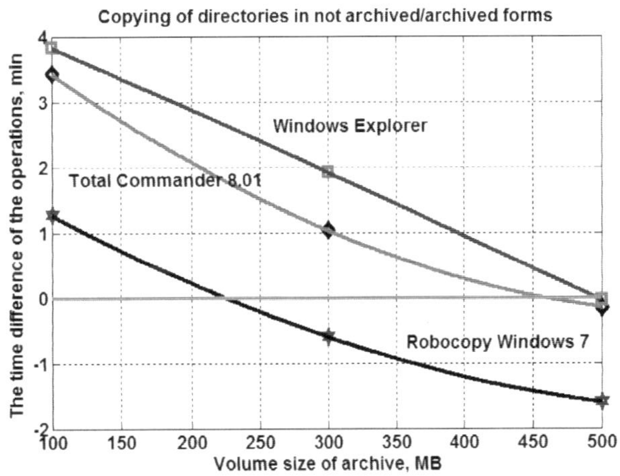

Figure 2. The difference in the time of copying the non-archived directories compared to the same process that contains the data-archiving step.

The data from the studies allow us to conclude that for networked storages it is more expedient to use the division of archives into 300 MB volumes. In the next step, we have begun to identify the factors influencing on the process of copying files into logical drives. We presented research results in Table 2-3 and in Figure 2.

Table 2. Time of copying (min) the test directory with a volume of 31.55 GB in non-archived form.

Media type information, that receives copy	File Manager		
	Total Commander 8.01	Windows Explorer	System Utility Robocopy
Internal hard drive, SATA	11	9.95	10
Deleting from hard drive, SATA	0.97	0.97	0.97
External hard drive, USB 2.0	26.83	27.05	29.63
Deleting from hard drive, USB 2.0	1.97	2	2
Total	40.77	39.97	42.6

During backup of a test directory in non-archived form on a SATA drive Windows Explorer failed to copy 132 files due to the lack of support for long filenames (over 260 characters), warning the user. However, this situation means that the program has not coped with the task of copying files, though, it showed the best time of copying (Table 2).

Table 3. Time of copying (min) the test directory with a volume of 31.55 GB in archived form.

Media type information, that receives copy	File Manager		
	Total Commander 8.01	Windows Explorer	System Utility Robocopy
Splitting into volumes 100 MB (301 pieces)			
Archive creation time		16.82	
Internal hard drive, SATA	6.23	5.67	5.67
Deleting from hard drive, SATA	0.02	0.02	0.02
External hard drive, USB 2.0	21.25	21.25	21.32
Deleting from hard drive, USB 2.0	0.05	0.05	0.05
Total	44.37	43.81	43.88
Splitting into volumes 300 MB (101 pieces)			
Archive creation time		15.05	
Internal hard drive, SATA	5.6	5.68	5.78
Deleting from hard drive, SATA	0.02	0.02	0.02
External hard drive, USB 2.0	21.12	21.12	21.13
Deleting from hard drive, USB 2.0	0.02	0.02	0.02
Total	41.81	41.89	42.0
Splitting into volumes 500 MB (61 pieces)			
Archive creation time		15.82	
Internal hard drive, SATA	5.63	5.65	5.75
Deleting from hard drive, SATA	0.02	0.02	0.02
External hard drive, USB 2.0	20.43	19.5	20.4
Deleting from hard drive, USB 2.0	0.02	0.02	0.02
Total	40.92	40.01	41.01

Total Commander 8.01 in a situation of copying without archiving showed good efficiency, but each time it met with a number of characters in the file name and asked about what to do next. This does not allow you to use it in the situation of automatic backup creation.

Robocopy [4] ideally behaved in the case of non-archived material, because without asking questions, copied everything as is and did not lose the files. In addition, this utility works in the command line mode, which gives great opportunities for its use in conjunction with the schedulers of OS tasks.

4. Discussion of results

If we compare the total time spent on the operation of creating a backup with non-archived form with the time of the backup process using archiving, we found the following patterns (Table 3 and Figure 2):

1. At first, the backup time of the non-archived directory is less than with the archive, but then the time of copying the multi-volume archive compares and even overtakes the time of copying the non-archived material.
2. The trend of decreasing the time of copying files from the volume size (segment) of a multi-volume archive clearly revealed.
3. There is a tendency to reducing the time of the archive preparation depending on the size of the archive volume.
4. Beginning with the size of the archive volume of 300 MB, there is a gradual decrease in the influence of the archive volume size on the time of copy process.
5. Robocopy showed the best results when creating a backup in combination with archiving, and demonstrated the high reliability of its work.
6. Most of all, the process of copying is influenced by the write operation on a portable disk. According to the specification (Table 3), USB 2.0 should provide a processing speed of 480 MB/s, but during the experiment, the speed changed from 280 to 380 MB/s. We assume that the reduction of the copying speed depends on the heterogeneity of the processed data such as their type, volume, etc.
7. Among the networked storages used, the most optimal in terms of price and data volume is Yandex Disk. The year of use of 1 TB of space in this storage costs approximately 3 times cheaper than in Google Drive. However, Yandex Disk does not provide the ability to control file versions, which dramatically increases the percentage of fatal errors that occur during work on projects. Therefore, in our opinion, it is better to use Google Drive as a network storage.
8. In terms of speed, network storage is directly related to the speed of the Internet channel and the amount of files. On average, with stable communication through an optical line, the amount of files equal to 31 GB should be synchronized in about 80 minutes.

5. Conclusion

Based on the results of our research, we propose a simple and economical version of the practical implementation of the backup process consisting of 4 components: the command line interpreter (CMD), the 7z archiver, the Robocopy utility, and network storage.

We also take into considering that the structure of the publisher's archive itself influences on the speed of archiving process. For example, in printing industry we can identify three parts of a common archive of the publishing house: rarely updated; constantly updated; almost not renewable.

Under the term of constantly updated, we mean the current projects of the publishing house, which ideally should be duplicated during the course of work, in the process itself, with any file change.

Rarely updated areas of the publisher's archive in our opinion are software archives and a graphic data bank (clipart, etc.) which are constantly used in the work. In our opinion, they should be updated by the administrator of the network and performed no more than once in 3 months.

Project directories already containing completed orders are the almost non-renewable part of the publisher's archive. Despite the fact that as the main criterion we determined the speed of the process of creating backup copies, with a high probability of creating archive that will not be updated frequently and consisting mostly of graphic materials, compression is possible at the maximum level. The speed of creating a backup will decrease at the same time, but this inconvenience will be compensated by a decrease in the total size of the archive.

As a result, we decided that the best option would be to create backup copies, consisting of three local copies of data and two network copies. We suggest creating the first and second copies on local disks - it is SATA (internal disk), and the third copy - USB (external hard drive). Copies on Google Drive and Yandex Disk are variants of network storage. However, these directories will be on different logical disks (SATA). Local disks and network storage will contain a multi-volume archive of 7z with a splitting into 300 MB.
In this case, the Yandex Disk and external disk (USB) will be used as additional data storage media with inconsistent synchronization with the general archive.
Google Drive and the second logical disk (SATA) will act as the main tools for saving data. The SATA disk is the only media in the described system with the maximum read/write speed, and Google Disk is a tool for creating and maintaining automatic backups of files (and their versions) in the process of employees' work.

6. The source of funding
The results of the research project are published with the financial support of Tula State University within the framework of the scientific project № 2017-34». Scientific Director N.E. Proskuriakov.

References
[1] Proskuriakov N E, Yakovlev B S 2016 The improving of the effectiveness of the technology of verification of websites in the Internet 2016 *Dynamics of Systems, Mechanisms and Machines, Dynamics 2016, art. no. 7819066 Conference Proceeding* (Institute of Electrical and Electronics Engineers Inc.)
[2] Mosk D Work with 7zip from the command line, available at https://www.dmosk.ru/miniinstruktions.php?mini=7zip-cmd
[3] Pakhomov S Comparison of 64-bit archivers WinRAR 4.2, WinZip 17.0 and 7-Zip 9.30, available at http://compress.ru/article.aspx?id=23664
[4] Malikov V We use Robocopy. Examples, available at http://vladimirmalikov.com/%D0%B8%D1%81%D0%BF%D0%BE%D0%BB%D1%8C%D0%B7%D1%83%D0%B5%D0%BC-robocopy-%D0%BF%D1%80%D0%B8%D0%BC%D0%B5%D1%80%D1%8B/

IOP Conf. Series: Journal of Physics: Conf. Series 998 (2018) 012027 doi:10.1088/1742-6596/998/1/

Smart transducer with radiomodem

V.N. Pugach[1] and E.L. Voronin[1]

[1]Research and Production Company Etalon JSC, 175, Lermontova ave, Omsk, 644009, Russia

E-mail: fgup@omsketalon.ru

Abstract. Systems for measuring different parameters enabling metering and wireless data transmission are an urgent problem in the industry. One of the most promising solutions is the developments of metering instruments enabling radio-link and GSM data transmission. The article describes a transducer operating with temperature sensors of different types as well as with the sensors of other physical values with the output signal represented as current or voltage with subsequent measurement data transmission from the transducer to the computer via radio-link. The article provides transducer measurement accuracy check. The work confirmed the claimed temperature measurement accuracy, noted a stable data transmission via radio link and convenience of work with the transducer and software.

1. Introduction

At small and large plants and industrial systems it is always necessary to supervise the processes by means of monitoring of physical values, such as temperature, pressure, flow rate etc. which entails the production rigging with the physical value data, auxiliary metering equipment and, consequently, communication lines for the signal transmission from the sensor to the meter and then to the control room etc.

The signal wired transmission has disadvantages, both engineering and economic:

- Signal transmission line purchase and assembly costs;
- Short signal transmission distance;
- Impossibility of connecting all the lines to one control room;
- High concentration of wired lines in the production premises, especially, if the signal transmission redundancy is required;
- Difficulties in the wire laying and assembly in hard-to-access and hazardous locations.

To eliminate these disadvantages it is appropriate to apply equipment with wireless data transmission links.

2. Problem description

The work objective is review of smart transducers with radio modem, discussion of its specifications and check. To attain the target set it is necessary to solve the following problems:

- Review the instrument and its specifications;
- Check the instrument measurement error;
- Check the signal instrument-to-PC transmission stability.

3. Theory

The smart transducer with radio modem (hereinafter referred to as the STRM) is designed for measuring the temperature and other physical values with subsequent data transmission to the computer via radio link. The transducer will enable continuous monitoring of the processes measured as well as supplying alarms of the critical values of a certain parameter to be measured.

The STRM is a small metal box with three cable inputs, two of which being measurement channels and the third one – for power supply. Outside the housing there is also a small antenna (Figure 1).

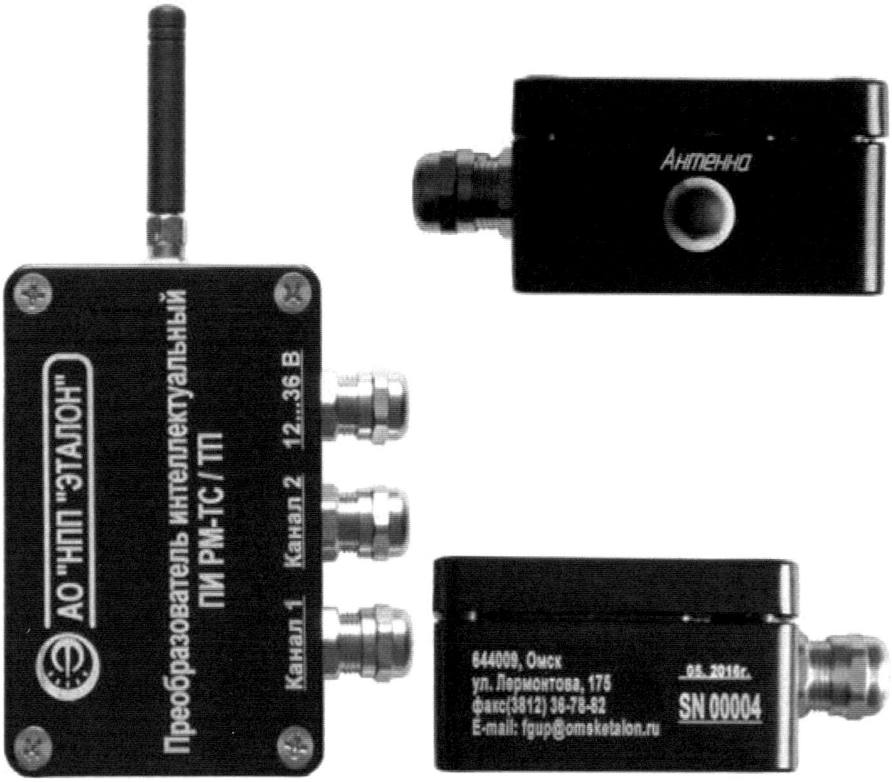

Figure 1. STRM Visual Appearance

Inside the transducer there are terminals of interchangeable measurement channels to which sensors are connected and a power supply terminal (Figure 2).

Figure 2. STRM Internal View

Both the transducer channels may operate with the following thermocouple types: ПП(S), ПР(B), ЖК(J), ХА(K), ХК(L) [1], and with the following resistance thermometer types: 50П, 100П, Pt50, Pt100, 50M, 100M [2], as well as with the current signal 4-20 mA and voltage 0-5 V and 0-10 V which makes a STRM a universal transducer capable of operating with any sensors of various physical values with a unified output signal.

Table 1. Basic specifications are provided in the table

Output Signal Type	Output Value Measurement Range	Unit of Measurement	Limits of the Basic Value Admissible Value, %
ПП(S)	From 0 to +1750 °C		0.5
ПР(B)	From +600 to +1700 °C		
ЖК(J)	From -100 to +1200 °C		0.2
ХА(K)	From -100 to +1300 °C		
ХА(L)	From -100 to +600 °C		
50П	From -100 to +750 °C	°C	
100П			
Pt50			
Pt100			
50M	From -100 to +200 °C		
100M			
0-5 В	Adjustable to the Required Sensor Type	Physical Value Measurement Unit	
0-10 В			
4-20 мА			

Data transmission from the transducer to the computer is performed by radio link at the distance up to 1.5 km.

STRM's may be joined in a uniform network with the number of up to 50 instruments (Figure 3). This function enables simultaneous monitoring of the parameters at the objects involved in various processes and warn the operator of the change in the physical value measured outside the user-set limits.

Figure 3. STRM Wireless Network

The main window of the software for operating МИРС Transducer (Measurement radio network manager) is a device manager (Figure 4), this window displays serial numbers of the instruments connected, their status, indication of the measured value excursion outside the limit set, the communication signal level and line for the comment input to each instrument.

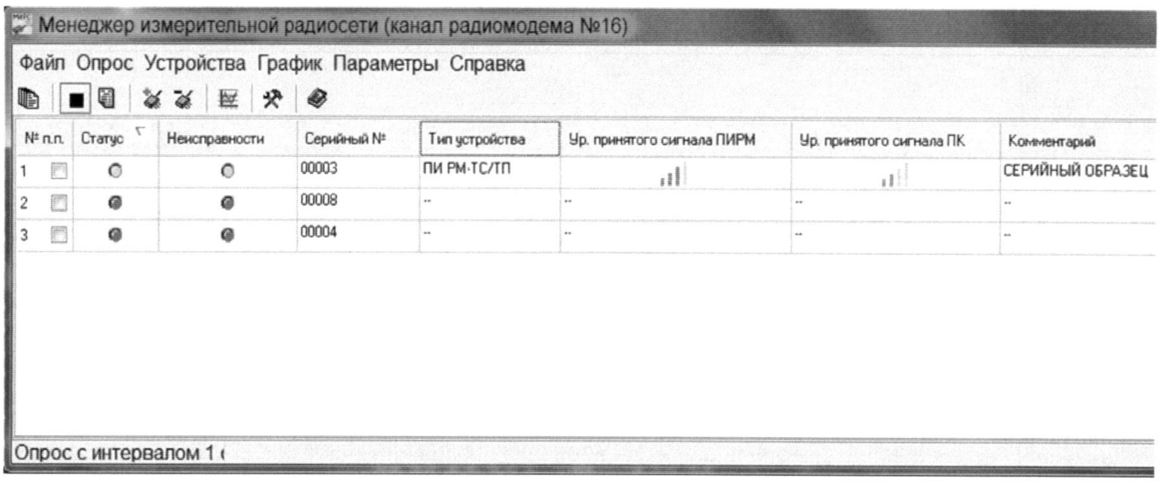

Figure 4. МИРС Device Manager

Then follows channel setting window (Figure 5) in which for each channel the sensor type connected to the channel is set or the ratio of the unified signal and value measured as well as values of the parameters for the emergency alarms are set.

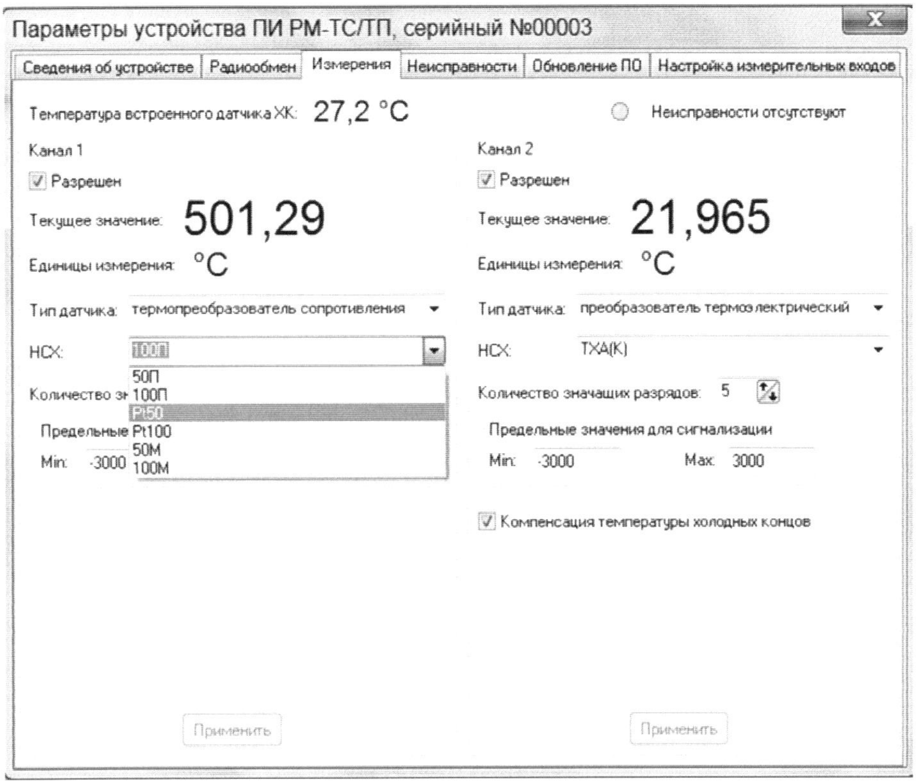

Figure 5. Channel Setup Window

There is also a function of displaying these data in real-time chart (Figure 6) which enables obtaining radically new information of the current processes.

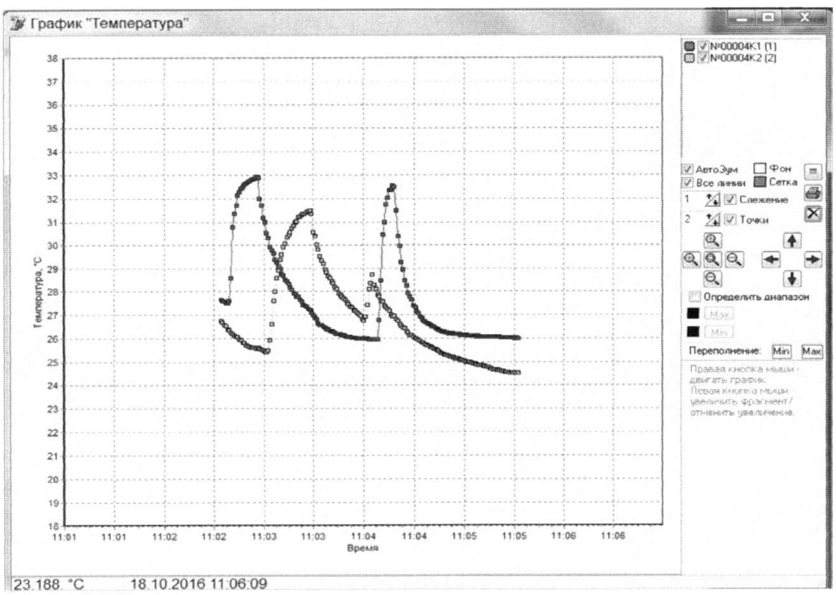

Figure 6. Data Display in the Chart

IOP Conf. Series: Journal of Physics: Conf. Series **998** (2018) 012027 doi:10.1088/1742-6596/998/1/012027

4. Experiment

SRTM measurement error was checked. The measurement error was checked during the operation with the thermocouple and resistance thermometers. The thermocouple measurement error was checked as per the diagram provided in Figure 7, resistance thermometer measurement error was checked as per the diagram provided in Figure 8.

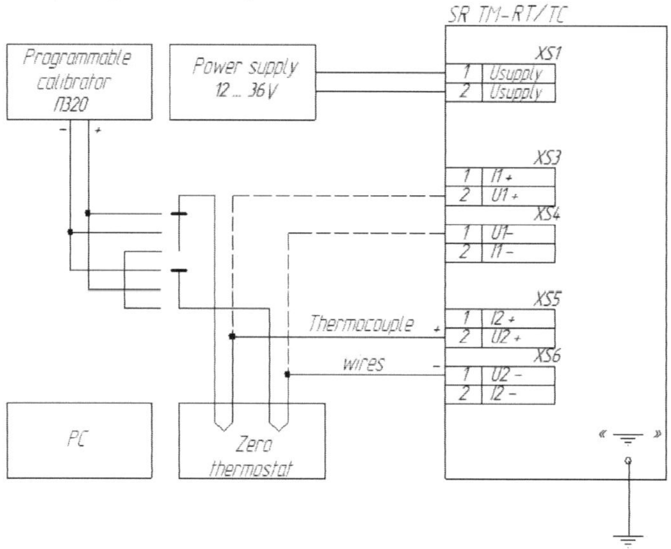

Figure 7. Check Diagram with the Thermocouple

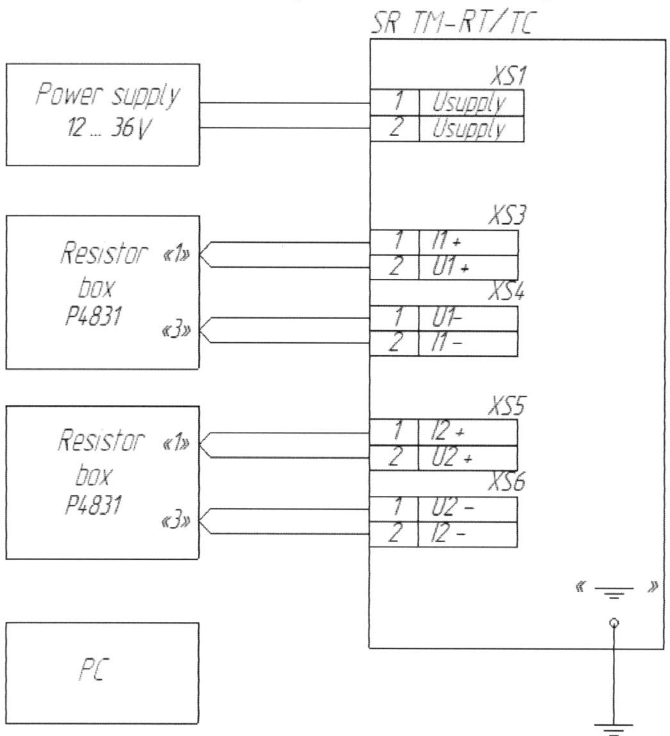

Figure 8. Check Diagram with the Resistance Thermometer

As per the diagram provided in Figure 7 thermal EMF was reproduced as per the rated static characteristic for ТЖК type thermocouple. As per the diagram provided in Figure 8 resistance boxes were used to reproduce the 100П type resistance thermometer rated static characteristic. After the measurements the measurement error was calculated as per equation 1.

$$\Delta T = ((T_{meas} - T_{in})/(T_{max} - T_{min}))\cdot 100\%. \quad (1)$$

5. Experiment Results

The measurement and calculation results are provided in Table 2 and 3.

Table 2. Error in case of thermocouple operation

T, °C	V_{init}, mV	STRM Readings, °C	ΔT, °C	ΔT, %
-100	-4.633	-98.54	1.46	0.11
0	0.000	1.08	1.08	0.08
100	5.269	100.88	0.88	0.07
200	10.779	200.92	0.92	0.07
300	16.327	300.96	0.96	0.07
400	21.848	401.00	1.00	0.08
500	27.393	501.04	1.04	0.08
600	33.102	601.08	1.08	0.08
700	39.132	701.12	1.12	0.09
800	45.494	801.15	1.15	0.09
900	51.877	901.28	1.28	0.1
1000	57.953	1001.40	1.40	0.11
1100	63.792	1101.46	1.46	0.11
1200	69.553	1201.50	1.50	0.12

Table 3. Error in case of resistance thermometer operation

T, °C	R_{init}, Ω	STRM Readings, °C	ΔT, °C	ΔT, %
-100	60.26	-100.02	-0.02	-0.0024
-50	80.31	-50.050	-0.05	-0.0064
0	100.00	-0.040	-0.04	-0.0052
50	119.40	49.95	-0.05	-0.0059
100	138.51	99.95	-0.05	-0.0059
150	157.33	149.94	-0.06	-0.0071

200	175.86	199.95	-0.05	-0.059
300	212.05	299.99	-0.01	-0.0012
400	247.09	399.98	-0.02	-0.0024
500	280.98	499.98	-0.02	-0.024
600	313.71	600.01	0.01	0.012
700	345.28	699.97	-0.03	-0.0035
800	375.70	800.02	0.02	0.0021

6. Discussion of Results

The transducer in question has the required measurement accuracy, it enables measurements from two sensors of different types and the data transmission to the PC without radio link communication. Therefore, this instrument may be used both as a redundancy system and a the main measurement system for the monitoring of different physical values.

7. Conclusions

1. The transducer presented enables operating with different temperature sensors as well as sensors of other physical values with a unified output signal.
2. The transducer enables measurements from a large number of sensors without signal transmission line laying and using one PC.
3. The instrument measurement error does not exceed the claimed accuracy ranges.
4. These transducers with wireless communication are becoming more and more popular due to their operation convenience and possibility of large-distance measurements without expenses for wired data transmission lines.

References

[1] GOST R 8.585-2001. 2010 State system for ensuring the uniformity of measurements. Thermocouples. Nominal static characteristics of conversion. Superseding GOST R 50431-92; made effective on 30.06.2002 (Moscow: Standartinform)

[2] GOST 6651-2009. 2011 State system for ensuring the uniformity of measurements. Platinum, copper and nickel resistive temperature transducers. General technical requirements and test methods. – Superseding GOST 6651-94; made effective on 01.01.2011. (Moscow: Standartinform)

Transient Oscilliations in Mechanical Systems of Automatic Control with Random Parameters

B Royev[1], **A Vinokur**[1,2], **G Kulikov**[1]
[1] Moscow Polytechnic University, 38, B. Semenovskaya str., Moscow, 107023, Russia
[2] National Research Nuclear University MEPhI (Moscow Engineering Physics Institute), Moscow, Russia

E-mail: kulikov.gb@mail.ru

Abstract. Transient oscillations in mechanical systems of automatic control with random parameters is a relevant but insufficiently studied issue. In this paper, a modified spectral method was applied to investigate the problem. The nature of dynamic processes and the phase portraits are analyzed depending on the amplitude and frequency of external influence. It is evident from the obtained results, that the dynamic phenomena occurring in the systems with random parameters under external influence are complex, and their study requires further investigation.

1. Introduction

Recently, in various fields of science and technology statistical methods have been widely used for solving the diverse challenges arising in machine building, control theory, transport etc. [1,4,5]. Among the application tasks solved with the methods of statistical dynamics a large place is occupied by the study of the behavior of mechanical systems with random parameters [1, 5].
The study of these systems presents serious mathematical difficulties [1, 5]. To solve this problem a modified spectral method was applied for the first time. [2, 3]. For other methods, such as method of moment functions, it is quite challenging to obtain a closed system of equations for moment functions [1]. The processes occurring in real systems are treated as extended Markov processes. To describe them, along with the equations of motion, additional relations (equation filters) are made in which the system is put into the Markov form. However, the empirical finding of the necessary filters is quite a complex and not always feasible task. The modified spectral method is free of most of these difficulties, which makes it possible to solve application problems [2, 3].

2. Statement of the problem

In this paper, a modified spectral method is used to study the transient oscillations in the mechanical automatic control systems with random parameters. The equation for the oscillations of such systems is as follows:

$$\ddot{y} + 2\varepsilon\dot{y} + \omega_0^2\left(1 - f(t)\right)y = q(t) \qquad (1)$$

Here y (t) is the displacement, ω_0^2 is the squared eigenfrequency, ε is the damping ratio, $f(t)$ and $q(t)$ are parametrical and external effects. Using the modified method of spectral representations [3, 4], we introduce the following integral representations for $y(t)$ and $f(t)$:

$$y(t) = y_0(t) + \int_{-\infty}^{\infty} \psi(\omega,t)\mathrm{Y}(\omega)e^{i\omega t}d\omega \qquad f(t) = \int_{-\infty}^{\infty} F(\omega)e^{i\omega t}d\omega \tag{2}$$

where $y_0(t)$ is the mathematical expectation of the process *y(t)*; *ψ(ω,t)* is an unknown deterministic function; *Y(ω), F(ω)* are random spectra that satisfy the conditions of orthogonality of the stochastic. The external influence is determined and given in the form q(t) = $q_0 \cos\Omega t$ (q_0, Ω is the amplitude and frequency of the external influence).

3. Theory

After substituting (1) into (2) and compiling the torque ratios we obtain the system of equations for the functions *y0(t)* and *ψ(ω,t)*

$$\begin{gathered} \ddot{y}_0 + 2\dot{y}_0 + \omega_0^2 y_0 - \omega_0^2 \int_{-\infty}^{\infty} \psi(\omega,t) S_{vf}(\omega) d\omega = q_0 \cos\Omega t, \\ \left[\ddot{\psi} + 2(\varepsilon + i\omega)\dot{\psi} + \left(\omega_0^2 - \omega^2 + 2i\varepsilon\omega\right)\psi\right] S_{yf}(\omega) - \omega_0^2 y_0 S_f(\omega) = 0 \end{gathered} \tag{3}$$

Here $S_{yf}(\omega), S_f(\omega)$ are the spectral densities of the processes *y(t)*and *f t)*. The solution to this system of equations is found under the following initial conditions

$$\begin{gathered} \langle y \rangle = s_0, \langle \dot{y} \rangle = \dot{s}_0, \\ \langle yv \rangle = 0, \langle \dot{y}v \rangle = 0, t = 0 \\ \langle y\dot{v} \rangle = 0, \langle \dot{y}\dot{v} \rangle = 0 \end{gathered} \tag{4}$$

Here s_0 and $\dot{s}_0$ are the initial values of the mathematical expectation of displacement and velocity. First we shall find the general solution to the homogeneous system of equations

$$\begin{gathered} \ddot{y}_0 + 2\dot{y}_0 + \omega_0^2 y_0 - \omega_0^2 \int_{-\infty}^{\infty} \psi(\omega,t) S_{yv}(\omega) d\omega = 0, \\ \left[\ddot{\psi} + 2(\varepsilon + i\omega)\dot{\psi} + \left(\omega_0^2 - \omega^2 + 2i\varepsilon\omega\right)\psi\right] S_{yf}(\omega) - \omega_0^2 y_0 S_f(\omega) = 0 \end{gathered} \tag{5}$$

The general solution to the mathematical expectation $y_0(t)$ and to the function *ψ(ω,t)* will be sought in the form

$$y_0(t) = Ce^{\lambda t} \qquad \psi(\omega,t) = D(\omega)e^{\lambda t} \tag{6}$$

where λ is characteristic indicators, *C* is constants, *D(ω)* is non - random spectra.
The characteristic equation relative to the values of λ has the form

$$(\lambda+\varepsilon)^2 + \varpi_1^2 - \omega_0^4 \int_{-\infty}^{\infty} \frac{S_v(\omega)d\omega}{(\lambda+\varepsilon+i\omega)^2 + \varpi_1^2} = 0 \tag{7}$$

Suppose the parametric effect is an ideal narrowband process with the spectral density

$$S_v(\omega) = \frac{\sigma_f^2}{2}\delta(|\omega| - \omega_*). \tag{8}$$

Here, σ_f^2 is the dispersion, ω_* is the carrier frequency of the process.
In this case, the characteristic equation is a sextic polynomial

$$\sum_{k=0}^{6} a_k p^{6-k} = 0, \tag{9}$$

the coefficients of which have the form:

$$a_0 = 1; a_1 = 6\delta; a_2 = 12\delta^2 + 2\gamma^2 + 3; a_3 = 4\delta\left(2\gamma^2 + 2\delta^2 + 3\right); a_4 = 12\delta^2\left(1+\gamma^2\right) + \gamma^4 + 3 - \sigma_f^2;$$
$$a_5 = 2\delta\left(3 + 4\gamma^2\delta^2 + \gamma^4 - \sigma_f^2\right); a_6 = \left(1-\gamma^2\right)^2 + 4\gamma^2\delta^2 - \left(1-\gamma^2\right)\sigma_f^2; p = \lambda/\omega_0; \delta = \varepsilon/\omega_0; \gamma = \omega_*/\omega_0.$$

Then we shall find a particular solution to the system (3) which describes the forced oscillations. Since the right part of the system is $q(t) = q_0 \cos\Omega t$, the solution will be sought in the following form

$$y_*(t) = A\cos\Omega t + B\sin\Omega t,$$
$$\psi_*(\omega,t) = Q(\omega)\cos\Omega t + G(\omega)\sin\Omega t \tag{10}$$

We shall find a particular solution to the displacement $y_*(t)$. After substituting (10) into (3) and excluding spectra $D(\omega)$ and $G(\omega)$ from the obtained ratio we shall derive a system of algebraic equations for constant A and B

$$\alpha_{11}A + \alpha_{12}B = q_0,$$
$$\alpha_{21}A + \alpha_{22}B = 0 \tag{11}$$

The solution to this system of equations has the form

$$A = \frac{\alpha_{22}q_0}{\Delta_1}, B = -\frac{\alpha_{21}q_0}{\Delta_1}, \Delta_1 = \alpha_{11}\cdot\alpha_{22} - \alpha_{12}\cdot\alpha_{21}, \tag{12}$$

where

$$\alpha_{11} = \alpha_{22} = \omega_0^2 - \Omega^2 - \omega_0^4\cdot I_1, \alpha_{12} = 2\varepsilon\Omega + \omega_0^4\cdot I_2, \alpha_{21} = -\alpha_{12}, \overline{\omega}_1^2 = \omega_0^2 - \varepsilon^2,$$
$$I_1 = \int_{-\infty}^{\infty}\frac{\left[(\varepsilon+i\omega)^2 + \overline{\omega}_1^2 - \Omega^2\right]}{\Delta_1}S_f(\omega)d\omega, I_2 = \int_{-\infty}^{\infty}\frac{2\Omega\cdot(\varepsilon+i\omega)}{\Delta_1}S_f(\omega)d\omega$$

After substituting (12) into the first expression (10) we obtain the following the expression for the displacement and velocity

$$y_*(\tau) = \frac{q_0}{\omega_0^2\cdot\Delta_1}\left(\alpha_{11}\cos\gamma\tau + \alpha_{12}\sin\gamma\tau\right),$$
$$\dot{y}_*(\tau) = \frac{q_0\cdot\gamma}{\omega_0^2\cdot\Delta_1}\left(-\alpha_{11}\sin\gamma\tau + \alpha_{12}\cos\gamma\tau\right),, \tag{13}$$

where

$$\alpha_{11} = \alpha_{22} = \omega_0^2\cdot\left(1-\gamma^2 - I_1\right), \alpha_{12} = \omega_0^2\cdot\left(2\delta\gamma + I_2\right), \alpha_{21} = -\alpha_{12}, I_1 = \sigma_v^2\frac{a_1A_1 + b_1B_1}{A_1^2 + B_1^2},$$

$$I_2 = \sigma_v^2\frac{c_1A_1 + d_1B_1}{A_1^2 + B_1^2}, a_1 = 1-\gamma^2-\gamma_1^2, b_1 = 2\delta\gamma_1, c_1 = 2\delta\gamma, d_1 = 2\gamma\gamma_1,$$

$$A_1 = \left(1-\gamma^2-\gamma_1^2\right)^2 + 4\cdot\left(\gamma^2\left(\delta^2-\gamma_1^2\right) - \delta^2\gamma_1^2\right), B_1 = 2\delta\gamma_1\cdot\left(1+3\gamma^2-\gamma_1^2\right),$$

$$\gamma = \frac{\Omega}{\omega_0}, \gamma_1 = \frac{\omega_*}{\omega_0}, \delta = \frac{\varepsilon}{\omega_0}, \tau = \omega_0 t.$$

IOP Conf. Series: Journal of Physics: Conf. Series **998** (2018) 012028 doi:10.1088/1742-6596/998/1/012028

For the constants C_j and spectra $D_j(\omega)$ we substitute the general solution of the system of equations (3) into the initial conditions (4).

4. Discussion of results

The analysis of the processes occurring in systems with random parameters was carried out numerically and is shown in Figure 1-6. All the results are presented in the dimensionless form. In Figure 1-3 the time variation of the mathematical expectation of the displacement is shown, as well as of.

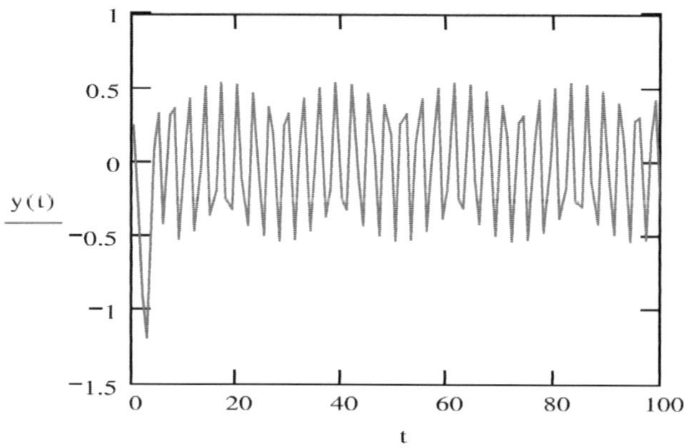

Figure 1. The expected displacement $y_0(t)$ with the external influence rating $q_0/\omega_0^2 = 2$ and $\gamma = 2$.

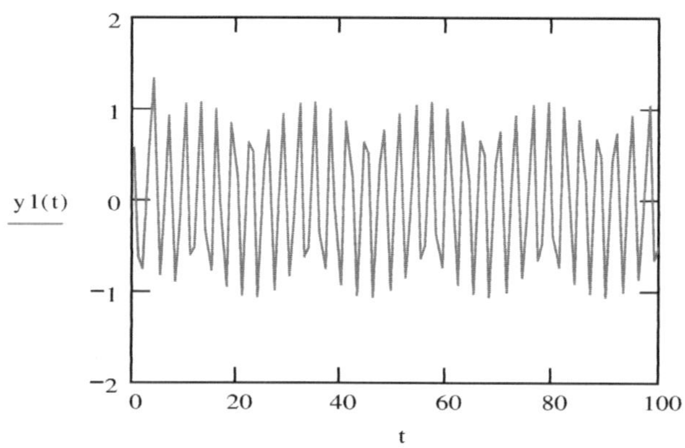

Figure 2. The expected velocity $y1(t) = \dot{y}_0(t)$ with the external influence rating $q_0/\omega_0^2 = 2$ and $\gamma = 2$

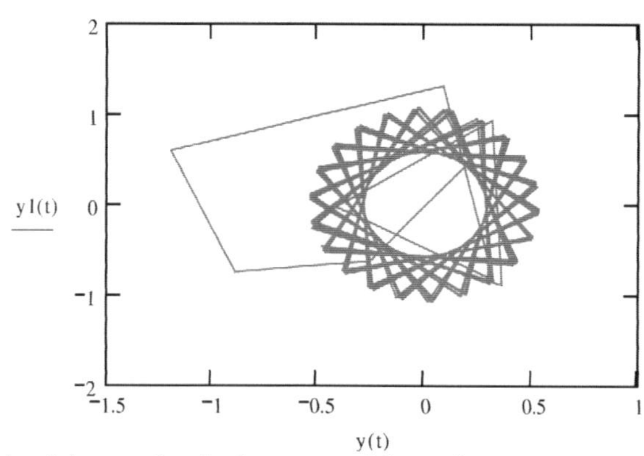

Figure 3. Phase portrait of the mechanical system with random parameters $y1(t) = \dot{y}_0(t)$ with the external influence rating $q_0/\omega_0^2 = 2$ and $\gamma = 2$.

The velocity with the carrier frequency $\gamma_1 = 1.5$, the parametric effect dispersion $\sigma_f^2 = 0.5$ and the damping parameter $\delta = 0.5$, which were chosen in accordance with the stability terms of the solution to the system of equations (5). The amplitude of external effect q_0/ω_0^2 was taken equal to 2.0 and the frequency $\gamma = 2$. From Figure 1-2 it is evident that the process of forced oscillations establishment is quite fast. In Figure 3 the phase portrait is shown that is characterized by a change in displacement and velocity along damped trajectories, with jumps from one trajectory to another. Next we shall consider the response of a mechanical system to a change in the frequency of the external influence γ with

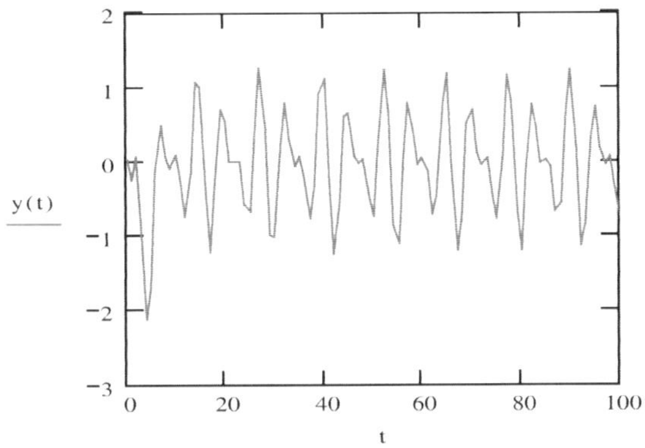

Figure 4. The expected displacement $y_0(t)$ with the external influence rating $q_0/\omega_0^2 = 2$ and $\gamma = 1.5$.

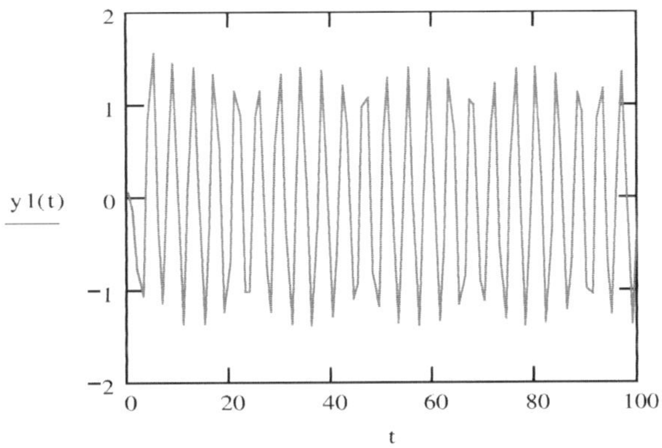

Figure 5. The expected velocity $y1(t) = \dot{y}_0(t)$ with the external influence rating $q_0/\omega_0^2 = 2$ and $\gamma = 1.5$.

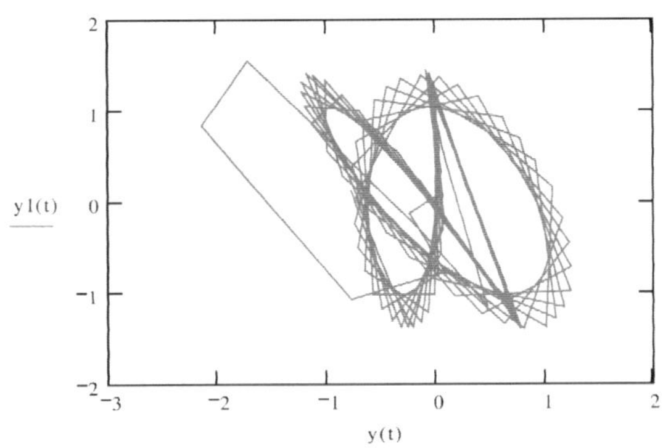

Figure 6. Phase portrait of the mechanical system with random parameters with the external influence rating $q_0/\omega_0^2 = 2$ and $\gamma = 1.5$.

the carrier frequency of the parametric effects $\gamma_1 = 1.0$ being in the area of the side parametric resonance, the dispersion $\sigma_f^2 = 0.5$, the damping parameter $\delta = 0.5$ and the amplitude of external influence $q_0/\omega_0^2 = 2$. Firstly, the frequency of the external influence changes significantly the nature of dynamic processes (Figure 4-5). Secondly, there is a change in the nature of the trajectories on the phase portraits (Figure 6). It is seen that when the carrier frequency $\gamma_1 = 1.0$ is in the area of the side parametric resonance, another group of damped trajectories occurs on the phase portraits.

5. Conclusion

Based on the obtained results, the following conclusions can be made. First, the mechanical systems with random parameters are very sensitive to their change. Second, the nature of the processes depends essentially on the external influences, which reduce the effect of parametric oscillations. Dynamic phenomena occurring in these systems under external influence are complex, and their analysis requires more detailed discussion and further study.

References

[1] Bolotin V V 1979 Slochasticheskye kolebaniya uprugikh sistem (Random Oscillations of Elastic Systems), Moscow: Nauka p 336
[2] Roev B A 2013 Forced and parametric oscillations. Technical application. LAP LAMBERT Academic Publishing p 128
[3] Roev B and Vinokur A 2016 On the analysis of forced oscillations of systems with two random sources of parametric effects. JVE International LTD *Vibroengineering PROCEDIA. OCT* **8** pp 147-152
[4] Whitney C 1990 Random Processes in Physical Systems. New York: John Willey p 320
[5] Srinamachchivaya N, Pavlyukevich I and Wedig W 2016 Random Perturbations of Periodically Driven Nonlinear Oscillators *Procedia IUTAM* **19** pp 91-100

Expanded uncertainty estimation methodology in determining the sandy soils filtration coefficient

A D Rusanova[1], L D Malaja[1], R N Ivanov[1], A V Gruzin[1] and V V Shalaj[1]

[1]Omsk State Technical University, 11, Mira Pr., Omsk 644050, Russian Federation

E-mail: ad.rusanova@mail.ru

Abstract. The combined standard uncertainty estimation methodology in determining the sandy soils filtration coefficient has been developed. The laboratory researches were carried out which resulted in filtration coefficient determination and combined uncertainty estimation obtaining.

Key-words: filtration coefficient, combined standard uncertainty, standard uncertainty, error, soil

1. Introduction

The metrology community of Russia is currently moving from the measurement results error calculation to the uncertainty estimation which is demonstrated by the standard [1]. The authors of such papers as [2, 3, 4] speak of the «error» and «uncertainty» concepts difference and emphasize the measurement uncertainty significance as a central concept in ensuring the measurement results accuracy. At the moment, many uncertainty estimation methodologies in various fields of science [6, 7, 8] have been developed abroad on the basis of [5]. Uncertainty estimation is important not only for the measuring instruments verification, instruments and auxiliary apparatuses qualification but also in conducting various physical characteristics laboratory studies.

2. Problem statement

Due to the development of new oil and gas fields located in adverse climatic and geological conditions characterized by the swelling and collapsible soils presence, the water-saturated soil foundations settlements rate correct prediction problem remains particularly relevant. This could be explained by the foundation soil mechanical properties changing at the humidity value variation [9, 10, 11] as well as by the soil consolidation, that is foundation soil compaction by means of water squeezing out of the pore space. Consolidation rate characterizing its dynamics depends on the filtration coefficient determining soil permeability. The consolidation process completion is followed by the foundation soil settlements stabilization.

On the basis of the above, the investigated soil filtration coefficient is necessary to be defined and the measurement results uncertainty estimation methodology is required to be developed.

3. Theory

In accordance with the requirements [12] for the filtration coefficient defining, three tests of the investigated soil in the device of Governmental Scientific Research Institute for Roads are required to be carried out. The test consists in defining the water level falling time in the piezometer from 0 to 50 mm when water flows through the filtration tube perforated bottom into which the prepared soil under study is placed.

The medium sized sandy soil of the Irtysh River floodplain was chosen for the study.

The sandy soil filtration coefficient K_f^{10} m/day reduced to the filtration conditions at the temperature of 10 °C is calculated by the formula:

$$K_f^{10} = \frac{h}{t} \cdot \varphi\left(\frac{S}{H_0}\right) \cdot \frac{864}{T}, \tag{1}$$

Where h is the soil sample height in the tube, cm;
t is the water level falling time, s;
S is the observed water level falling counted from the initial level in piezometer, cm;
H_0 is the initial pressure, cm;
$\varphi(S/H_0)$ is the dimensionless coefficient defined according to [12, annex 4];
T is the correction for reducing the filtration coefficient value to the water filtration conditions at the temperature of 10 °C calculated by the formula: $T = 0.7 + 0.03 \cdot T_f$;
T_f – the actual water temperature when testing, °C;
864 is the conversion coefficient (from cm/s in m/day).
The filtration coefficient defining results are represented in table 1.
When calculating the filtration coefficient, the mercury thermometer with the accuracy of ± 0.5 °C, stopwatch with the accuracy of 10 minutes ± 0.6 s, metal ruler with the accuracy of ± 0.15 mm were used.

Table 1. The log of the filtration coefficient of sandy soil

Test No	h, cm	S, cm	H_0, cm	t, s	T_f, ^{0}C	T	$\frac{S}{H_0}$	$\varphi\left(\frac{S}{H_0}\right)$	K_f^{10}, m/day	$\bar{K}_f^{10}$, m/day
1	2	3	4	5	6	7	8	9	10	11
1	17	0.5	20	109	17	1.21	0.025	0.025	2.78	
2	17	1.0	20	214	17	1.21	0.050	0.051	2.89	2.87
3	17	1.5	20	323	17	1.21	0.075	0.078	2.93	

The «combined standard uncertainty» concept representing the positive square root of the total dispersion obtained by the formula below is used to describe the indirect measurements [1]:

$$u_s^2 = \sum_{i=1}^{N}\left(\frac{\partial f}{\partial x_i}\right)^2 \cdot u^2(x_i), \tag{2}$$

where f is the function under consideration (1);
$u(x_i)$ is the input variable standard uncertainty estimated by A or B type

4. Results

To define the input variables partial differentials, the dependence type is determined as follows: all points from [12, Appendix 4] are plotted on the chart and approximated (figure 1). The obtained dependence is exponential:

$$\varphi = 0.0759e^{4.0346(S/H_0)}. \tag{3}$$

IOP Conf. Series: Journal of Physics: Conf. Series **998** (2018) 012029 doi:10.1088/1742-6596/998/1/012029

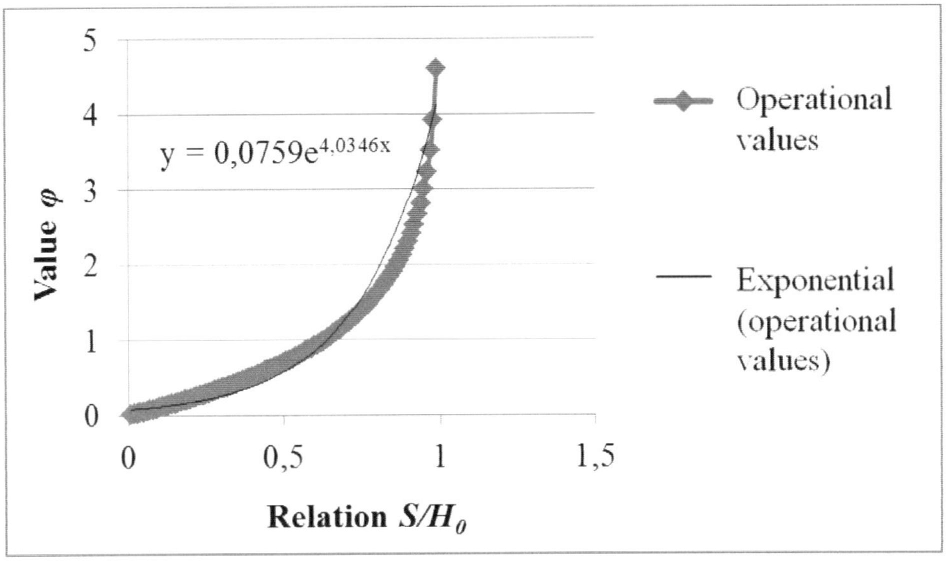

Figure 1. The operational values approximation.

Having plugged the equation (3) into the formula (1), the sandy soil filtration coefficient defining equation will be as follows:

$$K_f^{10} = \frac{65.5776 \cdot h \cdot e^{4.0346 \cdot (S/H_0)}}{t \cdot T}. \quad (4)$$

The partial derivatives of h, t, T, (S / H_0) are calculated correspondingly:

$$\frac{\partial K_f^{10}}{\partial h} = \frac{65.5776 \cdot e^{(4.0346 \cdot (S/H_0))}}{t \cdot T}, \quad (5)$$

$$\frac{\partial K_f^{10}}{\partial t} = -\frac{65.5776 \cdot h \cdot e^{(4.0346 \cdot (S/H_0))}}{t^2 \cdot T}, \quad (6)$$

$$\frac{\partial K_f^{10}}{\partial T} = -\frac{65.5776 \cdot h \cdot e^{(4.0346 \cdot (S/H_0))}}{t \cdot T^2}, \quad (7)$$

$$\frac{\partial K_f^{10}}{\partial (S/H_0)} = \frac{264.5794 \cdot h \cdot e^{(4.0346 \cdot (S/H_0))}}{t \cdot T}. \quad (8)$$

On the basis that input variables were not determined as a result of repeated observations, the standard uncertainty is defined by *B* type [1].

The uniform distribution law theory of the obtained experiment results is accepted as an assumption. Then the standard uncertainty of each input variable is calculated according to the formula:

$$u_{B_i} = \frac{\Delta x}{\sqrt{3}}, \quad (9)$$

where Δx is the absolute error presented in the technical documentations.

For the variable $\varphi(S/H_0)$ the absolute error is defined as the maximum values resolution $\varphi(S/H_0)$ [12, Annex 4].

Hence, the total standard uncertainty by type B is as follows:

$$u_B\left[K_f^{10}\right] = \sqrt{\left(\frac{\partial K_f^{10}}{\partial h} \cdot u_{B_h}\right)^2 + \left(\frac{\partial K_f^{10}}{\partial t} \cdot u_{B_t}\right)^2 + \left(\frac{\partial K_f^{10}}{\partial T} \cdot u_{B_T}\right)^2 + \left(\frac{\partial K_f^{10}}{\partial (S/H_0)} \cdot u_{B_{(S/H_0)}}\right)^2}. \quad (10)$$

The measurement result of the sandy soil filtration coefficient defining is written in the following form:

$$K_f^{10} = \overline{K}_f^{10} \pm k \cdot u_B \left[K_f^{10} \right], \tag{11}$$

where $\overline{K}_f^{10}$ is the filtration coefficient mean value:

$$\overline{K}_f^{10} = \frac{\sum_{i=1}^{n} K_{\phi_i}^{10}}{n}, \tag{12}$$

n is the number of measurements;
k is the sweep efficiency usually accepted to be 2.

5. Conclusions

Based on the results of laboratory studies presented in table 1, the investigated soil filtration coefficient was defined and the expanded uncertainty estimation was obtained:

$$K_f^{10} = 2.87 \pm 0.53 \text{ m/day}.$$

The developed methodology allows to estimate the expanded uncertainty for the sandy soil filtration coefficient.

References

[1] State Standard R 54500.3-2011/ISO/IEC Guide 98-3:2008 Uncertainty of measurement Part 3 Measurement uncertainty guidance 2012 (Moscow: Standartinform)
[2] Frolov S V and Grabarev S P 2010 *Fires and emergencies: prevention, elimination* **1** 86-94
[3] Frolov S V and Grabarev S P 2010 *Technologies of technospheric safety* **1** 1-9
[4] Kadis R L 2008 Factory laboratory. Diagnostics of materials **3** 61-5
[5] ISO/IEC Guide 98-3:2008 Uncertainty of measurement Part 3 Guide to the expression of uncertainty in measurement
[6] Dorozhovets M, Popovych I and Warsza Z L 2016 *Proc. Int. Conf. Advanced mechatronics solutions* **393** (Switzerland) p 421-30
[7] Bolotov M A, Pechenin V A and Murzin S P 2016 *Computer optics* **40** 360-9
[8] Hwang R J, Beltran J, Rogers C, Barlow J and Razatos G 2017 *Canadian society of forensic science journal* **50** 114-24
[9] Rusanova A D 2016 Investigation of the humidity impact on the tank foundation soil compaction of the type PBC-50000 *The relevant problems of the oil and gas complex development in Russia* (Moscow: Gubkin University) p 296
[10] Rusanova AD and Gruzin AV 2015 Oil storage tank foundation RBSPK-50000 soils settlement modelling *The Future Generation: A View of Young Scientists* (Kursk: Southwest State University) p 182-4
[11] Malaja L D and Rusanova A D 2016 Determination of the axisymmetric compression units impact on the experiment results *Proc. Int. Conf. Fundamental and applied researches in the modern world* (St. Petersburg: Strategija Budushhego) p 44-8
[12] State Standard 25584-90 Soils Filtration coefficient laboratory determination methods 2008 (Moscow: Standartinform)

The study of the stress - strain state of the tank with bottom water drainage during operation

Yu V Shchipkova[1], V V Tokarev[1]

[1]Omsk State Technical University, 11, Mira av., Omsk, 644050, Russia

E-mail: ylia_sipkova@mail.ru

Abstract. Bottom drainage from tank is a current problem in modern tank usage. This article proposes the use of the bottom drainage system from the tank with the shape of the sloped cone to the centre of it. Changing the bottom design alters the stress - strain state to be analyzed in the Ansys. The analysis concluded that the proposed drainage system should be applied.

1. Introduction

The tank for the storage of crude oil and oil products is a complex engineering and technical facility for the reception, storage and issuance of oil and oil products.

When oil and oil products are stored in tanks, the liquid densities stratification occurs. Due to this so-called bottom water, containing dissolved salts, is accumulated in the lower part of the tank. In addition, oil stored in the tank contains sulphur, i.e. the bottom water is a weak solution of sulphuric acid.

The prolonged contact of the bottom water with the inner surface of the tank, in the hold-up spots (when the gradual settlement takes place), causes the bottom and lower ring of the tank to be subjected to accelerated corrosion [1]. The need to protect the inner surface of the lower ring and the bottom from the corrosion occurs.

According to [1], it was proposed to equip tanks with a water header collector: a special water header collector with a diameter of about 1 m and a depth of 70-80 is cut into the bottom of the tank. Thus, the authors' intention is to localize the area where the bottom water is accumulated, and it is also possible to equip water header with level sensors and a phase switch. The disadvantage of such a system is the complexity of the bottom structure, the introduction of stress concentrator on the tank bottom - the water collector, as well as the creation of differential aeration corrosion in the section of phases in the water collector. This also reduces the corrosion reliability of the tank bottom.

Figure 1 shows a schematic version of the use of the water collector in the automated system for bottom water drainage from the tank [2], which includes: electrodes 1, globe valve 2, blocking relay 3, automatic control unit 4, display unit 5, valve 6, drainage pipe 7, pit-water controller 8.

Similar schemes can be found in the work [3-4].

Thus, the disadvantage of all existing schemes is the formation of hold-up spots of bottom water in the most loaded parts of the tank bottom.

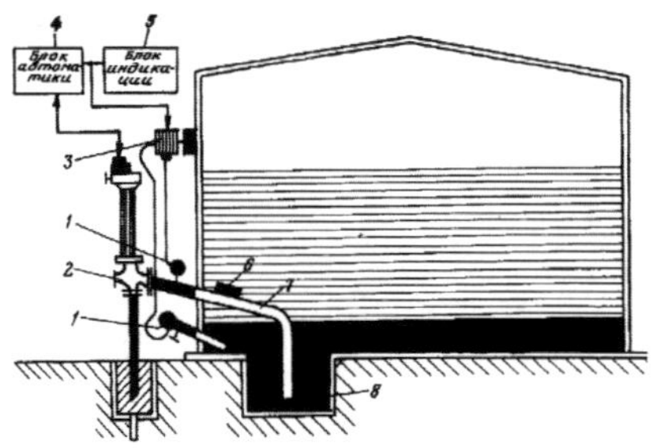

Figure 1. The tank equipped with the system for the automatic drainage of bottom water

According to [5] a gauging nipple 1 placement under the conic tank bottom 2 (figure 2) cannot be considered to be effective, since, despite the useful tank volume increase, its state monitoring deteriorates. Also, this increases leaks through the sealing surfaces and corrosive defects because of its ground placement.

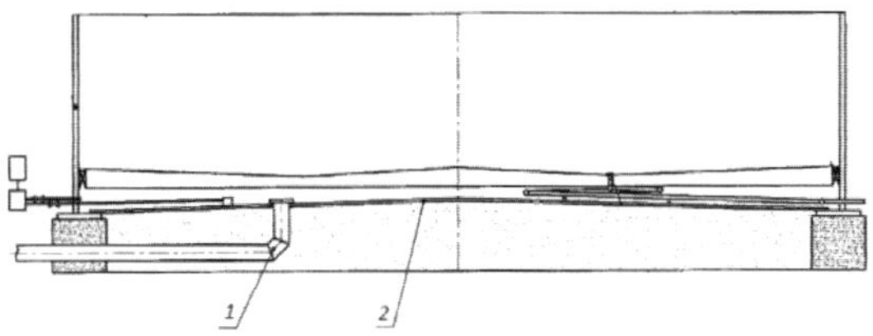

Figure 2. Schematic of the gauging nipple under the tank bottom

2. Problem statement

In analyzing the disadvantages of existing bottom water drainage systems, a perfect end result can be formulated to improve the bottom of the water drainage scheme:

- the bottom water evenly flows to one of the tank points;
- at the point of the selection of the bottom water there should be a swing pit for the efficient operation of the siphon drain;
- water drainage is done automatically.

This technical result can be achieved by changing the structure of the tank bottom: the bottom must be conical and have a slope to the center.

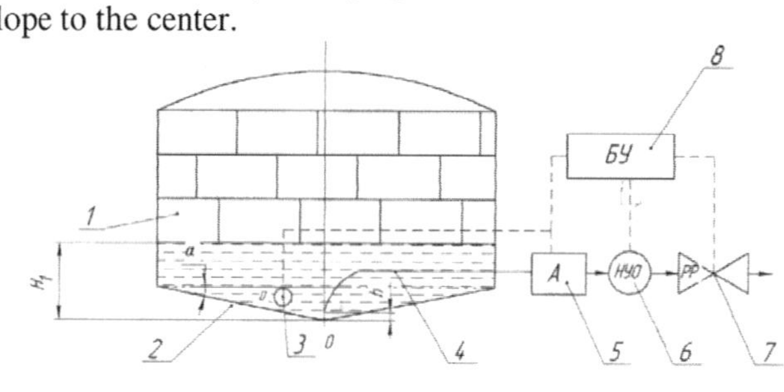

Figure 3. The device for bottom water drainage from the tank

The paper [6] proposes the oil and oil products tank bottom in the form of cone, with the cone vertex being the centre of the tank bottom and ensuring bottom water accumulation.

This assembly works as follows: in the lower part of the tank there are the hold-up spots of bottom water "H1", resulting from the storage of oil and oil products in the most loaded tank areas. These allow the bottom water to accumulate in its maximum in the lower part of the tank bottom, as oil gravity is less than water one. When bottom water appears in zone (H1), the sensor (D) is activated and sends a signal to the control unit (CU), which, in its turn, switches on the pumping unit (PU) draining bottom water out of the tank through the pipeline. One pipe end is located at the distance H from the center of the bottom, to ensure the suction of the bottom water from the zone H1. Another end is sealed in the tank sidewall and with analyzer (A) is connected with the pumping unit (PU), linked by pipeline with bottom flow regulator (FR). The bottom flow regulator (FR), the pumping unit (PU), the analyzer (A), and the interphase oil-water sensor (D) are electrically linked between themselves and with automatic control unit (CU). The form of piping 4 can be different, the main requirement is: to provide bottom dewatering from the accumulating area "H1". The drained bottom water flows through analyzer "A" 5 to determine its treatment quality and sends a signal to the control unit "CU" 8 in order to stop the pumping unit "PU" 6 and analyzes the presence of bottom water in pipe line 4.

The significant distinguishing features of the given tank for oil and oil products are:

- the bottom of the tank is conical with the slope from the periphery;
- the tank is equipped with a pipeline for the drainage of bottom water;
- the pipeline is equipped with: quality analyzer, pump, flow regulator;
- automatic control unit;
- sensor regulating the position of the interphase oil-water level.

Changing the direction of the bottom cone gives the opportunity to accumulate the bottom water directly in the swing pit, reducing the chance of commercial oil being pumped together with the bottom water. In its turn the interest in such bottom design has been increasing, and the experience with slope -bottom tanks has not been accumulated.

As a result, the study of the stress and strain state by numerical experiment is relevant.

3. Theory

The papers [7-8] contain an extensive experimental and numerical study of the stress - strain state of the vertical steel tank with floating roof. The shell and bottom stresses are measured by special sensors. To study the impact of key factors influencing on stress and strain state a finite element model is used. The selection and justification of finite elements is confirmed by computational and experimental data.

To analyze the stress and strain state, in this work we use the finite element method (FEM) allowing us to solve a wide range of physical tasks that are mathematically formulated by systems of differential equations. The system of equations obtained by FEM for the static linear-elastic model in question can be considered as a system of linear algebraic equations.

Study subject was a number of tanks from 5000 m^3 to 50000 m^3 with nominal sizes in accordance with [9].

As different variations, it was decided to use the slope angles of the tank bottom from 2 to 20 degrees.

The creation of VST mathematical model and design calculations are carried out by ANSYS Workbench 15.0.

The model geometry is very simplified, some elements are omitted because of their small influence on the main structure behavior (figure 4).

The model consists of the following elements:

- the wall of the tank;
- the bottom of the tank;
- soil bedding;
- clay bed.

The geometry of the tank bottom with 0 degree slope angle is shown in figure 4 a; 8 degrees -Figure 4 b.

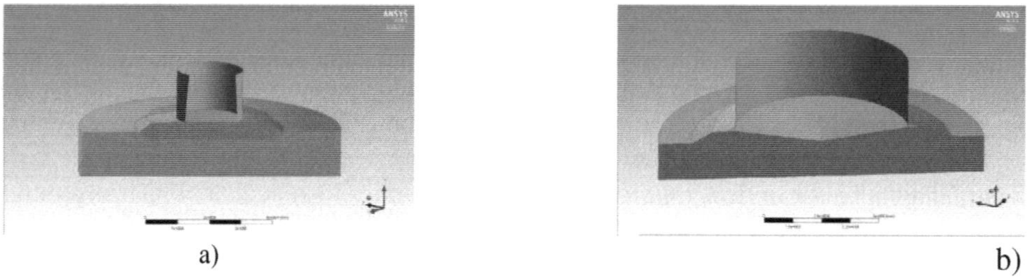

Figure 4. Tank bottom model geometry: a) with 0 degree angle; b) with 8 degree angle

We use steel 09Г2С as the most common in the industry and with sufficient strength and plasticity.
Medium-grained sand is used for bedding. Clay is a soil base.
The tank walls and bottom arc simulatcd by shcll clcmcnts.
The sand bedding and the soil base are modeled as solids.
Then the geometry is broken into finite elements.
The applied loads on the main structure and the boundary conditions consist of:

- fixed movement restriction, displacement is prohibited in all directions for soil base on the depth of 10 m;
- limiting the movement of the top tank ring of under the roof impact;
- force putting pressure on the upper ring of the tank and equal to total load from the weight of the spherical roof and the snow load;
- hydrostatic pressure of the product;
- gravity pull shall be indicated for the self-weight of the structure.

4. Experimental results

The analysis of the data obtained at the end of the main task calculation, in particular the tank, resulted in stresses and deformations appearing in the tank. The resulted calculations in the form of a graphical interpretation in the light gradation are shown in figures 5 (a) and 6 (a) for tanks with 0 degree slope angle, and 5 (b) and 6 (b), with 8 degree slope angle.

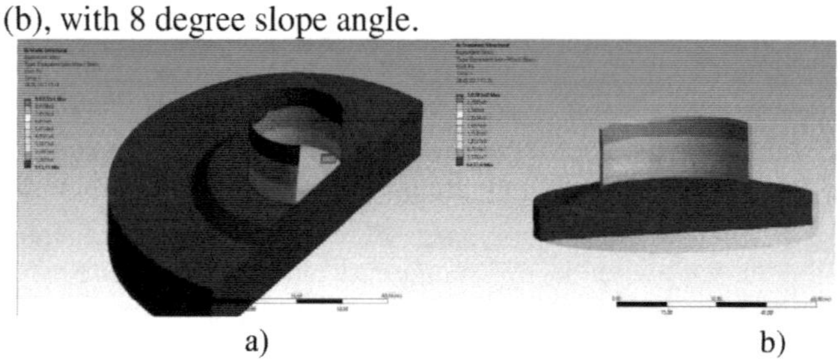

Figure 5. Von Mises stress in the tank: a) with 0 degree angle; b) with 8 degree angle

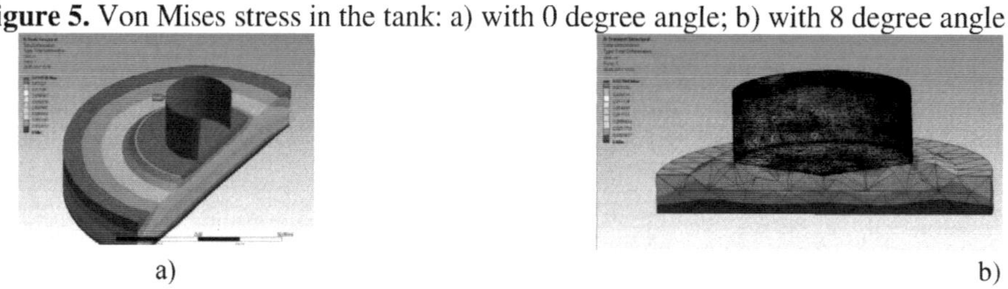

Figure 6. General deformations in the tank: a) with 0 degree angle; b) with 8 degree angle

In analyzing the stress and strain state of the flat bottom tanks, a slight stress change at the tank bottom has been detected during postprocessing stage. The main stresses are concentrated in the rim weld. In the case of gradual settlement, all the points of the tank structure have the same displacement.

The situation is quite different when calculating conical-bottom tank with slope to the center due to the bottom water drainage system. During the analysis of the stress and strain state, no concentrations of stress were found. However, the picture of the displacements will change dramatically. However, the maximum displacements are in the middle of the radius.

5. Results and discussion

Practical tests of the tank with 2-3 degrees bottom slope were carried out at oil pump station Ukhta-2, Transneft North, JSC.

The following features have been found during testing process:

- the selection of bottom water has been effectively carried out by the proposed system for 10 years;
- upon tank opening there were no paraffin deposits in practice;
- for effective selection of bottom water from TVS 5000 m^3 , it is sufficient to use one PDP located at the tank bottom.

The rim weld was diagnosed after the TVS test and cleaning. The results of the diagnosis showed that the rim weld had been deformed within the limits.

As the test of 2-3 degrees slope and 5000 m^3 TVS was successful, this design could be proposed for tanks with larger volumes.

6. Conclusion

During the study of the stress - strain state of the tank wall and bottom by ANSYS Workbench package there were created the geometric model and the model of loading the tank structure and calculated the stress - strain state of the tank, taking into account the elements of the bottom water drainage system.

During the analysis of the calculation results, it was found that the main load in the operation of such tank was in the central part of the bottom. However, the practical study of such design showed its run capability.

During the graduate qualification work, a new promising system for bottom water drainage was developed and proposed, based on the use of the conical bottom structure with slope to the center and the installation additional design in the problem area that provides greater bottom strength and more complete bottom water drainage, as well as the possibility of work automating.

During the tank operation it has been found that when draining bottom water, oil sludge discharge from VST also happens, thus reducing by half the cleaning time.

Consequently, the proposed structure of VST bottom is efficient, workable and can be implemented into tanks with 5000-50000 m^3 volume.

References

[1] Breiman M I 1978 Safe operation of equipment in open sites Chemistry
[2] Gubaidullin M M 1985 The reduction of air pollution at crude distillation unit TSNIITEneftrkhim
[3] Kuzmin S T 1987 Industrial and automation equipment in the oil and petrochemical industry Chemistry
[4] Dianov V G 1968 Automation of the production processes in oil and petrochemical industry Chemistry
[5] Shatsky AS , Lucyk A F, Gabelaja R D and Ivakin A V 2009 Cylinder storage tank for oil products Pat. 2348576 Russian Federation, MPK7 B65D 88/34 (2006.01), E04H 7/02 Newsletter. No. 7
[6] Potocari B V, Kucherenko M V and Pohlebaeva D 2015 The device for bottom water drainage

from the tank Pat. 151321 Russian Federation, MPK7 B65D 88/12 Newsletter. No. 9

[7] Lei Shi, Jian Shuai, Xiaolin Wang and Kui Xu 2017 Experimental and numerical investigation of stress in a large-scale steel tank with a floating roof **117** 25-34

[8] Rasiulis K, Šapalas A, Vadlūga R and Samofalov M 2006 *Journal of Constructional Steel Research* **62(12)** 1232-1237

[9] GOST 31385-2016 Vertical cylindrical steel tanks for oil and oil-products General specifications Standartinform 96

Comparative tests of bench equipment for fuel control system testing of gas-turbine engine

E V Shendaleva

Omsk State Technical University, 11, Mira av., Omsk, 644050, Russia

E-mail: shendalevaev@yandex.ru

Abstract: The relevance of interlaboratory comparative researches is confirmed by attention of world metrological community to this field of activity. Use of the interlaboratory comparative research methodology not only for single gages collation, but also for bench equipment complexes, such as modeling stands for fuel control system testing of gas-turbine engine, is offered. In this case a comparative measure of different bench equipment will be the control fuel pump. Ensuring traceability of measuring result received at test benches of various air enterprises, development and introduction of national standards to practice of bench tests and, eventually, improvement of quality and safety of a aircraft equipment is result of this approach.

1. Introduction

A quality and reliability of reference, calibrating, testing and measuring result in laboratories can be provided with laboratory participation in qualification test programs by means of interlaboratory comparison test during trueness and precision of characteristic measurement result of the same object will be defined in each laboratory [1].

Widespread introduction of interlaboratory comparative researches in modern practice was shown, first of all, in the field of substance and material characteristic measurements. But the methodology of comparative tests can be used also, for example, to other objects as gages established on places of their operation, or test benches. At the same time all compared and a control measuring instrument can be the identical level of accuracy [2].

2. Problem definition

We will consider tests of a fuel control pump (FCP) for fuel management systems (FMS) of gas-turbine engines (GTE). FCP tests and debugging are carried out at the semi-natural test bench [3, 4] containing the GTE imitating dynamic model [5], fuel-flow meter, electric drive and electropneumatic converter for providing FCP set characteristics (figure 1).

Measurements procedures are performed on the same object under identical circumstances don't yield, as a rule, identical results. Especially they differ if measurement conditions aren't identical. So results of tests of the same FCP at various stands often considerably differ from each other.

It is explained by existences of various factors which not always are giving in to control and inevitable casual errors inherent in each measuring procedure.

The test stand for FCP is much more difficult measuring tool, than single gages. Though for gages established at test benches, checking and calibration procedures are compulsory it doesn't solve a problem of test result traceability at different stands. Therefore it will be more correct when checking bench equipment characteristics as a measure of checking to choose a fuel control pump, his entrance and output technical characteristics.

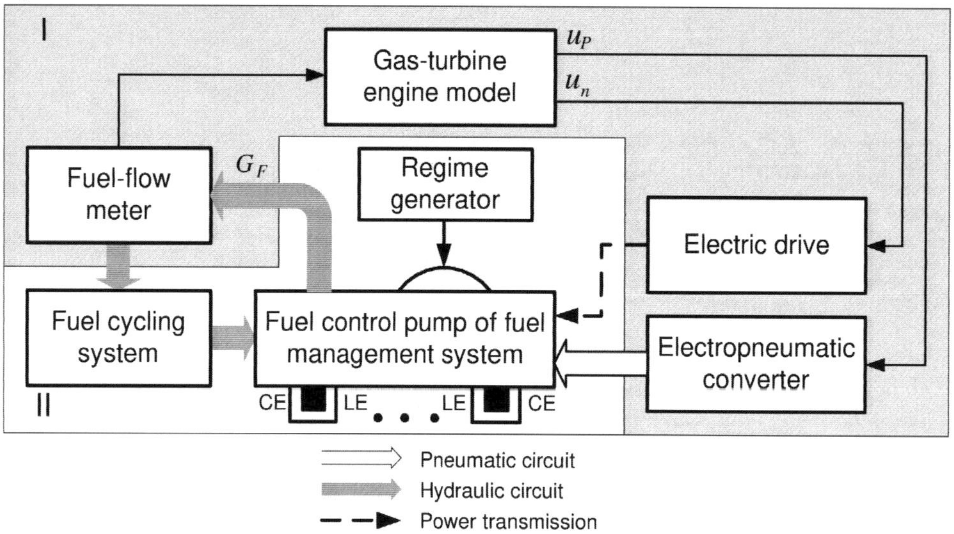

Figure 1. The scheme of the semi-natural stand for FCP
I – GTE semi-natural model; II – FCP of fuel management system
CE and LE – control and locking elements, u_n, u_P – control signals for electric drive and electropneumatic converter, G_F – fuel consumption.

3. Theory

Tests which are carried out in the course of development, operational development and mass FCP production can be united in groups [6]:

1) research experiments at a stage of preparation for development of FCP;
2) tests of separate blocks which are carried out in the course of development and operational development of FCP;
3) smoothing and special tests of pilot FCP;
4) tests of serial FCP.

By mass production of FCP carry out bench tests:

– demand;
– acceptance;
– qualification;
– periodic;
– standard (technological).

In the course of mass production each FCP copy is subjected short-term (demand and acceptance) and to periodic tests. After the last the dismantling, survey, identification of detail defects and repeated assembly of FCP are made.

The FCP debugging is corrected by the result of demand tests, and then cycles of FCP check and bench development is repeated on the FCP collaboration modes with the GTE semi-natural model. A final inspection and adjustment of the fuel control pump are carried out during acceptance tests.

FCP qualification tests are carried out after passed acceptance tests, for the purpose of held action assessments of earlier revealed shortcoming elimination and production transfer to the customer.

Periodic tests are carried out for monitoring of production and technological processes stability. The resource equivalent cyclic test providing the set quantity of start, acceleration and discharge cycles and also the set operating time on maximum modes according to established requirements to resource of the fuel control pump often act as periodic tests.

Thus, all life cycle of FCP is accompanied by different types of tests. At the same time use of «reference» FCP as a comparison measure for other fuel control pumps and identification of FCP with unsatisfactory characteristics doesn't allow speak with high degree of confidence about reliability of the received result.

Test stands for FCP represent a difficult constructions equipped with power equipment, equipment of a fuel feed system and air-gas communications, control systems, monitoring and measurements. The fuel equipment tests are accompanied by expenses of the electric power, fuel and other resources for which decrease it is necessary to increase informational content of tests and accuracy of measurement results.

In modern FMS GTE management programs are realized:

$$n_{LP} = f\left(\alpha_{ECL}, T^*_{IN}, P^*_{IN}\right),\ n_{HP} = f\left(\alpha_{ECL}, T^*_{IN}, P^*_{IN}\right),\ \pi^*_{C\Sigma} = f\left(\alpha_{ECL}\right),\ \pi^*_{GTE} = f\left(\alpha_{ECL}, T^*_{IN}, P^*_{IN}, M_N\right),$$

where n_{LP} – the rotor rotation frequency of low pressure compressor (LPC);

α_{ECL} – turning angle of the engine control lever (ECL);

T^*_{IN} – air stagnation temperature on an entrance to the engine;

P^*_{IN} – air pressure on an entrance to the engine;

n_{HP} – the rotor rotation frequency of high pressure compressor (HPC);

$\pi^*_{C\Sigma} = P^*_C / P^*_{IN}$ – extent of increase in the compressor air pressure;

$\pi^*_{GTE} = P^*_T / P^*_{IN}$ – extent of increase in the engine air pressure;

P^*_C – air pressure behind the compressor;

P^*_T – gas pressure behind the turbine;

M_N – Mach number.

The management of an operating mode in the reserve fuel control pump can be carried out on simplified programs:

$$n_E = f\left(\alpha_{ECL}\right),\ n_{HP} = f\left(\alpha_{ECL}\right),\ G_F = f\left(\alpha_{ECL}\right),\ G_F / P^*_C = f\left(\alpha_{ECL}\right),$$

where n_E – the rotor rotation frequency of the exhauster;

G_F – fuel consumption in the main combustion chamber.

Dynamic characteristics of fuel control pumps are defined on transitional modes: acceleration, dumping, partial acceleration and dumping, advanced acceleration.

Management of fuel consumption is carried out according to programs:

$$\dot{n}_{HP} = f\left(n_{HPs}, P^*_{IN}\right),\ \dot{n}_{LP} = f\left(n_{HPs}, P^*_{IN}\right),\ G_F / P^*_C = f\left(n_{HP}\right),\ G_F / P^*_C = f\left(n_{LP}\right),$$
$$G_F / P^*_C = f\left(n_{HP}, T^*_{IN}, P^*_{IN}\right),$$

where n_{HPs} – the specified rotation frequency of HPC rotor.

Reserve programs of management have an appearance:

$$n_{HP} = f(\tau),\ G_F = f(\tau),\ G_F = f\left(\pi^*_C\right),\ G_F = f\left(P^*_C\right),\ G_F = f\left(\tau, P^*_C\right),\ G_F = f\left(P^*_C, P^*_{IN}\right),$$

where τ – time.

At acceleration mode management programs of fuel consumption are applied as in intra motive parameters (P^*_C, T^*_C, n_{HP} or their combinations), and start on time. Management programs for intra motive parameters have an appearance:

$$G_F / P^*_C = const,\ G_F / P^*_C = f\left(n_{HPs}\right),\ G_F / P^*_C = f\left(n_{HP}, T^*_C\right),\ G_{Fs} / P^*_C = f\left(n_{HPs}\right),$$
$$G_F / P^*_{IN} = f\left(\Delta n_{HP}/\Delta t\right) \text{ at } \Delta n_{HP}/\Delta t = f\left(n_{HP}\right),$$

where T^*_C – gas temperature behind the compressor;

G_{Fs} – reduced fuel consumption in the main combustion chamber.

Management programs of fuel consumption for time has an appearance

$$G_F = f(\tau),\ dG_F / d\tau = const,\ dG_F / d\tau = f\left(n_{HP}\right),\ n_{HP} = f(\tau).$$

As directly set or prototype parameters at FCP of the GTE frequency rotation regulation contour tests consider:

– the rotor rotation frequency of the exhauster or LPC;
– the rotor rotation frequency of HPC;
– air pressure behind the compressor;
– extent of increase in compressor air pressure.

The FCP output parameter is fuel consumption G_F

Due to the need of various factor influences for correct understanding at compared test benches the same FCP has to be tested. As the purpose of comparative tests in this case will be «test» of test benches, but not the FCP, his design has no essential value. It is necessary to test, at least, two fuel control pumps of the same type on each of compared stands for more clear interpretation of the stand and FCP influence into repeatability and reproducibility of bench test results [7].

At a choice of comparison base for FCP output characteristics at different stands it is necessary to prefer complex characteristics, such as G_T/n_{HP}, G_T / P_C^*, $\pi_{C\Sigma}^*$, that is connected with high sensitivity of complex characteristics to parameter changes.

Among all bench measurement instrumentation flow meter indications [8] have the most considerable dispersion especially as flow meters of various dimension-types can be used at various measurement levels. Therefore along with the chosen measures of stand checking (FCP) it is necessary to use additional checking measures of fuel consumption gages [2], for example, reference mass flow meters.

FCP tests are carried out in reproducibility conditions under which test results receive by the same method, on one device under test, but at different stands, different operators, with various equipment using.

Under one FCP test the measured parameter dispersion [9] at the stand i

$$s_{ij}^2 = \left(n_{ij} - 1\right)^{-1} \cdot \sum_{k=1}^{n_{ij}} \left(\bar{y}_{ij} - y_{ijk}\right)^2, \quad (1)$$

where k – the single measurement index on j-level of i-stand;
j – measurement level (ECL);
n_{ij} – the number of measurements on j-level of i-stand;
$\bar{y}_{ij}$ – average value of measurement results on j-level of i-stand

$$\bar{y}_{ij} = n_{ij}^{-1} \cdot \sum_{k=1}^{n_{ij}} y_{ijk}. \quad (2)$$

Repeatability dispersion of the measured parameter

$$s_{rj}^2 = \left[\sum_{i=1}^{p} \left(n_{ij} - 1\right) s_{ij}^2\right] \cdot \left[\sum_{i=1}^{p} \left(n_{ij} - 1\right)\right]^{-1}, \quad (3)$$

where p – number of stands.

Interbench (interlaboratory) dispersion of test results

$$s_{Lj}^2 = \frac{s_{dj}^2 - s_{rj}^2}{\bar{n}_j}, \quad (4)$$

where

$$s_{dj}^2 = \frac{1}{p-1} \sum_{i=1}^{p} n_{ij} \left(\bar{y}_{ij} - \bar{y}_j\right)^2; \quad (5)$$

$\bar{n}_j$ – medium of measurements on j-level of all stands

$$\bar{n}_j = \frac{1}{p-1} \left[\sum_{i=1}^{p} n_{ij} - \left(\sum_{i=1}^{p} n_{ij}^2\right) \cdot \left(\sum_{i=1}^{p} n_{ij}\right)^{-1}\right]; \quad (6)$$

$\bar{y}_j$ – average value of measurement results on j-level of all stands

$$\bar{y}_j = \left(\sum_{i=1}^{p} n_{ij}\bar{y}_{ij}\right)\cdot\left(\sum_{i=1}^{p} n_{ij}\right)^{-1}. \tag{7}$$

Reproducibility dispersion of the measured parameter

$$s_{Rj}^2 = s_{Lj}^2 + s_{rj}^2. \tag{8}$$

In the presence of accepted reference value μ_{0j} for measured parameter (specifications, passport) the critical difference of reference value and average of parameter on each of stands [10]

$$CD_{ij} = \frac{2{,}8}{\sqrt{2}}\sqrt{s_{Rj}^2 - s_{rj}^2\left(1 - \frac{1}{n_{ij}}\right)}. \tag{9}$$

Critical difference of reference value and general average of parameter for several stands

$$CD_j = \frac{2{,}8}{\sqrt{2p}}\sqrt{s_{Rj}^2 - s_{rj}^2\left(1 - \frac{1}{p}\sum_{i=1}^{p}\frac{1}{n_{ij}}\right)}. \tag{10}$$

In the presence of two FCP it is possible to consider not only the accuracy of bench control and measurement systems, but also the errors brought by FCP [7].

For this purpose define:

a) arithmetic mean of test results and a divergence between test results (k = 1, 2) for each FCP on j-level of i-stand

$$\bar{y}_{ijt} = (y_{ijt1} + y_{ijt2})/2, \tag{11}$$

$$w_{ijt} = |y_{ijt1} - y_{ijt2}|; \tag{12}$$

b) arithmetic mean of test results and a divergence between average values of test results (t = 1, 2) for both FCP on j- level of i-stand

$$\bar{y}_{ij} = (\bar{y}_{ij1} + \bar{y}_{ij2})/2, \tag{13}$$

$$w_{ij} = |\bar{y}_{ij1} - \bar{y}_{ij2}|; \tag{14}$$

c) arithmetic mean of parameter and a standard deviation for both FCP on j-level of both stands

$$y_j = p^{-1}\sum_{i=1}^{p}\bar{y}_{ij}, \tag{15}$$

$$s_{yj} = (p-1)^{-1/2}\cdot\left[\sum_{i=1}^{p}(\bar{y}_{ij} - \bar{y}_j)^2\right]^{1/2}. \tag{16}$$

d) divergence squares sum between test results for each FCP on j-level of i-stand (t =1, g; i =1, p)

$$SS_{rj} = \sum_{i=1}^{p}\sum_{t=1}^{g} w_{ijt}^2; \tag{17}$$

e) divergence squares sum between arithmetic means of test results for both FCP on j-level of i-stand (i =1, p)

$$SS_{Hj} = \sum_{i=1}^{p} w_{ij}^2; \tag{18}$$

f) dispersions of repeatability and reproducibility

$$s_{rj}^2 = \frac{SS_{rj}}{4p}, \tag{19}$$

$$s_{Rj}^2 = s_{yj}^2 + \frac{SS_{rj} - SS_{Hj}}{4p}; \tag{20}$$

g) dispersions which are divergence measures between parameter values of different FCP

$$s_{Hj}^2 = \frac{SS_{Hj}}{2p} - \frac{SS_{rj}}{8p}. \tag{21}$$

During FCP tests at the modeling test bench the data on fuel consumption, rotor rotation frequency of the high pressure compressor (electric drive) and pressure behind the compressor (electric-pneumatic converter) provided in table 1 have been obtained.

4. Discussion of results of an experiment

The calculation executed for one FCP according to (1) – (10) (table 2) show that conditions on a critical difference $CD_{ij} > |\mu_{0j} - \bar{y}_{ij}|$ and $CD_j > |\mu_{0j} - \bar{y}_j|$ are executed not completely. As final results of tests it is possible to use average values of the FCP parameters on two stands except for fuel consumption for both stands at = 45 angular degree and rotation frequency for the first stand at = 83 angular degree.

Non-performance of conditions on a critical difference at FCP tests speaks about considerable influence of stand variability factors. For a research and elimination of this influence it is necessary to use multifactorial experiments.

Results of calculation for two FCP (11) – (21) (table 3) show that a standard deviation s_H for fuel consumption G_F at = 45 angular degree exceeds a reproducibility standard deviation s_R. It means that distinction influence of two FCP on test accuracy is higher than influence of other stand factors. Besides FCP distinction influence on the rotation frequency n_{HP} is comparable to influence of other factors operating at the stand.

5. Abstract and conclusions

Carrying out inter-bench comparative tests with one or several FCP as samples for comparison allows impartial estimating both trueness and precision of test benches, and statistical uniformity of FCP parameter values. Standardization of FCP comparative tests and adaptation of corresponding statistical methods for measurement results processing of various parameters will provide reasonable values of characteristics for fuel control pumps and test benches.

6. References:

[1] GOST R ISO 5275-1-2002 Russia *Accuracy (correctness and precision) of measurement methods and results. Part 1. Basic provisions and definitions* Moscow: Publ. house of standards

[2] MI 1832–88 Russia *State system of measurement ensuring unity. Group verification of checking means for identical accuracy level. Basic rules: Methodical instructions* Moscow: Publ. house of standards

[3] Shendaleva E V, Zhiltsov V V and Tetter V Y 2005 Russia The fuel apparatus adjustment technology for automatic control systems of gas turbine engines with model stands using. *Assembling in mechanical engineering and instrument making* **7** p 15-21

[4] Shendaleva E V 2017 Russia The regulation of gas turbine engine automatic control system in time of trials for technologic state forecasting *SibADI bulletin* **3 (55)** p 111 – 19

[5] Shendaleva E V 2012 Russia Gas turbine engine simulation in state space: dynamic aspect. *SibADI bulletin* **5 (27)** p 101 – 06

[6] Aleksandrovskaja L N, Kruglov V I, Kuznetsov A G, Kuznetsov V A and Kutin A A and Sholom A M 2003 *Testing theory and experimental processing of complex technical systems* Moscow: Logos p 735

[7] GOST R ISO 5275-5-2002 Russia *Accuracy (correctness and precision) of measurement methods and result. Part 5. Alternative methods of precision definition for standard measurement method* Moscow: Publ. house of standards

[8] German V M, Avgustovich V G, Arkhipov G N, Bereznjakov S V, Pipekin V I, Rakitin M M, Ryzhov I D, Sakhautdinov V N and Smolko V V and Tchervanjuk V V 1990 *Optimization methods for tests and simulation of control systems for gas turbine engines* ed Vasilchenlo K J, Dolgolenko G P, Znamenskaja A M, Kljachko V D, Mahonkin Y E, Mironov A D, Hejfets M I under the editorship of . Dedesh V T (Moscow: Mechanical engineering p 160)

[9] GOST R ISO 5275-2-2002 Russia *Accuracy (correctness and precision) of measurement methods and results. Part 2. Main repeatability and reproducibility procedure for standard measurement method* Moscow: Publ. house of standards

[10] GOST R ISO 5275-6-2002 Russia *Accuracy (correctness and precision) of measurement methods and results. Part 6. Use of accuracy values in practice* Moscow: Publ. house of standards

IOP Conf. Series: Journal of Physics: Conf. Series **998** (2018) 012031 doi:10.1088/1742-6596/998/1/012031

Table 1. Data of experiments.

Technical characteristics	α_{ECL}, angular degree	Stand 1		Stand 2		Parameter values on specification
		Test 1	Test 2	Test 1	Test 2	
Fuel control pump 1						
Fuel consumption G_F, kg/h	45	1426	1428	1429	1431	1420
	83	2973	2981	2983	2979	3000
	120	3396	3387	3378	3388	3400
Air pressure behind the compressor P_C^*, kg/cm^2	45	4.63	4.68	4.71	4.76	4.7
	83	8.87	8.96	9.05	8.99	9.0
	120	11.92	12.03	11.96	11.91	12.0
Rotor rotation frequency of high pressure compressor n_{HP}, min^{-1}	45	4141	4139	4122	4132	4130
	83	4762	4753	4752	4753	4750
	120	4881	4896	4894	4895	4890
Fuel control pump 2						
Fuel consumption G_F, kg/h	45	1429	1427	1419	1421	
	83	3012	3011	3006	3008	
	120	3398	3407	3405	3409	
Air pressure behind the compressor P_C^*, kg/cm^2	45	4.71	4.73	4.72	4.79	
	83	9.07	9.06	8.95	9.04	
	120	12.18	12.12	12.13	12.07	
Rotor rotation frequency of high pressure compressor n_{HP}, min^{-1}	45	4131	4134	4127	4129	
	83	4753	4755	4747	4745	
	120	4895	4886	4893	4896	

Table 2. Computational results for one fuel control pump.

Characte-ristics	α_{ECL}	Stand	$\bar{y}_{ij}$	$\bar{y}_j$	μ_{0j}	s_{ij}^2	s_{rj}^2	s_{Rj}^2	$\lvert\mu_{0j}-\bar{y}_{ij}\rvert$	CD_{ij}	$\lvert\mu_{0j}-\bar{y}_j\rvert$	CD_j
G_F, kg/h	45	1	1427	1428	1420	2.00	2.00	2.00	7.00	1.98	7.50	1.40
		2	1428			2.00			8.00	1.98		
	83	1	2977	2994	3000	32.00	16.25	607.31	23.00	48.47	5.75	34.27
		2	3012			0.50			11.50	48.47		
	120	1	3392	3397	3400	40.50	40.50	90.88	8.50	16.64	3.00	11.77
		2	3403			40.50			2.50	16.64		
P_C^*, kg/cm^2	45	1	4.66	4.69	4.70	0.0012	0.0007	0.00	0.045	0.095	0.013	0.067
		2	4.72			0.0002			0.020	0.095		
	83	1	8.92	8.99	9.00	0.0041	0.0021	0.01	0.085	0.215	0.010	0.152
		2	9.07			0.0000			0.065	0.215		
	120	1	11.98	12.06	12.00	0.0060	0.0039	0.02	0.025	0.253	0.063	0.179
		2	12.15			0.0018			0.150	0.253		
n_{HP}, min^{-1}	45	1	4140	4136	4130	2.00	3.25	30.56	10.00	10.65	6.25	7.53
		2	4133			4.50			2.50	10.65		
	83	1	4758	4756	4750	40.50	21.25	22.06	7.50	6.70	5.75	4.73
		2	4754			2.00			4.00	6.70		
	120	1	4889	4890	4890	112.50	76.50	76.50	1.50	12.25	0.50	8.66
		2	4891			40.50			0.50	12.25		

Table 3. Computational results for two fuel control pumps.

Characteristics	α_{ECL}	Stand	FCP	Test 1	Test 2	$\bar{y}_{ijt}$	w_{ijt}	$\bar{y}_{ij}$	w_{ij}	$\bar{y}_j$	SS_{rj}	SS_{Hj}	s_{Rj}	s_{Hj}
G_F, kg/h	45	1	1	1426	1428	1427	2	1428.5	4.5	1426	16.0	73.0	1.78	4.15
			2	1429	1431	1430	2							
		2	1	1429	1427	1428	2	1424.0						
			2	1419	1421	1420	2							
	83	1	1	2973	2981	2977	8	2979.0	30.3	2994	85.0	36.3	21.53	1.94
			2	2983	2979	2981	4							
		2	1	3012	3011	3012	1	3009.3						
			2	3006	3008	3007	2							
	120	1	1	3396	3387	3392	9	3387.3	17.5	3396	278.0	92.5	13.28	2.4
			2	3378	3388	3383	10							
		2	1	3398	3407	3403	9	3404.8						
			2	3405	3409	3407	4							
P_C^*, kg/cm^2	45	1	1	4.63	4.68	4.66	0.05	4.70	0.043	4.72	0.001	0.008	0.035	0.001
			2	4.71	4.76	4.74	0.05							
		2	1	4.71	4.73	4.72	0.02	4.74						
			2	4.72	4.79	4.76	0.07							
	83	1	1	8.87	8.96	8.92	0.09	8.97	0.063	9.00	0.003	0.016	0.050	0.003
			2	9.05	8.99	9.02	0.06							
		2	1	9.07	9.06	9.07	0.01	9.03						
			2	8.95	9.04	9.00	0.09							
	120	1	1	11.92	12.03	11.98	0.11	11.96	0.170	12.04	0.003	0.004	0.129	0.000
			2	11.96	11.91	11.94	0.05							
		2	1	12.18	12.12	12.15	0.06	12.13						
			2	12.13	12.07	12.10	0.06							
n_{HP}, min^{-1}	45	1	1	4141	4139	4140	2	4133.5	3.2	4132	117.0	189.3	11.35	1.52
			2	4122	4132	4127	10							
		2	1	4131	4134	4133	3	4130.3						
			2	4127	4129	4128	2							
	83	1	1	4762	4753	4758	9	4755.0	5.0	4753	90.0	89.0	7.64	3.55
			2	4752	4753	4753	1							
		2	1	4753	4755	4754	2	4750.0						
			2	4747	4745	4746	2							
	120	1	1	4881	4896	4889	15	4891.5	1.0	4892	316.0	52.0	7.81	5.79
			2	4894	4895	4895	1							
		2	1	4895	4886	4891	9	4892.5						
			2	4893	4896	4895	3							

Information about the author:

Elena V. Shendaleva (Omsk, Russian Federation) – Ph. D. in Technical Sciences, Ass. Professor, Department of Oil&Gas Busyness, Standardization and Metrology, Omsk State Technical University (644050, Mira prospect, 11, Omsk, Russian Federation, e-mail: shendalevaev@yandex.ru)

E. V. Shendaleva

Omsk State Technical University, Russian Federation, Omsk

UDC 001.891.5 : 629.7.018.4
***Keywords*: comparative tests, fuel control system equipment, gas-turbine engine, test bench.**

Quality Control of the Print with the Application of Statistical Methods

K V Simonenko, G S Bulatova, L B Antropova and L G Varepo

Omsk State Technical University, 11, Mira av., Omsk, 644050, Russia

E-mail: cirywka@gmail.com

Abstract. The basis for standardizing the process of offset printing is the control of print quality indicators. The solution of this problem has various approaches, among which the most important are statistical methods. Practical implementation of them for managing the quality of the printing process is very relevant and is reflected in this paper. The possibility of using the method of constructing a Control Card to identify the reasons for the deviation of the optical density for a triad of inks in offset printing is shown.

1. Introduction

The quality of products (services) is crucial in assessing the performance of each enterprise or organization. The release of effective and high-quality products allows the company to obtain more additional profits and ensures rapid production development. In order to achieve high-quality performance of printing products, it is required to combine the creative potential and practical experience of many specialists from different divisions.

An important role in solving the problem of quality improvement is played by consumers who dictate their demands and requests to producers of goods and services and evaluate the results of the organization's activities.

The task of statistical methods is to ensure the production of products that meet the requirements of consumers at the lowest cost. For this purpose, the analysis and research of operations in the enterprise are carried out. Due to the use of statistical methods of quality control of products, it is possible to achieve an improvement in the quality of both the purchased materials and the products produced and the reduction in the quantity of rejects, as well as the increase in the service life of the goods.

In addition, statistical methods of quality control help to more competently and rationally build the technological process, exercising control at its various stages.

2. Statement of the problem

Various theoretical principles and practical methods include the Total Quality Management TQM. Quality are achieved through the involvement of all personnel in the work improvement activities. The aim of quality improvement is to meet the needs of users and benefit all interested parties and society as a whole.

It is known that the statistical analysis of product quality is the study of conditions and factors that influence the quality of the analyzed object [1-5]. Manufacturing of quality products at minimum costs is the main goal of each enterprise.

The standard of printing is the determination of optimal reference tolerances and conditions for a particular technology and production conditions. They should ensure the optimal result in accordance with the recommendations.

The work on standardized methods is not limited to obtaining the results corresponding to the requirements of the standard. Strict control, designed to ensure the highest quality for print production, is not limited to a single calibration test run - the results of printing must be checked constantly. Therefore, the daily issue of quality control is more and more relevant and determines the main task of this work.

3. Methods of research

To solve this problem, statistical methods of control are used in the work: control chart of Shewhart [6-7]. Control charts are plotted on the chart time series with the indicated upper and lower boundaries, allowing understanding the ongoing process. The horizontal lines on the map represent the upper control limit (UCL), the central line (CL), and the lower control limit (LCL). Using these lines, you can trace the following dependencies:

- if too many experimental points are located above the UCL (below the BCL), the quality parameters do not correspond to the norms;
- if a number of experimental points are located between the CL and the UCL (or CL and BCL), the process requires intervention;

The main components of the printing system are the following: Speedmaster SM-102 printing machine, offset paper, ToyoInk.
Procedures were carried out to acclimatize paper (cardboard), maintain climate control and stabilize the printed image.

4. Results and discussion

The analysis of the widely used quality control for multi-color products of ISO 12647-2 standard [8] showed that the following standardized indicators are presented in it:

- reference paper types with CIELAB coordinates;
- CIELAB coordinates for basic colors (CMYK) and tolerances for reference paper types;
- secondary coordinates (RGB) for reference paper types;
- tonal values and tolerances for reference paper types;
- definition of dot gain in midtones;
- CMYK values for the gray balance scales;

Variation tolerances for base colors and dot gain when printing runs

The normalized values of optical density are presented in Table 1 [8].

Table 1. Optical density values according to GOST ISO 12647-2

Paints	Optical density and tolerances		
	Coated glossy	Coated matt	Uncoated
Blue	1.55± 0.1	1.45± 0.1	1.00± 0.1
Purple	1.50 ±0.1	1.40 ±0.1	0.95 ±0.1
Yellow	1.45± 0.1	1.25± 0.1	0.95± 0.1
Black	1.85 ±0.15	1.75 ±0.15	1.25 ±0.15

IOP Conf. Series: Journal of Physics: Conf. Series **998** (2018) 012032 doi:10.1088/1742-6596/998/1/012032

Since in the standard ISO 12647-2 the above indicators are given only for standard paper types, the number of which is limited to five, then when working with other types of paper, the evaluation of the quality of multicolour products will be approximate. The assortment of paper in the modern market is quite large. This causes certain difficulties in choosing a reference group for them, the indicators of which are taken as standard.

The standard defines measurable results, which need to be addressed, but does not disclose the methodologies and recommendations for their achievement.

Modern methods of quality management allow us to monitor the quality of the product at all stages of its life cycle. The main function of the quality control system is the production of high-quality competitive products. During the control all possible deviations from the established norms are analyzed. At the same time, statistical methods of quality control used at all stages of the product life cycle are an important condition for the rational construction of the technological process.

Figure 1 shows the results of studies of changes in the optical density for all four CMYK ink.

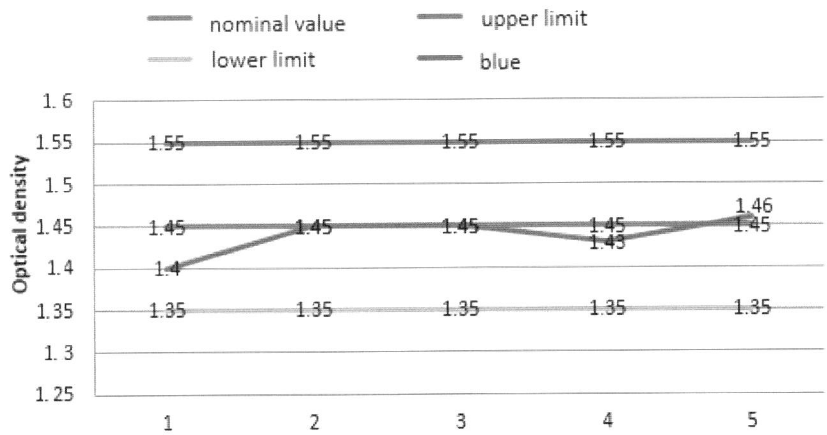

a) blue ink

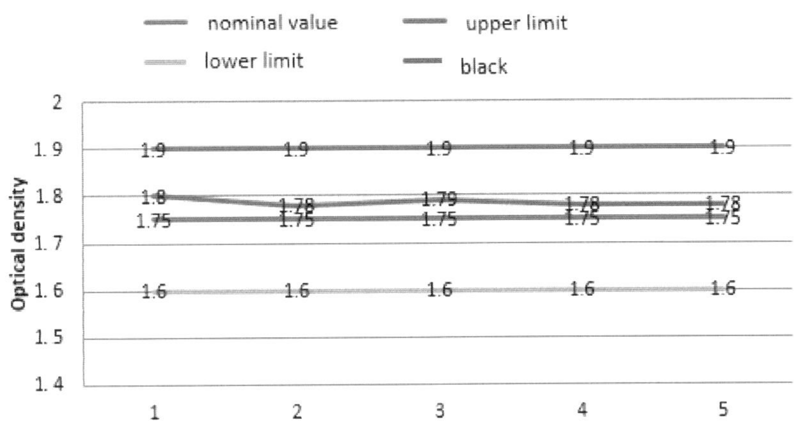

b) black ink

Figure 1. Beginning. Control charts of the dependence of the optical density on time

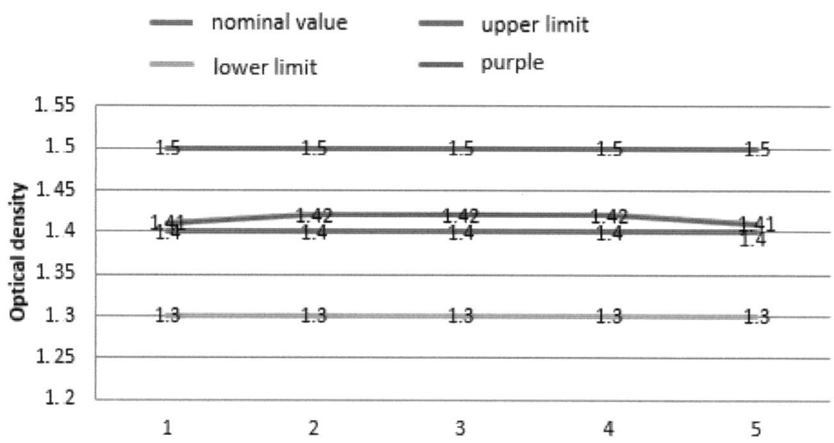

c) purple ink

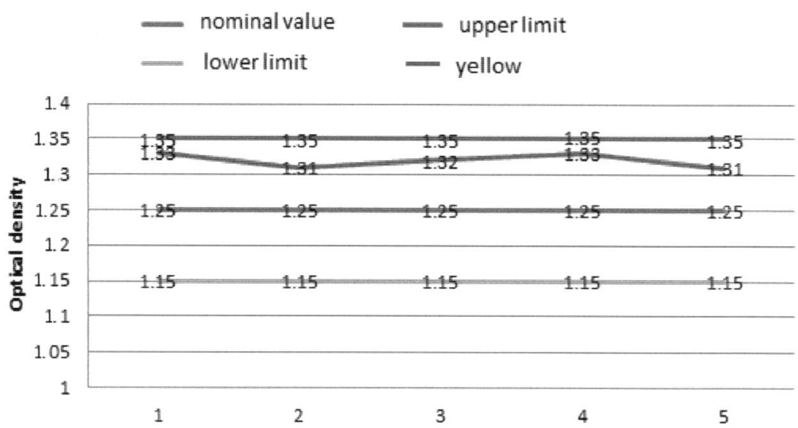

d) yellow ink

Figure 1. End. Control charts of the dependence of the optical density on time

For comparison, the charts also show normalized optical densities for the recommended reference paper. The value of the optical density of the printing impressions was determined with an interval of 10 minutes.

As it can be seen from the figure, in almost all cases the changes in the indices are within the permissible limits ± 0.1. The maximum deviation is observed in yellow ink. But since it is the least noticeable to the human eye, this does not greatly affect the quality of the print. The most pronounced is the contour (black) ink. In this study, deviations in black ink are minimal.

Therefore, one of the leading places in the management and control system should be the construction of control charts.

Also, in the work to predict print quality based on the structuring of the quality function, the degree of tightness of pair interactions between consumer requirements and the characteristics of the quality of multicolour products was assessed. The House of Quality has been built (Fig. 2). This allowed us to identify the following relationship: customer requirements and controlled characteristics of the printed matter; customer requirements and print quality indicators.

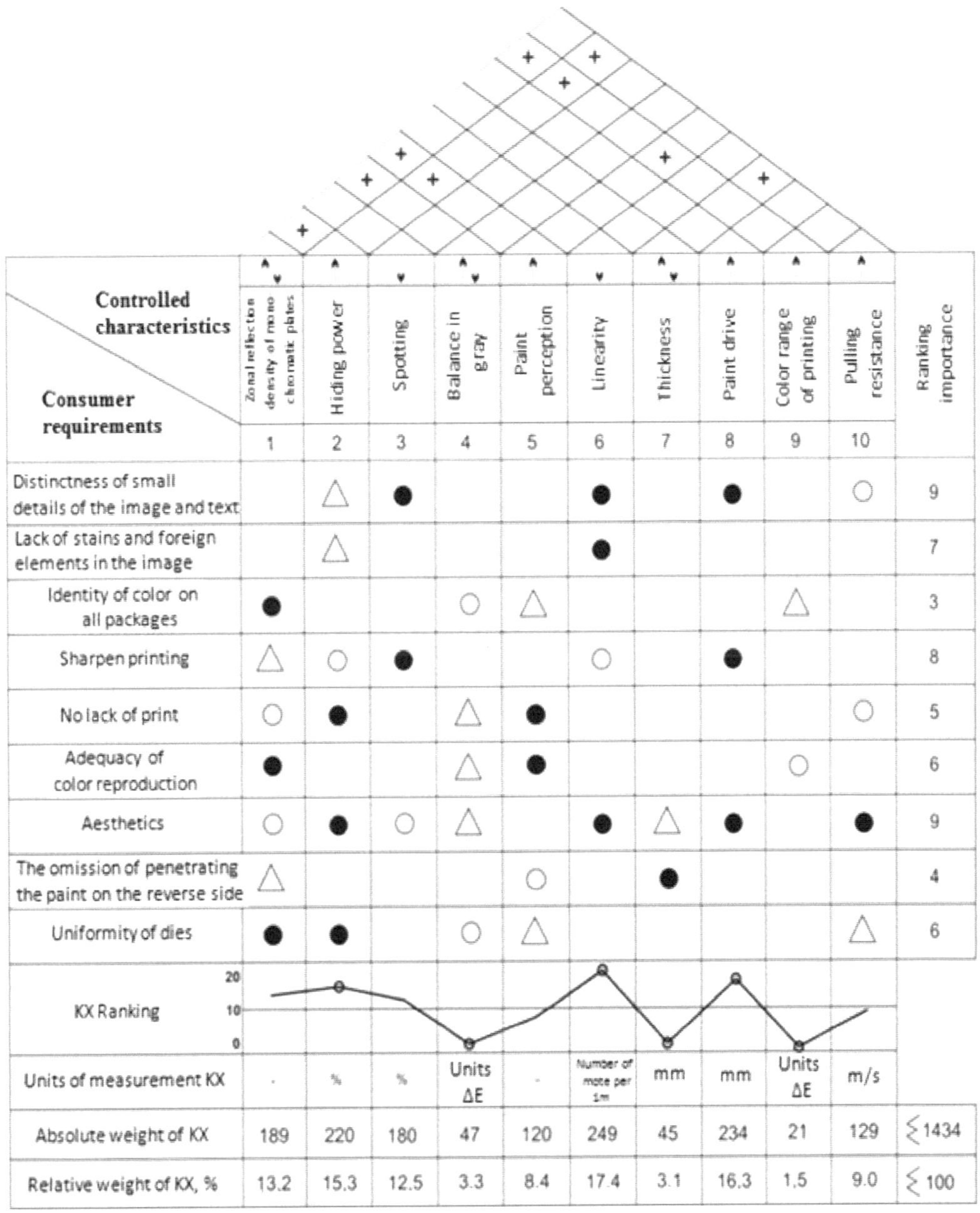

Figure 2. Quality House for predicting print quality parameters (symbols of the degree of tightness of communication are placed in the matrix cells: • - strong, ○ - medium, Δ - weak).

5. Conclusion

Thus, the possibility of using the method of constructing the Shukart Control Chart for revealing the reasons for the deviation of the optical density for a triad of inks in offset printing is shown. Analysis of the graphs of the optical densities depends on all four colors on time showed that the greatest deviation from the norm is observed in yellow paint.

The analysis of the research results using quality structuring functions showed that the most important requirements of consumers are the discernibility of the image and text small details, the clarity of the print, the absence of stains and extraneous elements in the image, and its aesthetics.

The approaches considered allow us to determine the weakest place in the printing system, in particular, paper - printing ink, to assess the quality level and to give appropriate recommendations for the management of the technological process and the achievement of standardized quality indicators.

References

[1] Borisova A S and Varepo L G 2011 *Izvestiya vysshih uchebnyh zavedeniy. Problemy polygrapfii i izdatelskogo dela* **4** p 27-35 Russian

[2] Borisova A S and Varepo L G 2011 *Quality, standardization, control: theory and practice: mater 11th Intern. scientific-practical. Conf.* ATM of Ukraine) p 13-15

[3] Varepo L G, Borisova A S and Golunov A V 2010 *Papers of the 42-th conference: international circle of educational institutes for graphic arts technology and management* (Moscow) p 145-151

[4] Goodilin D 2007 Offset printing on cardboard: problems of standardization *Compuart J.* **6** p 28-31 Russian

[5] Vernhes P Jean-Francis B Mercier C and Pineaux A B-B 2008 *Applied Surface Science* **254 (22)** 7431-7437

[6] ISO 7870-1: 2014 Control charts Part 1: General guidelines

[7] ISO 7870-2: 2013 Control charts Part 2: Shewhart control charts.

[8] ISO 12647-2: 2004(E) Graphic technology Process control for the production of half-tone color separations, proof and production prints Part 2: Offset lithographic processes.

Revisiting measuring colour gamut of the color-reproducing system: interpretation aspects

I A Sysuev[1], L G Varepo[1] and O V Trapeznikova[1]

[1]Omsk State Technical University, 11, Mira, av., Omsk, 644050, Russia

E-mail: sia1960@mail.ru

Abstract. According to the ISO standard, the color gamut body volume is used to evaluate the color reproduction quality. The specified volume describes the number of colors that are in a certain area of the color space. There are ways for evaluating the reproduction quality of a multi-colour image using numerical integration methods, but this approach does not provide high accuracy of the analysis. In this connection, the task of increasing the accuracy of the color reproduction evaluation is still relevant. In order to determine the color mass of a color space area, it is suggested to select the necessary color density values from a map corresponding to a given degree of sampling, excluding its mathematical calculations, which reflects the practical significance and novelty of this solution.

Keywords: color gamut body measuring, integrated index, color accuracy, printing quality, coordinate system L*a*b*.

1. Introduction

It is known that the task of evaluating the quality of color reproduction is rather complicated. Approaches to evaluating the quality of color printing are very diverse. There is a gap between instrumental measurements and visual evaluation. Color management systems ensure the correct color control throughout the entire production cycle, starting with the prepress stage, integrating all the devices into one common color space.

According to some authors [1], the color reproduction of a seal is determined by the properties of the printed material, in this case - of paper, and by the type of printing equipment. A similar opinion is shared by the authors of [2] who also note that variations in the quality of offset printing are connected with numerous parameters of printing equipment and paper. Since in most cases quality evaluation is carried out manually, the investigators [2] propose a new system for evaluating the print quality. This system is based on the use of conventional printed labels. So-called virtual sensors are used to evaluate print attributes using double mesh images. The main feature of this solution is the connection between the estimated print quality and the known process conditions.

In [3], a method for calibrating a printing machine based on regression models is developed. According to this research, the method is advantageous because it allows us to obtain a mathematical model describing the printing machine characteristics, which forms the basis for the current colorimetric correction of the printing.

However, to test complex multi-colour printouts, the resources of many known systems are limited. In addition, many well-known models mainly reflect the relationship of the evaluated optimization parameter (the color reproduction index) to the variables being investigated, for example, the paper characteristics.

An indicator that meets all the requirements of the ISO standard (i.e., describing the number of reproduced colors) is the color gamut body volume of the color-reproducing system, evaluated, including the printing system, in units of the color space ΔEE CIE L*a*b* [4, 5].

T e specified volume describes the number of colors that are in a certain area of the color space (for example, belonging to the printing system color gamut body) and differentiated by the typical observer. Obviously, other things being equal, a larger color gamut body means more colors reproduced by the printing system. The color in such a system is represented as one brightness coordinate and two color-difference chromaticity coordinates.

The determination of the volume of the color-reproducing system gamut body consists of two main parts: the description of the gamut body in the L * a * b * space and the determination of its volume. In [6], it is noted that in this case it is appropriate to use numerical integration methods.

The description of the printing system color gamut body and the calculation of the color gamut body volume in units of ΔE and ΔE_{00} is discussed in detail in [7-10]. A graphic interpretation of the color gamut body is given there.

Nevertheless, the task aimed at improving the accuracy of the color rendering index is not fully solved.

2. Statement of the problem

The more precisely the color differences are calculated, the more the result of calculating the color gamut body volume will be consistent with the color perception. The search for solutions to the problem under consideration remains relevant.

For an objective evaluation of the printing system, in addition to the existing ones, it is necessary to use an integral numerical indicator expressing its color gamut. On the one hand, this indicator is required in order to objectively compare the printing systems color gamut with a minimum error; on the other hand, as a criterion used to solve problems of determining the optimal color characteristics of the triad stimuli used by the system being evaluated. The replacement of the color-difference function ΔE by another, more perfect, is an up-to-date task.

3. Problem solution

As it was noted above, L*a*b* system is not a contrast one, as it is usually named.

An ideally equal-contrast system is homogeneous in terms of a person color perception. This means that an equally contrasting color space has a constant color density that is, the location density of the points of the colors differentiated by man in the color space. In a truly color-equal (isotropic in color perception) space, the indicator of its color density is the parameter $\Delta E = \text{const}$. In this case, by calculating the volume of the color gamut body in units of ΔE (as the distance between the points of colors in an equally contrasting, uniformly dense, color space where the same distances between the points of the colors being compared correspond to the same visual differences), one can determine its color mass.

In this application, the color mass is an indicator characterizing the amount of colors distinguished by a person in a certain area of the color space, for example, within the body of the color gamut of the color reproducing system and the printing system in particular. In the L*a*b* color space ΔE parameter cannot be an indicator of the color density, since the system is largely unequal and not isotropic from the point of view of the color difference. In other words, the color space in its different areas and in different directions is characterized by a different color density.

As an indicator of the color density, parameter ΔE_{00}, which to the greatest extent (nowadays) corresponds to the coloration of the standard observer can be used. Indicator ΔE_{00} takes into account the heterogeneity of the L*a*b* color space in different areas and directions ($\Delta E_{00L^*} \neq \Delta E_{00a^*} \neq \Delta E_{00b^*}$).

Integral index Θ of the color gamut evaluation is the indicator characterizing the number of colors, reproduced by a certain color-reproducing system and differentiated by a typical observer.

It was calculated as the volume of the unequally-contrasting (non-isotropic from the point of view of color differentiation) color space area. At the same time, the unevenness of the color space when calculating the area volume was taken into account using the refined color separation function ΔE_{00} at the stage of the elementary volume determination [6].

Actually, the essence and the algorithm for calculating the index Θ was a natural development of the indicator of the color gamut body volume of the color-reproducing system (or an random color space area) as an integral indicator of their color gamut: evaluation of the number of colors reproduced by the printing system (or contained in a certain area). Namely: if we assume that the color space is equally contrasting, i.e. the volume of a certain closed region of this space serves as an indicator of the amount of colors that are detected by the standard observer in it.

If the color space is equally contrasting, the volume of a closed area is calculated taking into account the description of this inhomogeneity (function ΔE_{00}). The result is the interpretation of the index Θ as the volume of the color space area calculated using the refined color function ΔE_{00}.

By introducing the color density and color mass concepts index Θ can be interpreted much more accurately by its nature and in accordance with the physical analogy. If it is assumed that the function ΔE_{00} is a description of the color space L*a*b* density, then the elementary volume V_{el} (ΔE_{00}) is, in fact, not specified from the point of view of color perception by the volume of the micro-area, but is an indicator of the number of colors distinguished by a typical observer that is the elementary volume color mass. At that, a certain color space area, composed of several volumes, will have a color mass M_{CSA} that is an index, characterizing the number of its colors, distinguished by atypical observer.

Color mass of a certain area of the color space (Fig. 1) can be found by using the numerical integration method:

$$M_{CSA} = \sum_{L^*} \sum_{a^*} \sum_{b^*} M_{el}, \tag{1}$$

where M_{CSA} is the colour mass of the colour space area under investigation,
L*, a*, b* are the numerical integration limits, describing the area boundaries,
M_{el} is the colour mass of the elementary volume $V_{el} = \Delta L^* \times \Delta a^* \times \Delta b^*$, where $\Delta L^*=\Delta a^*=\Delta b^*$).

$$M_{el}(\Delta E_{00}) = \Delta E_{00L^*} \times \Delta E_{00a^*} \times \Delta E_{00b^*}, \tag{2}$$

where $\Delta E_{00L^*} = \Delta E_{00}\,(L^*_1, a^*_1, b^*_1;\ L^*_2, a^*_1, b^*_1)$,

$\Delta E_{00a^*} = \Delta E_{00}\,(L^*_1, a^*_1, b^*_1;\ L^*_1, a^*_2, b^*_1)$,

$\Delta E_{00b^*} = \Delta E_{00}\,(L^*_1, a^*_1, b^*_1;\ L^*_1, a^*_1, b^*_2)$,

L^*_1, a^*_1, b^*_1 и L^*_2, a^*_2, b^*_2, are the colour coordinates of the elementary volume reference points, where $\Delta L^*=L^*_2-L^*_1$, $\Delta a^*=a^*_2-a^*_1$, $\Delta b^*=b^*_2-b^*_1$.

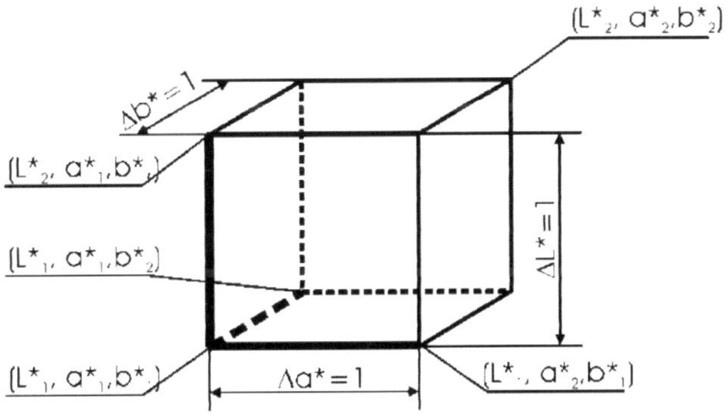

Figure 1. Elementary volume (ΔL*×Δa*×Δb*) in L*a*b space (ΔL*=Δa*=Δb*)

In this interpretation of Θ, as the authors suggest, a complete analogy with the physical concepts of "density - volume - mass" will take place.

The specified values of colour density P(L*,a*,b*) are found as the ratio of the color mass of an elementary volume to this volume:

$$P(L^*, a^*, b^*) = \frac{Mel}{V_{el}} \tag{3}$$

where $V_{el} = \Delta L^* \times \Delta a^* \times \Delta b^*$, having ΔL*=Δa*=Δb*.

Such map of colour density can be calculated for all the colour space L*a*b* at the specified value of its sampling ΔL*=Δa*=Δb*, and will include the values of P (L*, a*, b*) for every elementary volume of the colour space.

4. Conclusion

Based on the above, we can state the following.

- When calculating the color weight of the color space area, you do not need to calculate the color density each time, you only need to select the desired values from the map corresponding to the specified degree of sampling.
- The determination of the color weight of a certain area of the color space in this case will be reduced only to determining the occurrence of an elementary volume in the color space area and summing the corresponding values of the color masses of elementary volumes.

However, we believe that the presented interpretation of the indicator Θ as the color mass, as well as the very introduction of the concepts of color density and color mass, requires further discussion.

References

[1] Maryam Ataeefard 2014 Investigating the effect of paper properties on color reproduction of digital printing *J. Progress in Organic Coatings* vol 77 Issue 9 p 1376-1381

[2] Undström J, Verikas A, Tullander E, Larsson B 2013 Assessing, exploring, and monitoring quality of offset colour prints *J. Measurement* vol 46 Issue 4 p 1427-1441

[3] Olejnik-Krugły A, Różewsk P, Zaikin O, Sienkiewicz P 2013 Approach for color management in printing process in open manufacturing systems *J. IFAC Proceedings* vol 46 Issue 9 p 2104-2109

[4] ISO 15076 1(2005) Image technology colour management – Architecture, profile format and data structure, 2005

[5] ISO 12647-2: 2004 Graphic technology – Process control for the production of half-tone colour separations, proof and production prints – Part 2: Offset lithographic processes, 2012

[6] Pozharsky A O, Sysuev I A 2006 Evaluation of the printing system color gamut by color gamut body volume, calculated by using color space heterogeneity *J. Educational Institutions. Problems of printing and Publishing* No 4 p 3–12 Russian

[7] Pozharsky A O and Sysuev I A 2005 Evaluation of the printing system color gamut by color gamut body volume, calculated by using color differences precise function *J. Omsk scientific bulletin* No 4 pp 180 Russian

[8] Pozharsky A O 2006 Plotting of color gamut body, reproduced by a printing system as a part of its volume calculation task *J. Omsk scientific bulletin* No 2 p 136–138 Russian

[9] Varepo L, Golunov A, Golunova A, Trapeznicova O and Nagornova I 2015 Method of calculation volume of the color gamut body *Testing and Measurement: Techniques and Applications* p 69-71

[10] Varepo L, Trapeznicova O, Golunov A, Panchuk K and Lyashkov A 2017 Computer visual representation of color gamut of the multicolor image reproduction systems *J. Geometry and graphics* vol 5 Issue 3 p 86-91 Russian

Accuracy Improvement of Multi-Axis Systems Based on Laser Correction of Volumetric Geometric Errors

V I Teleshevsky, V A Sokolov, and Ya I Pimushkin

Moscow State Technological University "Stankin", 1, per. Vadkovskiy, Moscow, 127055, Russia

E-mail: vitel@stankin.ru

Abstract. The article describes a volumetric geometric errors correction method for CNC-controlled multi-axis systems (machine-tools, CMMs etc.). The Kalman's concept of "Control and Observation" is used. A versatile multi-function laser interferometer is used as Observer in order to measure machine's error functions. A systematic error map of machine's workspace is produced based on error functions measurements. The error map results into error correction strategy. The article proposes a new method of error correction strategy forming. The method is based on error distribution within machine's workspace and a CNC-program postprocessor. The postprocessor provides minimal error values within maximal workspace zone. The results are confirmed by error correction of precision CNC machine-tools.

1. Introduction

CNC-controlled multi-axis systems are widespread in contemporary industry. Both machine-tools and measuring systems (i.e. CMM's), used to produce and measure complex parts, include numerical controlled moving components [1-2].

There are several up-to-date trends in modern industry:

- Integration of measuring and machining operations;
- Constantly increasing accuracy requirements;
- Usage of complex shaped parts.

Thus, the requirements increase not only for positioning accuracy of machine's parts, but also for volumetric accuracy of the whole machine – it's ability to precisely reproduce complex surfaces.

2. Statement of the problem

Volumetric accuracy is characterised by volumetric error – vector between desired and actual position of cutting edge or measuring probe ending in any workspace point. The "Volumetric accuracy" term was first introduced in [3] and is actively studied in recent papers [4-6].

Volumetric accuracy depends primarily on technology of machine`s parts production and assembly. There are limited opportunities to increase volumetric accuracy by means of technology, making accuracy increase quite expensive or impossible [5]. Thus a new method is being developed, one based on error measurement and automated compensation [6-7].

The two especially strong trends benefit this method. First, CNC-control system evolve greatly, for both machining and measuring applications. Second, there is certain progress in the field of error measurement. Instead of microscopes, standards and strings, versatile laser interferometers are used, allowing measurement of not only linear, but also angular and straightness errors for wide range of machine dimensions and with high precision. Moreover, recent advances in laser measurements allow simultaneous measurement of several error components.

These trends, one in the field of control, and the other in the field of observation, provide new opportunities to increase volumetric accuracy. By means of software error correction. These opportunities are based on Kalman's concept of "observation-control" duality [7]. This concept is used in MSTU "Stankin" in the field of multi-axis systems accuracy [5-7]. The concept unites processes of observation and control.

As mentioned in [5-7], the information stream consists of control – system's adjustments - and observation – creation of system's state model based on measurements of spatial error functions listed in [7]. [7] states that in order to map volumetric errors a number of error functions should be measured, i.e. 21 function for 3-axis machine, and more than 30 for 5-axis one. It is possible to collect a number such amount of data quite rapidly using a versatile laser interferometer.

Thus an increased volume of observation provides more informative state model and more effective error correction strategy. The error map can be not only created once, but, using a rapid measuring interferometer, can also be regularly updated and corrected.

3. Theory

Any machine-tool or CMM consists of several CNC-controlled moving parts. The parts provide linear movement and rotation of workpiece along or around coordinate axes. Errors of these movements and axes misplacement make up volumetric error of the machine.

Parts are machined and measured within the three-dimensional machine`s workspace. Machining and measurement errors are caused by linear and angular errors that distort the workspace. Thus the task is to determine the part's distortion by means of measuring spatial error functions.

The versatile interferometer is used as an observer in order to create the machine's state model. This kind of models are frequently described in papers. Many methods are used for this purpose, i.e. Denavit-Hartenberg method [8], rigid body kinematics method [9], multiple-body kinematics method [10], matrix simulation method [11]. Some papers consider not only systematic errors, but random errors as well.

Volumetric geometric error can be calculated as follows:

$$\boldsymbol{XYZ} = R_X \left[R_Y \left(R_z^{-1} \boldsymbol{T} + \boldsymbol{Z} - \boldsymbol{Y} \right) - \boldsymbol{X} \right] \tag{1}$$

where X, Y, Z are the linear error vectors; R_X, R_Y, R_Z are the rotation errors matrices; T is the tool offset vector.

Hence, the equations for the axes errors are:

For X axis:

$$\begin{aligned} \Delta X &= \delta_{XX}(X) + \delta_{XZ}(Z) + \delta_{XY}(Y) + Y\left[\varepsilon_{ZY}(Y) + \varepsilon_{ZX}(X)\right] + Y\alpha_{YX} - Z\left[\varepsilon_{YY}(Y) + \varepsilon_{YX}(X)\right] - \\ &- Z\alpha_{ZX} + X_T - Y_T\left[\varepsilon_{ZX}(X) + \varepsilon_{ZY}(Y)\right] - Z_T\left[\varepsilon_{YZ}(Z) + \varepsilon_{YY}(Y) + \varepsilon_{YX}(X)\right]; \end{aligned} \tag{2}$$

For Y axis:

$$\begin{aligned} \Delta Y &= \delta_{YY}(Y) + \delta_{YX}(X) + \delta_{YZ}(Z) - X\varepsilon_{ZX}(X) - Z\left[\varepsilon_{XY}(Y) + \varepsilon_{XX}(X)\right] - Z\alpha_{ZY} + \\ &+ X_T\left[\epsilon_{ZY}(Y) + \epsilon_{ZX}(X)\right] + Y_T - Z_T\left[\varepsilon_{XY}(Y) + \varepsilon_{XX}(X) + \varepsilon_{XZ}(Z)\right]; \end{aligned} \tag{3}$$

For Z axis:

$$\begin{aligned} \Delta Z &= \delta_{ZZ}(Z) + \delta_{ZX}(X) + \delta_{ZY}(Y) - X\varepsilon_{YX}(X) - Y\left[\varepsilon_{XY}(Y) + \varepsilon_{XX}(X)\right] + \\ &+ X_T\left[\varepsilon_{YX}(X) + \varepsilon_{YZ}(Z) + \varepsilon_{YY}(Y)\right] + Y_T\left[\varepsilon_{XY}(Y) + \varepsilon_{XZ}(Z) + \varepsilon_{XX}(X)\right] + Z_T, \end{aligned} \tag{4}$$

Where X, Y, Z are the current workspace point coordinates.;

X_T, Y_T, Z_T are the vector T components;

IOP Conf. Series: Journal of Physics: Conf. Series **998** (2018) 012034 doi:10.1088/1742-6596/998/1/012034

Others are the error functions from part II.
The absolute value of volumetric error vector can be calculated as:

$$\Delta = \sqrt{\Delta X^2 + \Delta Y^2 + \Delta Z^2} \tag{5}$$

High volume of observation is required in order to obtain information on volumetric error, i.e. 21 error function must be measured for a 3-axis machine, and more than 30 for 5-axis machine. Thus, the accuracy control can only be realized by means of versatile contactless and rapid interferometers.

4. Experiments

A HAAS "VF3" machine was used as the observation object. It has a commonly used alignment of parts, with horizontal X-table and vertical spindle.

a) b)

Figure 1. View and scheme of HAAS VF3 alignment.

Rigid body kinematics method was used in order to map errors of this machine. Renishaw XL-80 interferometer was used as observer. Following workspace points were used to collect data:

- X axis: 50 mm. – 950 mm. step 100 mm – 10 points.
- Y axis: 25 mm. – 475 mm. step 50 mm – 10 points.
- Z axis: 25 mm. – 520 mm. step 55 mm – 10 points.

Special CNC programs were made in order to move machine's elements to preset points.
Figure 2 presents the resulting error map.

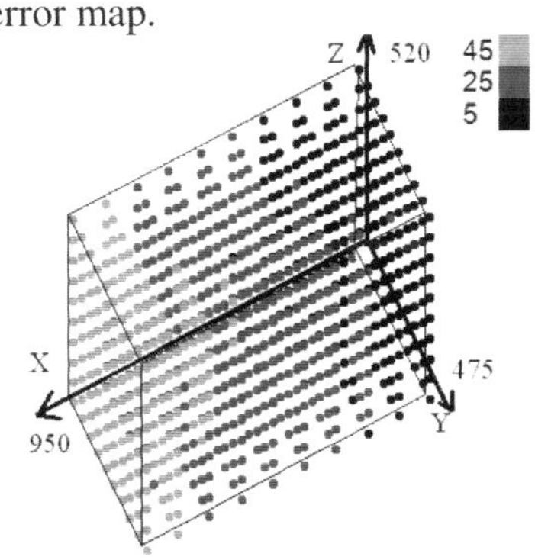

Figure 2. Volumetric error distribution within machine's workspace. Color scale in mm.
There are about 20% points with volumetric error less than 5 micron.

Contemporary industry is based on interaction between CAD/CAM systems and CNC systems in order to transform a drawing into CNC program and then into cutting edge path. Thus a method of error control was used, based on changing coordinates of path's points within CNC program. Figure 3 describes the method.

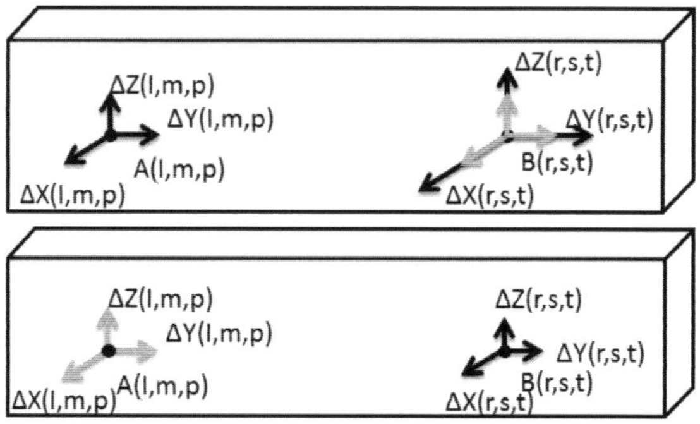

Figure 3. Error correction vector. Black – errors for each workspace point, grey – workspace point A errors, subtracted from all the other points.

For any workspace point A numbered (l,m,p) and point B numbered (r,s,t) error vectors are determined as $\Delta x_{(l,j,k)}, \Delta y_{(l,j,k)}, \Delta z_{(l,j,k)}$ and $\Delta x_{(r,s,t)}, \Delta y_{(r,s,t)}, \Delta z_{(r,s,t)}$. If errors for point A are subtracted from all workspace points (i.e., by means of coordinates origin shifting), then compensated errors for point A will be zero ($\Delta x_{(l,j,k)} = 0, \Delta y_{(l,j,k)} = 0, \Delta z_{(l,j,k)} = 0$), and compensated errors for other points will be calculated as:

$$\Delta x^{ck}{}_{(r,s,t)} = \Delta x_{(r,s,t)} - \Delta x_{(l,j,k)} \tag{6}$$

$$\Delta y^{ck}{}_{(r,s,t)} = \Delta y_{(r,s,t)} - \Delta y_{(l,j,k)} \tag{7}$$

$$\Delta z^{ck}{}_{(r,s,t)} = \Delta z_{(r,s,t)} - \Delta z_{(l,j,k)} \tag{8}$$

Thus, absolute value of volumetric error vector for point A is zero:

$$\Delta A = 0 \tag{9}$$

Absolute value of volumetric error vector for point B:

$$\Delta B = \sqrt{\left(\Delta x^{ck}_{(r,s,t)}\right)^2 + \left(\Delta y^{ck}_{(r,s,t)}\right)^2 + \left(\Delta z^{ck}_{(r,s,t)}\right)^2} \tag{10}$$

By picking all points within the workspace and subtraction their errors from all the other points, a point can be found so that sum of corrected errors will be minimal:

$$\sum_i \Delta F_i \to \min \tag{11}$$

Where i=1,… N; N is the number of workspace points; F is the random point in machine's workspace.

These changes can be introduced: 1) during CNC-program creation via CAM system; 2) via CNC system after the program is loaded; 3) after program creation but before loading it into the machine. The last method was chosen because it is not uncommon that access to CNC system is limited or restricted. In order to introduce changes into CNC program a spectial software postprocessor was developed.

The postprocessor is integrated into technological system as shown on figure 4.

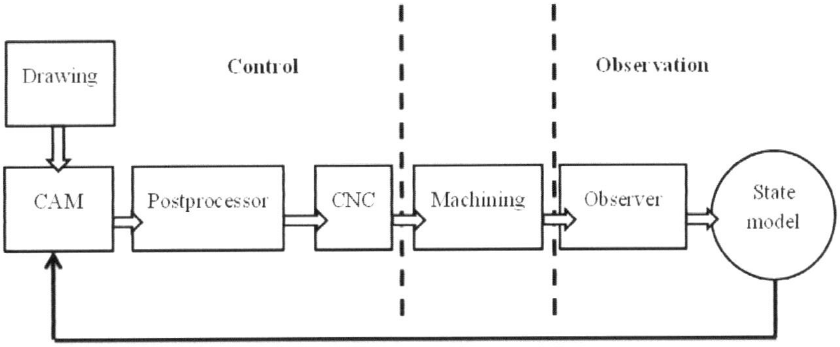

Figure 4. Technological streams interaction

Figure 5 presents result of error correction by means of a postprocessor. First, an optimal origin point was found. Then corrections were introduced into CNC system. Finally, all error functions were measured again. As result, error zonw with smaller errors (less than 5 micron) make up to 80% of machine's workspace (compared to 20% before correction).

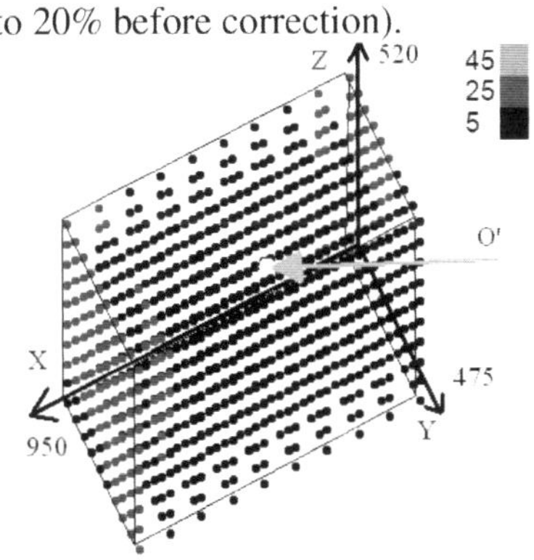

Figure 5. Error distribution after correction. Color scale in microns. White point O' is the optimal origin point.

The postprocessor works as follows:

1. Error measurements results are loaded into postprocessor;
2. Error vectors is being calculated according to (2)-(5) for all workspace points;
3. Coordinates origin shifts by calculated error value for a certain workspace point in order to correct measured errors;
4. Error vector for the given point is subtracted from all other points so that errors in other points decrease (or increase);
5. Steps 3-4 repeat for each workspace point so that an optimal point is found.
6. Condition (6) provides minimal errors within largest workspace area.
7. A CNC program created by Cam system is loaded into postprocessor;
8. The postprocessor analyses G-code lines and adjusts coordinates according to the correction values;
9. The corrected program is loaded into machine's CNC system.

5. Discussion

Error map gives opportunities to:

- visualize error distribution within the machine's workspace;
- determine workspace areas with required error value;
- determine workpiece's distortion due to machining errors;
- calculate corrections to load into CNC system in order to minimize errors.

6. Conclusion

1. The article presents an algorithm and software developed for rapid error mapping.
2. A strategy of "optimal origin" selection is produced based on error mapping in order to provide minimal errors within largest workspace area.
3. A rapid origin search method is proposed. method.

The proposed method of error correction is based on finding "optimal origin point" considering volumetric errors of a machine and providing minimal errors within largest workspace area.

Usage of interferometer provides this method with an important reference point – point of beam splitting within interferometer's optics. This point is machine-independent and can be used as a basis of machine's origin point search.

The method's effectiveness was tested by correcting real machine errors.

References

[1] Grigoriev S N, Masterenko D A, Teleshevskii V I and Emelyanov P N 2013 *Measurement Techniques* **55** (11) p 1311-1315
[2] Grigoriev S N and Teleshevskii V I 2011 *Measurement Techniques* **54** (7) p 744-749
[3] McKeown P A and Loxham J. 1973 *Ann. CIRP* **22** (1)
[4] Schwenke H. E. A. 2008 *Ann. CIRP* **57** p 660–675
[5] Grigoriev S N, Teleshevsky V I and Sokolov V A 2013 *Int. Conf. on Competitive Manufacturing «COMA`13» 30 Stellenbosch, South Africa* (Organized by the Department of Industrial Engineering Stellenbosch University) p 301-306
[6] Grigoriev S N, Teleshevsky V I and Sokolov V A 2013 *Int. Conf. on Competitive Manufacturing «COMA`13» Stellenbosch, South Africa* (Organized by the Department of Industrial Engineering Stellenbosch University) p 301-306
[7] Teleshevskii V I and Sokolov V A 2012 *Measurement Techniques* **55** (5) p 535-541
[8] Lamikiz A, López de Lacalle L N, Ocerin O, Díez D and Maidagan E 2008 *Int. J. of Advanced Manufacturing Technology* **37** p 122–139
[9] Rahman M, Heikkala J and Lappalainen K 2000 *Int. J. of Machine Tools & Manufacture* **40** p 1535-1546
[10] Okafor A C and Ertekin Y M 2000 *Int. J. of Machine Tools & Manufacture* **40** (8) p 1199–1213
[11] Lin Y and Shen Y 2003 *Int. J. of Advanced Manufacturing Technology* **21** p 243–248
[12] Ahn K Gee and Cho D W 2000 *Int. J. of Machine Tools & Manufacture* **40** p 2235–2248

Structural fabrication quality as a factor of industrial facilities safety

E V Tishkov, E M Kardaev, S Yu Stolbova and O S Shishova

Omsk State Technical University, 11 Mira ave., Omsk, 644050, Russia

E-mail: ossh2013@yandex.ru,

Abstract. In the conditions of industrial facilities high wear degree, it is very important to ensure the possibility of their safe operation in order to avoid various kinds of accidents and catastrophes. As practice shows, industrial plant collapses can occur suddenly under normal operating conditions. Usually, such accidents can take place at different stages of structures life cycle. One of the reasons for this is the initially low quality of reinforced concrete structures fabrication. The article considers the factors contributing to the collapse of reinforced concrete structures of water purification tanks located on the territory of the Omsk Region. The main surveys results on tank structures after collapse with the use of ultrasonic and physical methods of investigation are presented. On the basis of the obtained data analysis, it was found that the main cause of the accidents was the insufficient load-bearing capacity of typical reinforced concrete structures, caused by defects in their fabrication in the factory conditions because of exceeding the standard displacement from the design position of the working reinforcement. Recommendations are given on the identification of defective structures and the prevention of similar accidents when operating similar tanks at manufacturing plants constructed from standard designs.

Key words accident, reinforcement, caving, reinforced concrete structures, tanks, bearing capacity.

1. Introduction

The current stage of development of Russia is characterized by active construction and reconstruction of objects of predominantly civilian use. Industrial construction is developing to a lesser extent. A similar situation is observed in the analysis of infrastructure facilities construction in cities and industrial enterprises, in particular, water treatment facilities. We have to state that most such objects were built in the middle or the second half of the last century. To date, the structures of old treatment facilities have accumulated a great deal of physical wear, but they continue to function and serve city enterprises. Moreover, the intensity of exploitation increased with the growth in the cities number and the sewage volume. However, in some cases, the dysfunction of tanks, including their collapse, is not resulted from the accumulated normal wear and tear, but from initially hidden defects designs, accepted at the stage of fabrication and is not amenable to visual identification of the construction and normal operation of the facilities. Typical features of existing reservoir structures are:

- application, for the construction of variously designed tanks, of unified prefabricated reinforced concrete structures made according to standard series;

- a large useful volume of tanks, i.e. in case of failure in the event of an accident, it is necessary to redirect the water to the reserve storage, which is extremely difficult in conditions of congestion;
- practical impossibility of complete dehumidification of certain types of treatment facilities due to their continuous operation.

In this regard, the identification of possible defective structures for the purpose of preventing their strengthening and preventing the accident of storage tanks and water treatment is an urgent task.

2. Statement of the problem

In the process of tanks structures fabrication, there is a number of tolerances, within which displacements from design decisions are permissible. It is believed that if the tolerances are observed, the influence of the initial imperfections on the strength and operational properties of the structural elements is negligible. A number of parameters, such as overall dimensions, position of embedded parts, etc. can be determined using simple measuring instruments. However, to assess other parameters of the quality of reinforced concrete products (concrete strength class, class of reinforcement, etc.), more complex measurements are required, which, as practice shows, has not always been carried out in the manufacturing plants.

The problems of reinforced concrete structures fabrication and installation quality were investigated by leading research organizations and scientists both in Russia and abroad. In particular, in 1993, on the basis of a research, a program was developed to determine the destructive loads for the calculation of cylindrical ferro-concrete tanks [1]. The problems of reinforced concrete structures progressive collapse with the development of recommendations for designers [2, 3] are investigated. The experience of research indicates that the violation of the technology of reinforcement and concrete operations in the structures manufacturing and erection can lead to facility collapse, while the development of defects can occur as both controlled and sudden. Thus, the violation in the performance of reinforcement work led to the collapse of the silo building in 1954. [4]. Displacement from the requirements of the project for the reinforcement of the aero-tank bottom in Togliatti led to the destruction in the bottom of the wall panels sealing and the subsequent collapse of 11 wall panels [5]. In addition to the initial defects, specific aggressive conditions of maintenance facilities operation lead to accelerated aging of structural elements. N. K. Rosenthal points out that the presence of air in air tanks containing oxygen and carbon dioxide, a high rate of washing of the aerot-ank structures surface with water, potentially creates the conditions for the occurrence of concrete and reinforcement destruction in the underwater zone [6].

The investigation of the inadequate reinforcement influence on the load-bearing capacity of reinforced concrete tank panels was carried out at two sites, which identified wastewater reservoirs in the city of Omsk and Omsk Region. These objects are selected for research due to massive sudden collapse of building structures. The practical significance of the chosen topic is to substantiate the causes of man-made accidents of structures associated with poor-quality reinforcement of reinforced concrete structures and to develop recommendations for identifying and preventing similar situations.

3. Results of on-site inspections

Inspection of the facility number 1 was carried out according to the requirements [7] in connection with the large-scale collapse of the bearing wall and partitions. The waste water reservoir is located in sewage treatment plants for biological treatment of household and related industrial wastewater in the territory of the Soviet administrative district of Omsk. The construction is 4-corridor with a distributed intake of waste water and variable volume of the regenerator. The object surveyed is a buried rectangular open top structure with dimensions in the axes of 108x72 m and a depth of about 5.0 m relative to the ground level. The structure is interlocked from 2 aero-tanks (left in axes 1-5 / A-B and right in axes 5-9 / A-B), divided by a continuous bearing wall along axis 5. Each aeration tank consists of 4 sections with a width of 9 m, separated by partitions with technological apertures for water passage. The structure was built in 1974; at the time of collapse, the service life was 40 years. Overhaul of reservoir structures throughout the life of the plant was not carried out.

The main constructional elements of the structure are:

- bearing walls, located along the perimeter of the structure and an internal separation wall along axis 5. The walls are made of prefabricated reinforced concrete panels according to the standard series 3.900-2. The panels are cantilevered into the bottom grooves. Between the individual wall panels there are also monolithic reinforced concrete inserts;

- loading beds made of prefabricated reinforced concrete structures of individual fabrication, installed on partitions in prefabricated reinforced concrete beams.

According to the initial design, the joints of the dividing bearing wall panels between each other and the corner panels are splined and made by injecting the interlit weld with a cement-sand mortar. The joints of the corner panels and dividing partitions are designed to be rigid, located on the welding of horizontal reinforcement bars. The project provides for shotcrete monolithic sections of the walls with thickness of 20 mm.

During the examination, the following basic defects and structural damages were recorded:

- The wall panels of the dividing bearing wall section along axis 5 collapsed (figures. 1, 2);

- the internal dividing partitions collapsed together with the supplying reinforced concrete beds at different sites along the length;

- there were cracks in the panels of the bearing wall and partitions adjoining the destroyed sections;

- almost complete absence of the shotcrete concrete layer of on the panels surfaces (residual layer thickness up to 3 mm at the design value of 20 mm);

- corrosion of working reinforcement of wall panels;

- in separate wall panels there were no reinforcing bars.

Figure 1. Overview of the sewage water reservoir collapse (North side).

Figure 2. Overview of the sewage water reservoir collapse (South side).

The results of the object inspection and the destruction nature allow us to reconstruct the picture of the accident occurred. Immediately before the accident, the right aero-tank was not filled with water, while the left one was in operation. After filling the left aeration tank with water at the full calculated depth, a sudden destruction of 14 panels (a section 39 m long) of the bearing dividing wall occurred along axis 5. The panels collapsed as a result of strength loss of the normal section at the jammed place under the action of unilateral water pressure from the filled aeration tank. After the collapse of the bearing dividing wall section, there was a sharp outflow of water from the filled left aerotank, which resulted in the consequent destruction of the partitions in an unfilled aeration basin (along axis 6 coaxial with the collapsed section, a 39 m long section collapsed, along axis 7 panels collapsed over a length of 21 m) from the hydraulic shock of the formed wave and simultaneously in the filled part (first along axis 4 at a length of 56 m, then along axis 3 at a length of 39 m) due to asymmetric pressure provoked by a sharp outflow of water into the break. After the water level in the section in axes 6-7 / A-B of the right aero-tank increased, the partition wall was destroyed at a length of 27 m in 7 / A axes because of the remaining water pressure; thereon, the destruction ceased.

During on-site surveys, the strength characteristics of concrete were determined, the diameters and the pitch of the reinforcement were measured, and the reinforcement class was identified. In order to determine the possible causes of collapse, a comparative analysis of on-site surveys data and design solutions was performed, as a result of which the following structural features of the wall panels were found:

- the actual strength of bearing wall panels concrete corresponds to class B20, which is higher than the strength class B15 provided by the project;
- working reinforcement of load-bearing wall panels has ∅22 mm AIII with an average pitch width of the panel 140 mm (according to the project, it has ∅20 mm AIII with 200 mm pitch);
- the actual thickness of the bearing dividing wall panels of 200 mm corresponds to the design;
- working reinforcement of partitions panels has ∅12 ... 14 mm AIII with a pitch of 180 mm, distribution reinforcement has ∅6 mm AI and ∅8 mm AIII. According to the project, working reinforcement has ∅12 mm AIII, distribution reinforcement -∅8 mm AI with a pitch of 130 mm;
- in three load-bearing wall, the panels are set up in a row and located in the middle of the collapsed section along axis 5, the actual distance from the edge of the panel to the working reinforcement is from 80 mm to 110 mm (figure 3).

Figure 3. The collapse reason – excess permissible displacement of working reinforcement.

Inspection of facility No. 2 - a drinking water purification tank - was performed in connection with the collapse of the bearing wall and the slabs. The drinking water purification tank is located in the purification plant of a water purification station near Omsk. The surveyed object is a structure, grounded along the contour, with a coating of ribbed slabs and a heated roof, rectangular in shape with dimensions in the axes of 84x66 m and a height of 4.95 m to the bottom of the slabs. The structure is interlocked from 11 water sewage tanks separated by solid bearing walls. The construction was erected in 1969, by the time of the collapse the service life was about 45 years. The capital repeir of the structure was not carried out throughout the plant life.

Structurally, the facilities are frameless (except for sewage tank No. 5, mounted with an incomplete frame) with vertical bearing structures in the form of prefabricated reinforced concrete panels fabricated according to the standard series 3.900-2. Axis 5 has steel columns. As a coating, prefabricated reinforced concrete ribbed cover plates are used, supported by steel beams or located directly on wall panels.

During the examination, the following basic defects and structural damages were recorded:

- the load-bearing slabs of the sewage tank No. 4 covering in the axes 6-13 / GD and the wall panels of the bearing wall in the axes 6-13 / D collapsed. The total area of thecollapse was 130 m^2;
- there was a loss of stability of the coating steel beam along the D axis in the plane and from the plane;
- the unbroken wall panels of sewage tank No. 1 are inclined from the vertical plane to 100 mm in the direction of T axis;
- violation of structural requirements for the construction of a of sewage tank No. 5 steel structure;
- there was no any anticorrosive protection of steel structures; in some areas, the degree of corrosion of beams and columns metal was up to 60% of the cross section;
- working reinforcement corrosion damages in sewage tanks wall panels and slabs;
- overestimation of up to 80 mm on one side and underestimation of the protective layer of concrete of the of the bearing wall collapsed panels on the other side, inadmissible displacement of the reinforcement when the panels embedding in concrete.

Based on the survey results and the damage nature analysis, the overview of the accident that occurred was restored. Immediately before the destruction, sewage tank No. 4 was not filled with water. At the same time, the sewage tank No. 5 adjacent to it was 80% full. At a certain point in time, as a result of strength loss of the wall panels normal section, a part of the bearing wall partitions and the coating in axes 6-13 / G-D under the effect of unilateral hydrostatic water pressure on the side of the filled sewage tank broke down at the point of pinching (figure 4). The situation was aggravated by the fact that the ribbed slabs of the coating did not actually unlock the top of the wall panels, because the corrosion caused complete destruction of the welds and embedded parts.

Figure 4. Overview of the sewage water reservoir structures collapse

The analysis of the actual constructive wall panels made it possible to find out that:

- the actual thickness of the collapsed panels of the bearing wall is 200 mm;
- the longitudinal working reinforcement of the panels is made with periodic reinforcement ∅18 A-II with an average pitch of 100 mm along the width of the panel;
- The transverse reinforcement is made by different reinforcement of the periodic profile (∅10 A-II, ∅14 A-II) and plain bars (∅12 A-I) with a pitch of about 200 mm;
- the protective layer of concrete is within 25 ... 85 mm (excess allowance of the reinforcement from the design position);
- the strength class of concrete corresponds to class B20.

4. Calculated analysis of the failure reasons

As the survey has shown, the reservoirs had numerous defects and damages, which are mainly resulted from non-compliance of the operational regulations, maintenance and major repairs of structures. However, the immediate cause of the accident in both cases is the sudden failure of the load-bearing wall panels under the effect of unilateral hydrostatic water pressure. A series of calculations was performed to assess the actual bearing capacity and crack resistance of a wall panels section. The calculations took into account the actual characteristics of concrete and reinforcement of the collapsed panels, as well as the displacement of the working reinforcement from the design position.

The calculations were performed according to the requirements of two groups of limiting states from the condition:

- Strength check, i.e. force from external loads and impacts in the section F under consideration should not exceed the ultimate force F_{ult};

- width of crack opening, i.e. the crack opening width from the external load a_{crc} should not exceed the maximum allowable crack opening width $a_{crc, ult}$ [8].

The main results of the calculation analysis of the influence of displacement on the bearing capacity and crack opening in reinforced concrete panels are shown in the graphs (figures 5 and 6, respectively).

The comparison of load-bearing capacity of panels with efforts from unilateral hydrostatic water pressure, as well as with the Norms, regulating the requirements for reinforcement of reinforced concrete structures, indicates the following:

1. During wall panels fabrication, the requirements [9] for placing the longitudinal working reinforcement (Δ = 10 mm) and the value of the protective layer of concrete (+8 mm, -5 mm) were grossly violated. The main reason for the defect appearance is a violation of the reinforcement technology: the upper mesh in the case of panel horizontal concrete casting was not fixed and was displaced by the weight of the poured concrete to the center of the panel. The fact of several defective panels setting on the object No. 1 in a row is a typical one, which led to the collapse of the defective site.

2. For the panels of the object No. 1, the reinforcement displacement within the permissible [9] limits did not lead to the wall panels section overload at normal stresses. The limit of the condition for ensuring the strength of the normal section of the wall panels is the working reinforcement displacement by 40 mm relative to the design position. The normal section strength of object No. 2 wall panels is initially provided under the condition of the design placement of the reinforcement and with the obligatory unfastening of the panels tops with ribbed plates. In fact, in conditions of complete destruction of welded seams and embedded parts, the displacement of the working reinforcement, even within the limits of the standard allowance for object No. 2, was critical and led to overloading of the section with loss of strength and structures collapse.

3. By the condition of the cracks opening, the working reinforcement displacement should not exceed 47 mm. At the same time, violation of the maximum permissible crack opening condition can eventually lead to loss of concrete and working reinforcement adhesion, especially taking into account the aggressive influence of the medium, which in turn will affect the load-bearing capacity of the structure. In exploited reinforced concrete structures under the influence of aggressive gas-air and liquid chloride-containing media, one of the most common types of damage is the failure of reinforcement and concrete adhesion as a result of reinforcement corrosion [10, 11].

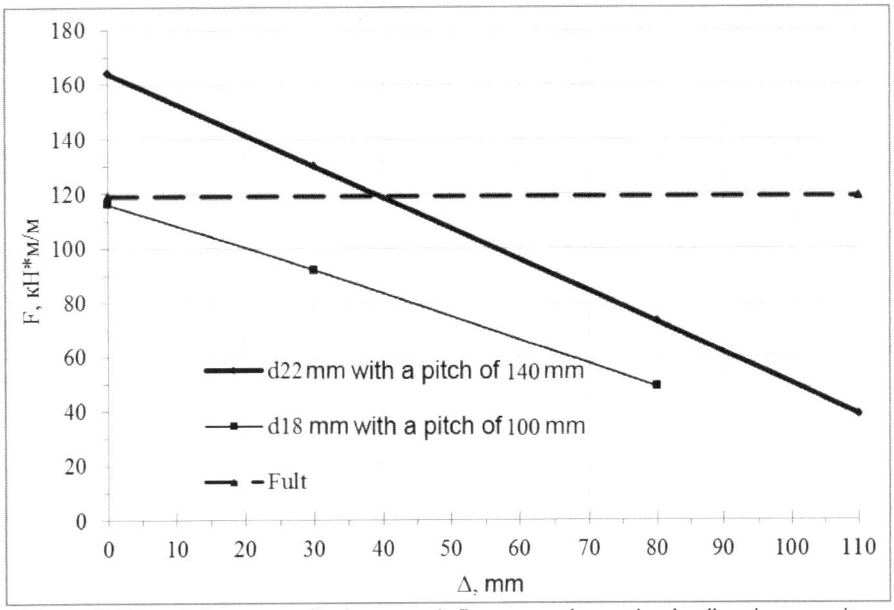

Figure 5. The reinforcement displacement influence on the section loadbearing capacity.

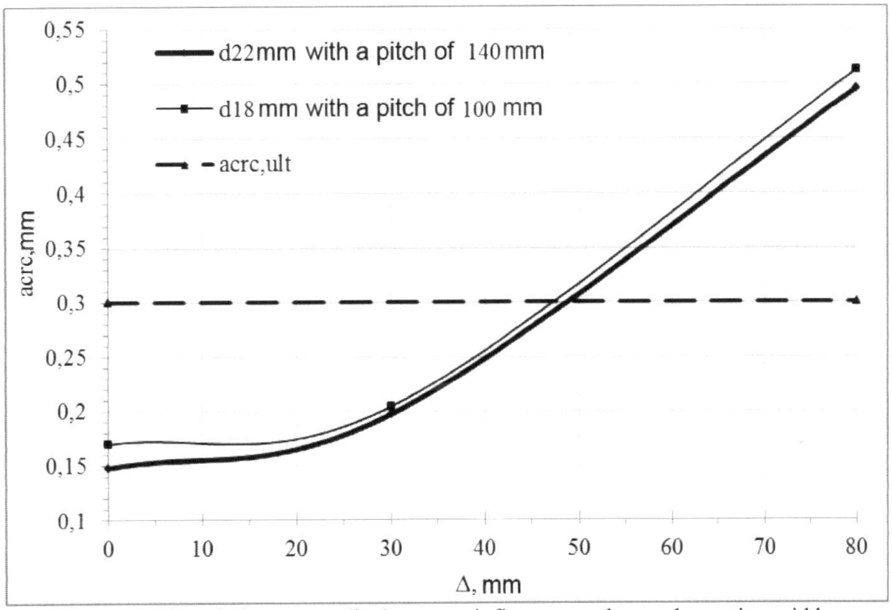

Figure 6. The reinforcement displacement influence on the crack opening width.

Thus, the occurrence of an accident at the facility number 1 resulted from violations in the fabrication of wall reinforced concrete panels. For facility number 2, in addition to the reinforcement displacement up to the collapse, there were violations of the repair mode during the facility operation, when the measures to restore and protect the welded joints of the ribbed plates were not carried out in time. According to the classification of A. G. Tamrazyan [12], the accident refers to the categories of accidents caused by poor-quality work and violation of the operation regulations.

5. Conclusion

1. All cases of accidents considered in the article occurred without human casualties. In terms of

the scale and degree of consequences, the examples given refer to the second group [13]. Nevertheless, due to the unsatisfactory quality of the wall panels and the subsequent accident, significant economic damage was caused as a result of the work suspension for up to six months and the need to restore the structures.

2. For the existing aero-tanks structures, a hidden defect of the non-project working reinforcement location is very dangerous, because, in the case of unilateral loading of the tank (for example, during cleaning or reconnaissance of the adjacent one), it can result in sudden collapse of the bearing wall panels.

3. For newly erected buildings with the use of prefabricated reinforced concrete wall panels, it is necessary to strengthen control over the products fabrication, and at the construction site, that over the acceptance inspection. In existing facilities, it is advisable to conduct full-scale inspection of wall panels with the detection of the reinforcement position, its corrosion damage and evaluation of structures bearing capacity. Ultrasonic methods for nondestructive testing can be used to determine the actual position of the working reinforcement.

References

[1] Ramanjaneyulu K and Gopalakrishnan S 1993 *Computers & Structures* **48 (2)** p 205-217

[2] Wang Hongyu, Su Youpo and Zeng Qingshen 2011 *Systems Engineering Procedia* **1** p 48-54

[3] Brunesi E, Nascimbene R, Paris F and Augenti N 2015 *Engineering Structures* **104** p 65-79

[4] Shkinev A N 1984 *Accidents in the construction* 4 th ed.(Moscow: Stroiizdat)

[5] Dobromyslov A N 2008 *Diagnostics of damage to buildings and structures* (M .: Publishing house of the Association of Construction Universities)

[6] Rosenthal N K 2011 *Concrete and reinforced concrete* **2** p 78-85

[7] GOST 31937-2011. Buildings and constructions. Rules for inspection and monitoring of technical condition. - Moscow: Standartinform, 2014.

[8] SP 63.13330.2011. Concrete and reinforced concrete structures. Basic provisions. Updated version of SNiP 52-01-2003 (with Changes No. 1, 2). - Moscow: Standartinform, 2013

[9] JV 70.13330.2012. Bearing and enclosing structures. Updated version of SNiP 3.03.01-87 (with Amendment No. 1) Moscow: Standartinform, 2013.

[10] Alekseev S N, Rosenthal N K 1976 *Corrosion resistance of reinforced concrete structures in an aggressive industrial environment* (Moscow: Stroiizdat)

[11] Pinus B I 1978 *Concrete and reinforced concrete* **11** p 37-38.

[12] Tamrazyan A G, Bulgakov S N et al. 2012 *Reducing risks in construction in emergency situations of natural and man-made nature* (Moscow: Publishing House of the ASV)

[13] MDS 12-4.2000 Regulations on the procedure for investigating the causes of accidents of buildings and structures, their parts and structural elements on the territory of the Russian Federation. [Order of the Ministry of Construction of Russia No. 17-48 from 06.12.94] 2001.

The information protection level assessment system implementation

E V Trapeznikov[1]

[1]Omsk State Technical University, 11, Mira av., Omsk, 644050, Russia

E-mail: evtrapeznikov@yandex.ru

Abstract. Currently, the threat of various attacks increases significantly as automated systems become more widespread. On the basis of the conducted analysis the information protection level assessment system establishing objective was identified. The paper presents the information protection level assessment software implementation in the information system by applying the programming language C #. In conclusions the software features are identified and experimental results are represented.

Key words: information security, neural network, development environment, legislation

1. Introduction

Modern automated systems are a set of various software and hardware technologies. Due to the abundance of technologies being used both at the time of establishing and during the operation of automated systems, threats and vulnerabilities can occur absolutely at any time.

Any automated system malfunction can lead to the irreversible consequences. A demonstration of this is the statistics of Kaspersky Lab ICS CERT for the first half of 2017. According to the mentioned statistics the number of automated control systems unique computers attacked by encrypting Trojans has increased that is shown in figure 1.

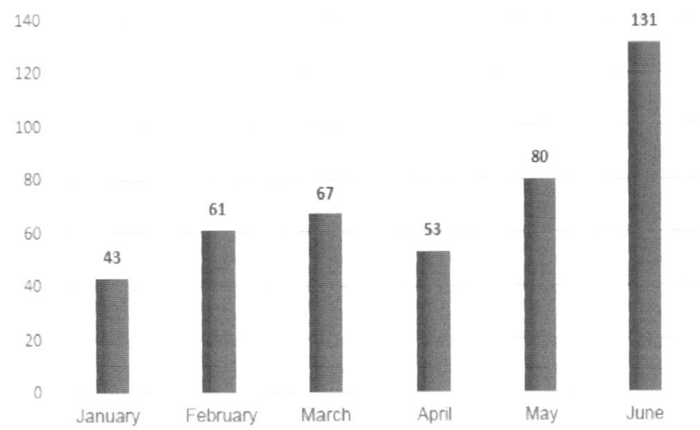

Figure 1. The statistics by the number of attacked computers.

In the given type of attack there is a disfunction of two information properties at once, namely integrity and accessibility. The information is no longer available to its owner. The chance to return the information becomes minimal. According to the same statistical data, the most vulnerable data transmission channel is the Internet and removable media. The detection and prevention of these attacks has become an important part of the security. The number of attacks that use the sophisticated methods has recently increased. The main purpose of these attacks is to filter to the destination and remain unnoticed. The intrusion detection system is one of the important ways of computer networks high security ensuring and is used for various attacks preventing [1, 2, 3, 4].

Most threats also occur due to the user errors [3, 4]. Paper [3] states that the attacker often applies real user accounts and standard administration tools to hide the stay inside the system. According to the authors, the most detection systems use only signature-based analysis and cannot identify malicious activities. In addition, it is impossible to analyze user actions manually owing to the task complexity. In the study the authors apply neural networks for detecting the user suspicious behavior. Besides, the authors draw the conclusions showing the qualitative improvement in detecting the suspicious users with an accuracy of 98%.

The authors [6] suggest the network packets analysis and certain attributes collection for neural network training material creating. The created classifier based on the neural network provides the fake packets identifying performance of 96%.

One of the topical problems common to the modern systems is the «hardware Trojans» emergence [7, 8, 9]. The authors of the presented papers use neural networks in their studies for analyzing the various features obtained from the equipment to determine the information extracting attempt fact.

It should be noted that most authors do not consider the comprehensive approach for assessing the information protection quality. However, this approach is be prioritized. The comprehensive approach should provide the complete assessment of the whole protection system from all alleged threats. The complete preliminary analysis of the protection system will ensure the correct assessment in accordance with the legislation.

2. The problem statement

To increase the vulnerabilities detection level most authors use various smart tools. One of the most frequently used are neural networks. Neural networks can be used for different problems solving. In recent years neural networks have been applied to provide security for the problems as attacks classification, prevention, detection. The general concept of neural networks is that neurons are able to correct weights in such a way as to predict the results correctly. Neural network different parameters are corrected at the training stage and subsequently the studied model is applied to the test data for assessment. The research objective is the information protection system quality assessment methodology implementation in accordance with the information security assessment model in the information system based on the neural network [10, 11].

3. Theory

For implementing the software, the programming language C # and development environment Microsoft Visual Studio 2015 were used. The local program database is formed by using SQLite, it is the embedded cross-platform database supporting the sufficiently complete SQL command set.
The general flow diagram of the software is shown in figure 2.

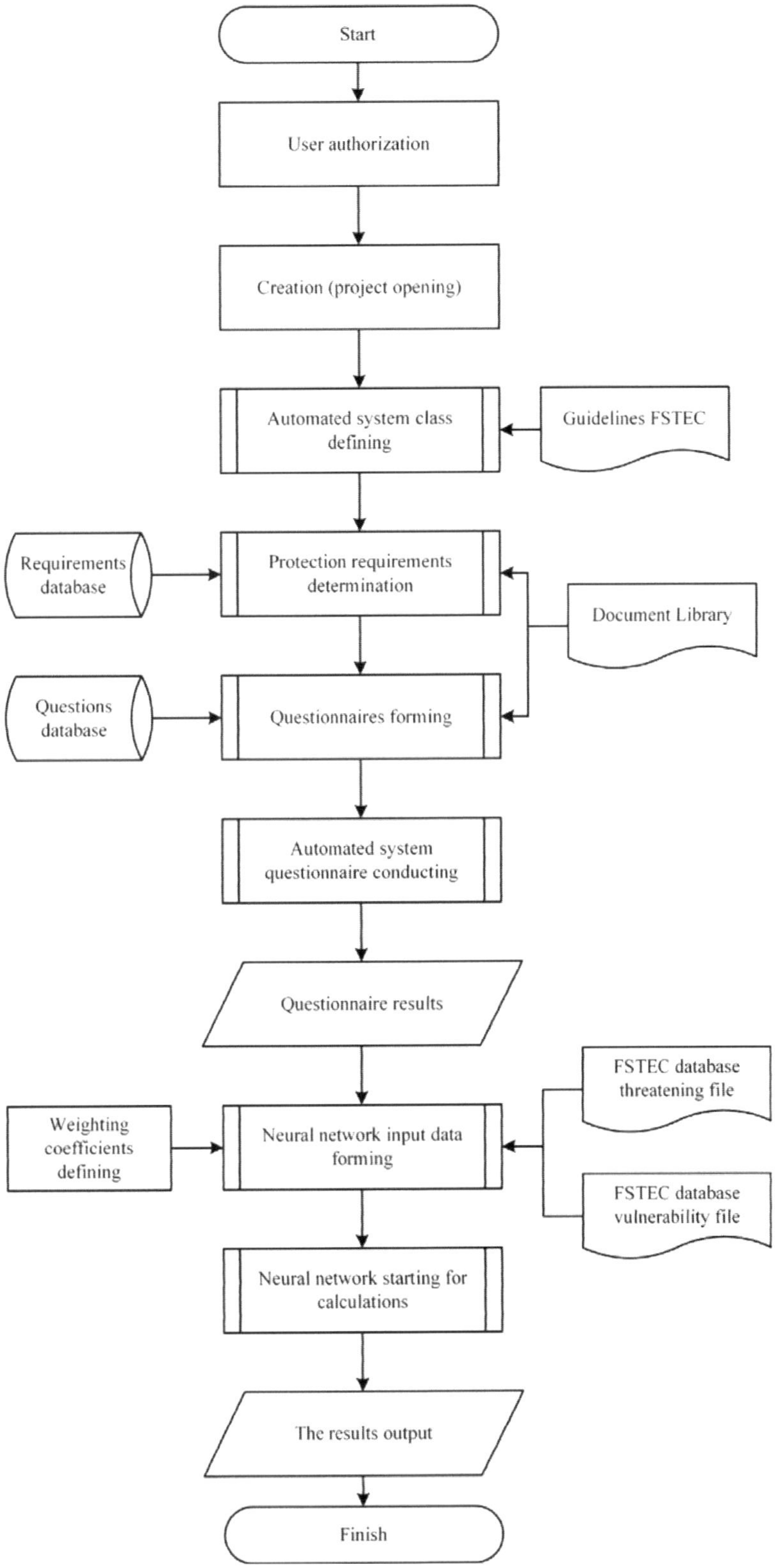

Figure 2. The program flow diagram.

To achieve the objective the object-oriented programming method was applied.

For the software development, the MVC architecture (a scheme for application data, user interface and control logic partitioning into three separate components, namely the model, view and controller) is used in a manner that each component can be modified independently. The program interface is shown in figure 3.

Working with the program is carried out on the basis of authentication data, namely the login and password. For ensuring the data storage security, the user password is stored in the encrypted form. The authentication is performed via the entered password encryption and comparison with the database password by the specified login.

To continue working with the software, it is necessary to initialize the project opening, or creation. The project is given its name and date. All the information about projects is stored in the database corresponding table (ProjectStructure).

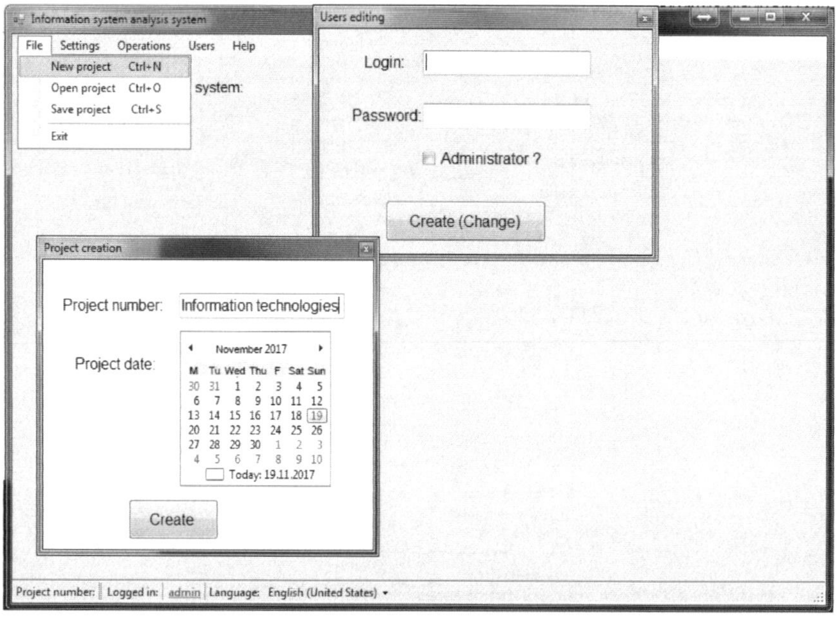

Figure 3. The program interface.

The automated system class identifying is necessary for information protection requirements compiling. The system class shows the processed information access level in the system, the users access, etc. The class definition is performed according to the guideline document of the Federal Service for Technical and Export Control «Automated Systems. Protection against unauthorized access to information. Automated systems classification and information security requirements».

The questionnaire conducting will make possible to define the information security system current level, to determine whether the information is present. The questionnaire is carried out by using the State Standard. The questionnaire results are the neural network input data set.

In addition to the questionnaire, the input parameters are FSTEC databank threats and vulnerabilities. Loading is performed in the corresponding local database tables, namely FstecThreatsTable and FstecVulnerabilityTable The data source are Excel files with threats and vulnerabilities. The program performs the transformation for the further work. After determining all the parameters, the neural network is run for calculations. A separate stage is neural network training according to the "with the teacher" method, in which the initial data decision is predetermined and neural network changes the weighting factors for achieving the solution.

The final stage is all the results output for the user.

4. Results and discussion

For the experiments conducting the hypothetical enterprise was determined. At the given enterprise there is the automated system in which the data to be protected are processed. The automated system class was defined as 2A: users have the same access rights (authority) to all AS information processed and (or) stored on various confidentiality level media.

Further, for determining the protection system current state, the survey was conducted by using GOST R ISO/IEC 15408 Information technology (IT). Security methods and tools. Information technologies security assessing criteria.

The results revealed the protection system actual state data. These data are the part of the neural network input data. In addition, the input data used the threats and vulnerabilities FSTEC bank data.

The neural network calculations found out the results about possible threats, for example, IST threat 030 «The threat of using default identifying /authentication information» was typical for the system as the most common passwords were used to authenticate the users.

However, for improving the neural network performance, it is necessary to conduct its training additionally by using data on real enterprises.

5. Conclusion

The studied researches demonstrate the software development relevance for the automated systems protection level assessment. Each study focuses on the identification and elimination of a certain type of threats, as well as the comprehensive approach.

The developed software makes possible:

- To differentiate the access rights to the program features according to the identification data.
- To create, edit users for working with the software
- To log all users actions.
- To create and save the current projects.
- To download the information from the FSTEC data bank
- To classify the investigated automated system.
- To conduct the automated system survey for determining the protection system current state.
- To form the list of possible threats based on the survey results and the destabilizing factors list by using the neural network.

References

[1] Akashdeep, Manzoor I and Kumar N 2017 *Expert Syst Appl* C **88** 249–57

[2] Saied R Overill E and Radzik T 2016 *Neurocomputing* **172** 385–93

[3] Ussath M et al 2017 *Identifying Suspicious User Behavior with Neural Networks* IEEE CSCloud (New York) p 255–63

[4] Banik D and Chowdhury R R 2016 *Offline signature authentication: A back propagation-neural network approach* Sixth ISED (Patna) pp 334–39

[5] Amruta M and Talhar N 2016 *Effective denial of service attack detection using artificial neural network for wired lan* SCOPES p 229–34

[6] Hasegawa K, Yanagisawa M and Togawa N A 2017 *IEICE T Fund Electr* E **100** 1427–38

[7] Ni L. et al 2016 *A method of noise optimization for Hardware Trojans detection based on BP neural network* 2nd IEEE ICCC (China: Chengdu) p 2800–04

[8] Li J. et al 2016 *A novel hardware Trojan detection based on BP neural network* 2nd IEEE ICCC (China: Chengdu) p 2790–94

[9] Danilova O T and Trapeznikov E V 2015 *Information Counteraction to the Terrorism Threats* **24** 24–9

[10] Trapeznikov E V 2017 *Proceedings of the Tula State University. Technical sciences* **2** 312–18

Information security system quality assessment through the intelligent tools

E V Trapeznikov[1]

[1]Omsk State Technical University, 11, Mira av., Omsk, 644050, Russia

E-mail: evtrapeznikov@yandex.ru

Abstract. The technology development has shown the automated system information security comprehensive analysis necessity. The subject area analysis indicates the study relevance. The research objective is to develop the information security system quality assessment methodology based on the intelligent tools. The basis of the methodology is the information security assessment model in the information system through the neural network. The paper presents the security assessment model, its algorithm. The methodology practical implementation results in the form of the software flow diagram are represented. The practical significance of the model being developed is noted in conclusions.

Key-words: information, intelligent tools, neural network, information system, guideline documents

1. Introduction

The technology development stage defines the concept of information as the information sphere important unit. Information processing is carried out by means of the automated systems resources including various information systems. Any automated system processing personal data in particular requires protection in accordance with the legislation. Protection is necessary for preserving three basic properties of information, namely confidentiality, integrity, accessibility.

In many studies the particular attention is paid to the intelligent tools. Currently, the most frequently used approaches are:

- genetic algorithms,
- neural networks,
- immune systems technologies,
- Bayesian network, etc.

Protection is necessary due to the increasing number of threats to data processing automated systems. At the moment, increasingly sophisticated and effective access violation detection strategies both by hardware and software are being developed and implemented. However, these systems need to be improved for preventing the illegal access. The authors of [1] propose their own solution based on the decision tree as a means of users behaviour models obtaining, neural networks as intrusion detection identifying classifiers. The paper points out that the given intelligent techniques combination makes possible to reduce the access violation risk. In [1], only user behaviour analysis for tracking unauthorized access is used.

Since network technologies are applied for information processing and storing speed increasing, at this level a large number of threats and vulnerabilities being used by the intruders occur. Network technologies form part of the automated system. The authors of [2] in their study constructed the

attacks protection and defence system theory model for network security based on the game theory. According to the model the authors developed the training algorithm on the genetic algorithm basis. These steps allow authors to evaluate human and system suspicious actions. The authors of [2] conclude with experimental data demonstrating the system effectiveness.
The automated system is a complex of components, for this reason the information security system quality is worth to be assessed in full.
The adaptive system based on the neural network and intelligent agents is given in paper [3]. The developed system will make possible to implement the information security adaptive control, the system provides the timely response to threats and performs the decision-making process in real time. In paper [3] the system is said to allow attacks identifying, classifying and preventing. The adaptive system algorithm is represented in figure 1.

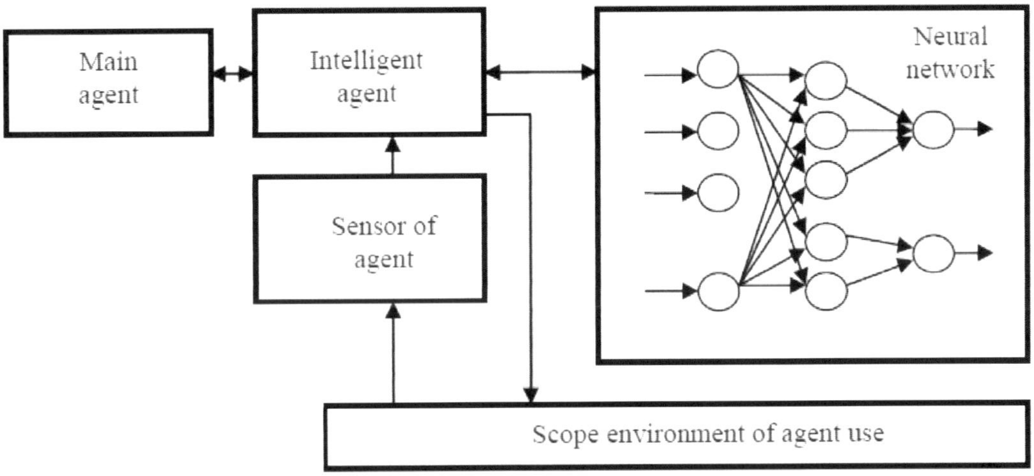

Figure 1. The adaptive system of the authors [3].

Some authors consider the neural networks application for creating the virus detection system. Computer viruses have become a serious threat to the information system. Due to the virus codes complexity and behavioural uncertainty, as well as the encryption emergence the traditional detection methods are becoming less effective. The viruses detecting approach application based on the artificial intelligence has become a major problem of the current antivirus researches [4, 5]. In paper [4], the approach on the basis of the dynamic and static viruses detection integration is proposed. The detection system uses the vector support mechanism as the elements classifier for dynamic virus behaviours model constructing, as well as applies the probabilistic neural network as the elements classifier for static behaviour modelling. The authors refer to the detection system effectiveness.
It should be noted that most authors examine the individual aspects of information security. One of the most frequently distributed attacks is DDoS (distributed denial-of-service attack.) The network resource fails as a result of a number of requests to it sent from different points).

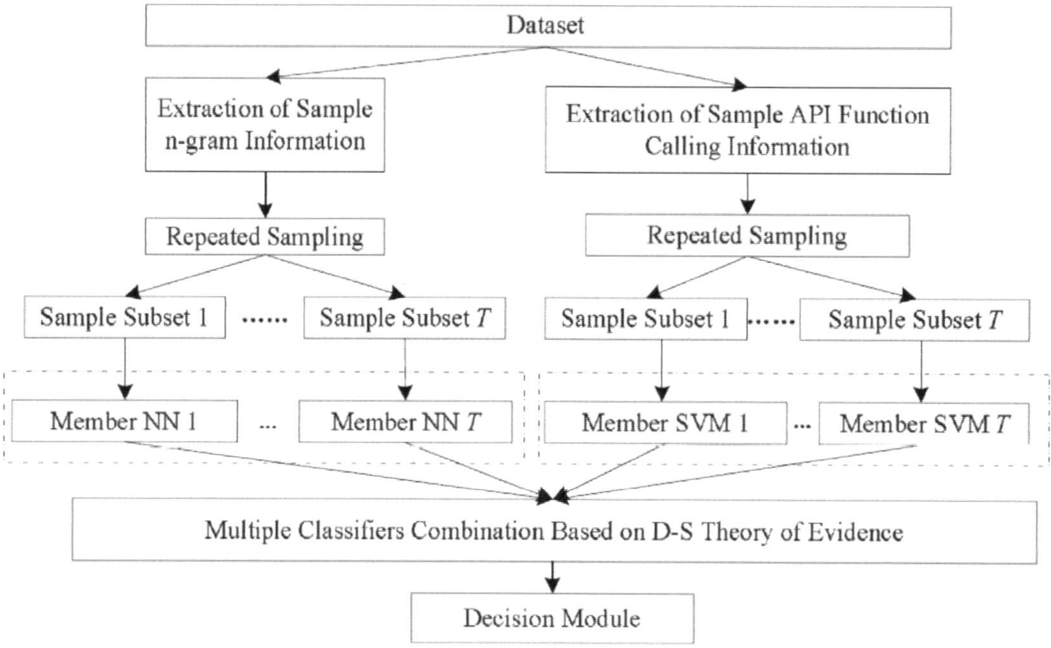

Figure 2. Viruses detection intelligent system of the authors [4].

In papers [6, 7] the IDS method is proposed. It is based on the neural networks delays. The method is based on the multi-tiered architecture, each host is stored in the network. Hosts are hidden in the first and second layers of the firewall. If a suspicious packet is detected, similar to the DDoS template, the system will automatically call the security module. Those packets that did not get over the security module will be isolated and then removed from the transfer.

2. Problem Statement

The various authors studies analysis shows the information security system quality assessment problem to be quite relevant under the increasing threats. The given research objective is to develop the information security system quality assessment methodology by applying the intelligent tools. One of the methodology components is the information security assessing model in the information system based on the neural network [8].

3. Theory

The information security system quality assessment methodology development objective by using the intelligent tools is improving the automated system protection level quality to counter the emerging threats. The methodology procedure algorithm is presented in figure 3.

According to the algorithm:

Step 1. In order to properly identify the particular automated system necessary protective actions, AS class is necessary to be determined. The automated system class is identified according to the guidance document of the Federal Service for Technical and Export Control «Automated Systems. Protection against unauthorized access to information. Automated systems classification and information security requirements».

Automated systems are classified into the following categories: 3B, 3A, 2B, 2A, 1D, 1G, 1V, 1B, 1A. For class defining the following set of parameters is applied:

1. The information availability of different confidentiality level in the automated system.
2. The automated system access subject privilege to the confidential information access.
3. The automated system data processing mode: collective or individual.

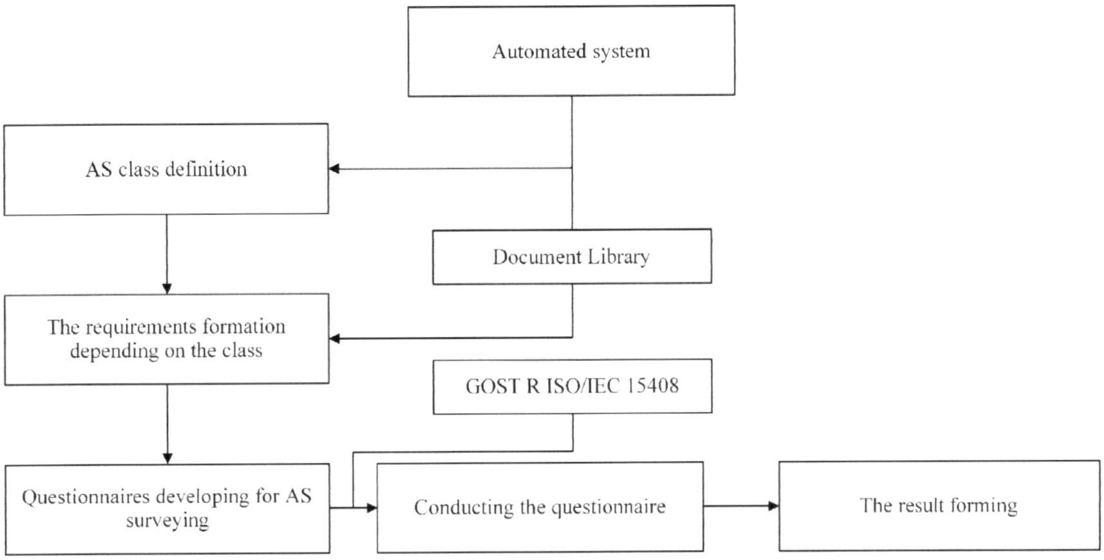

Figure 3. The methodology algorithm.

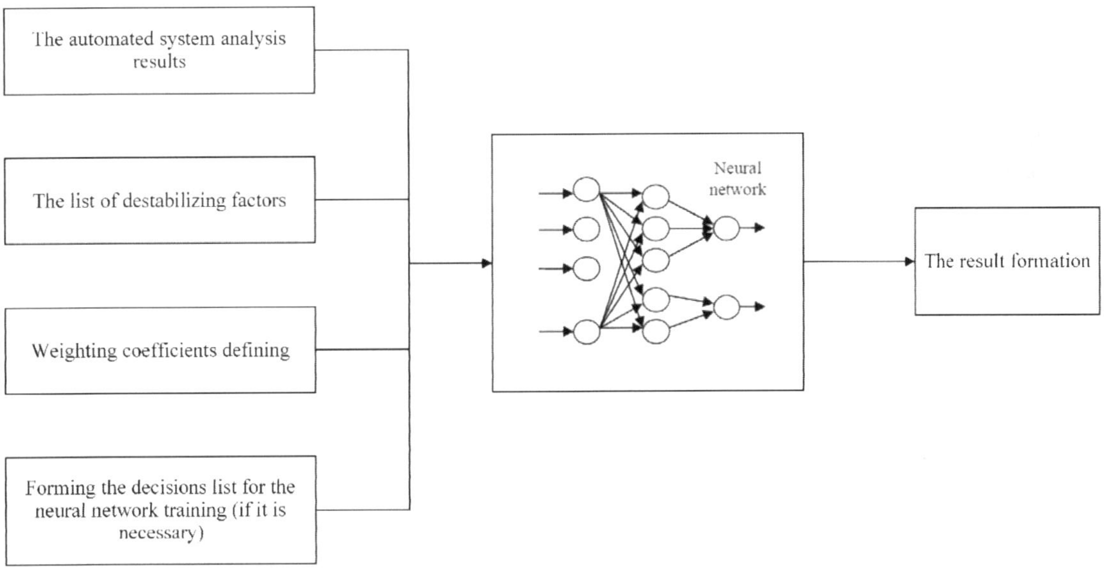

Figure 4. The security assessment model based on the neural network.

Step 2. For defining the protection requirements, the guideline document on the basis of FSTEC Order of Russian Federation No. 17 of February 11, 2013, «On the Requirements Approval for the Information Protection not Constituting the State Secret and Contained in the State Information Systems» is also used. According to this order, depending on the AS class the information system security class is defined as (1):

The protection class (C) = [information relevance level; system scale] (1)

where the information relevance level is defined as (2):

RL = [(confidentiality, damage degree) (integrity, damage degree) (accessibility, damage degree)] (2)

Step 3. The AS protection system current status determining questionnaires list is formed on the basis of the specified requirements for the particular automated system. For the questionnaires compiling and the survey conducting, «GOST R ISO/IEC 15408 Information technology (IT). Security features and methods. Information technologies security assessment criteria of information technology» is used.

The conducted survey results make possible to define the information security system current status, to determine its presence or absence. The questionnaires are drawn up depending on the automated system class for providing the information security necessary level.

The methodology first stage generated results are the input parameters for the information security assessment model. For identifying the possible threats and vulnerabilities, the FSTEC threat and vulnerability database is used as the model destabilizing factors [8]. The neural network training necessity is defined during the model formation process. The trained neural network allows to get the more reliable result and possible threats that may occur under the actual system state.

4. Results and discussion

The software presented in figure 5 was developed for testing the methodology.

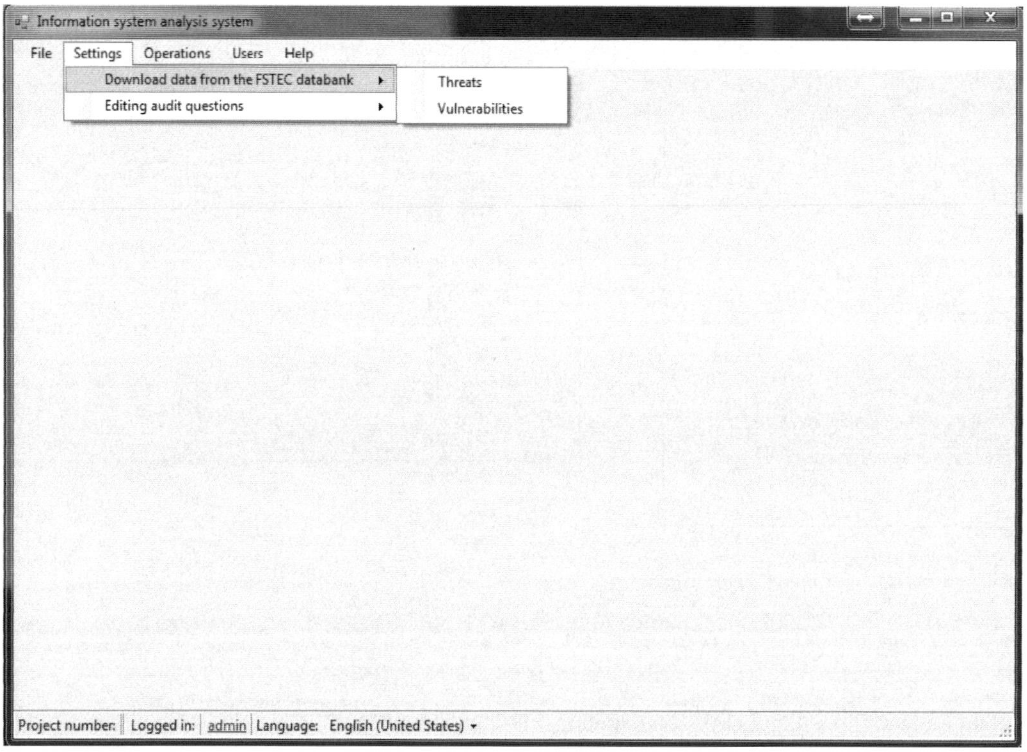

Figure 5. The software main window.

The software makes possible downloading the data from the FSTEC databank, assessing the automated system protection system status, conducting the surveys, etc.

The software provides the access rights to the software features differentiation and actions logging conducting. In operation all the information is stored in the local database. SQLite is the basis of the database.

To carry out the experiments according to the presented methodology, the automated system, a hypothetical enterprise having the existing protection system that is predetermined not to be satisfying the regulatory requirements, was used. By using the methodology, the automated system complete analysis was carried out. Subsequently, the analytical data and destabilizing factors list were applied as input parameters of the pretrained neural network. As a consequence of the methodology complete cycle performing, the results indicating the attacks that occur in the given automated system at the protection system current level were obtained.

5. Conclusions

In the context of attacks number and their complexity increasing, the information systems protection requires the reasonable approach which would help making the decisions of possible threats and attacks on the information system in accordance with the certain characteristics. Intellectual tools make possible to create the trained system which is able to make independent decisions with a high level of confidence. The developed methodology based on the neural networks system will allow to identify and classify the possible attack or threat.

6. References

[1] Guevara C, Santos M and López V 2017 *Intrusion detection with neural networks based on knowledge extraction by decision tree* Int. Joint Conf. SOCO'16-CISIS'16-ICEUTE'16 (Spain: Springer International Publishing) pp 508-17
[2] Ru-jun Y 2016 *Int. J. Secur. Appl.* **10** 317–24
[3] Mukhin V et al 2016 *Adaptive security system based on intelligent agents for distributed computer systems* Int. Conf. on Development and Application Systems (DAS) (Suceava) pp 320-25
[4] Zhang B et al 2014 *Neurocomputing* **137** C 24–33
[5] Mao W et al 2017 *Computers & Security* **68** C 47–68
[6] Aljumah A 2017 *Int. J. Adv. Appl* **4** 78–83
[7] Aljumah A and Ahamad T 2016 *Int. J. Comput. Sci. Netw. Secur* **16** 132–38
[8] Trapeznikov E V and Danilova O T 2016 The information protection analysis model on the basis of the neural network *Dynamics* **1** (Russia) p 302–08

IOP Conf. Series: Journal of Physics: Conf. Series 998(2018) 012038 doi:10.1088/1742-6596/998/1/012(

Printing quality control automation

O V Trapeznikova

Omsk State Technical University, 11, Mira av., Omsk, 644050, Russia

E-mail: ol-trapeznikova@yandex.ru

Abstract. One of the most important problems in the concept of standardizing the process of offset printing is the control the quality rating of printing and its automation. To solve the problem, a software has been developed taking into account the specifics of printing system components and the behavior in printing process. In order to characterize the distribution of ink layer on the printed substrate the so-called deviation of the ink layer thickness on the sheet from nominal surface is suggested. The geometric data construction the surface projections of the color gamut bodies allows to visualize the color reproduction gamut of printing systems in brightness ranges and specific color sectors, that provides a qualitative comparison of the system by the reproduction of individual colors in a varying ranges of brightness.

Key words: the distribution of ink layer, accuracy, software

1. Introduction

The quality ratings of multicolor image obtained by the methods of offset printing can be determined by the optimal matching of main components of printing system. Among the priority quality ratings of printed sheet the factor characterizing the distribution of ink layer on the printed substrate is chosen . The distribution of ink layer on the sheet has a significant effect on changes in optical density, color gamut, and other factors.

The surface characteristics of a printed material are also included in the number of estimated prior factors of substrates. Increasing the accuracy of the result of measuring the above-mentioned factors due to the development of modern high-precision methods and software is up-to-date.

2. Statement of the problem

The earlier grapho-analytical method for determining the ink layer on the sheet is based on an analysis of the characteristics of ink transfer from the plate to paper and the equation of ink balance in a contact with paper during the printing process [1] does not give a complete objective estimation for a number of reasons. From the scientific and practical point of view, the investigations by [2-4] are interesting but they illuminate the problem solving only by qualitative methods. When solving such problems, it is necessary to overcome significant mathematical difficulties.

Despite the printing process is preceded by the stages of prepress process and time on the printing press makeready, the process of offset printing today is sufficiently automated to produce a high quality circulation. Nevertheless, the improvement of existing methods of estimation and the development of new approaches in solving this problem is of both scientific and practical interest. To achieve this goal, the methods of calculus mathematics and mathematical analysis elements are applied.

3. Results and Discussion

Taking into account that the stock of paper is wide enough, and its characteristics are of an advertising nature, the existing methods of evaluating the quality rating of printing do not provide the required accuracy and the adequacy of estimation. Standard MedienStandart Druck 2008 classifies papers both by surface type and by color parameters, however, the ISO 12647-2:2004 limits the types of model papers that causes difficulties in assessment of the quality of color reproduction.

Taking into account that the printed substrates have a natural surface microrelief making the distribution of ink film on the sheet along its thickness irregular, an algorithm for its evaluation has been developed [5]. For the software development the high-level programming languages Visual C # and .NET Framework 4.0 were used; and the software environment of Microsoft Visual Studio 2010 was served. The functional of a program is removed under the tabs created with the TabControl component. To display the result of the program run an element of the StringBuilder class is formed into which the final result is written. To build the plots (sections) the standard component, i. e. Chart is used. To reduce the miscalculations the program provides an error control module. A fragment of program module for calculating the paper surface characteristics is represented on Fig. 1.

```
// Coordinate output on the screen
//=========================================================
    private void openFileBackgroundWorker_RunWorkerCompleted(object sender,
RunWorkerCompletedEventArgs e)
    {
      xNach.Text = Convert.ToString(nwx);
      yNach.Text = Convert.ToString(nwy);
      zNach.Text = Convert.ToString(nwz);
      zmax.Text = Convert.ToString(z_max);
      profil.Text = Convert.ToString(zn_profil);
      nachStep.Text = Convert.ToString(Math.Abs(Math.Round(h, 3)));
      znachN.Text = Convert.ToString(N);
    }
//=========================================================
// Calculation of Fourier polynomial elements
//=========================================================
// A number of Fourier polynomial elements is set by user.
//=========================================================
    private void countbackgroundWorker_DoWork(object sender, DoWorkEventArgs e)
    {
      int N1;
      int N2 = 0;
      int N3 = 0;
      int N4 = 0;
      int N5 = 0;
      int main_inch = 0, ich = 0, ich2 = 0, ich3 = 0, ich4 = 0, ich5=0;
      double fsin, yk, ts, fin, argsin, a, TempPopravka;
      double y0N, yN1, yN2;
```

Figure 1. A fragment of program module for calculating the paper surface characteristics

A fragment of program module for calculating the ink perception of a paper is represented in Fig. 2.

```
// Ink perception computing
//=========================================================
    private void GetKraska_Click(object sender, EventArgs e)
    {
      double kraska_got = 0;
      double kraska_1 = 0;
      double kraska_2 = 0;
      kraska_1 = 0;
      for (int lep = 0; lep != zn_profil; lep++)
      {
        kraska_2 = 0;
        for (int lep2 = 1; lep2 > i - 1; lep2 = 2 * lep2 + 1)
        {
          kraska_2 = kraska_2 + (cof_b[lep, lep2] / lep2);
        }
        kraska_1 = kraska_1 + kraska_2;
      }
      kraska_got = (0.00015 * (zn_profil * z_max - ((2 / Math.PI) * kraska_1))) / 1000;
      textBox21.Text = Convert.ToString(kraska_got);
    }
    private void tabControl1_SelectedIndexChanged(object sender, EventArgs e)
    {
      if (tabControl1.SelectedIndex == 5)
      {
```

```
if (cof_b == null)
{
  tabControl1.SelectedIndex=1;
  MessageBox.Show("Произведите расчеты");
```

Figure 2. A fragment of program module for calculating the ink perception of a paper

The irregularity of the ink layer surface on printed substrate is described by the sum of two independent functions that take into account macrogeometric deviations (nonplanarity, areas with a greater or lesser ink layer thickness) and a stochastic microgeometric component (reflects the surface roughness of the ink layer on the substrate being printed).

In the case, the macrogeometric deviation was defined as the distance from some point of the real surface to the ideally flat abutting surface, i. e. up to the plane, in accordance with the principle of abutting surfaces for measuring the macrogeometric deviations.

The algorithm for estimating the color reproduction quality of the system is offered in Fig. 3.

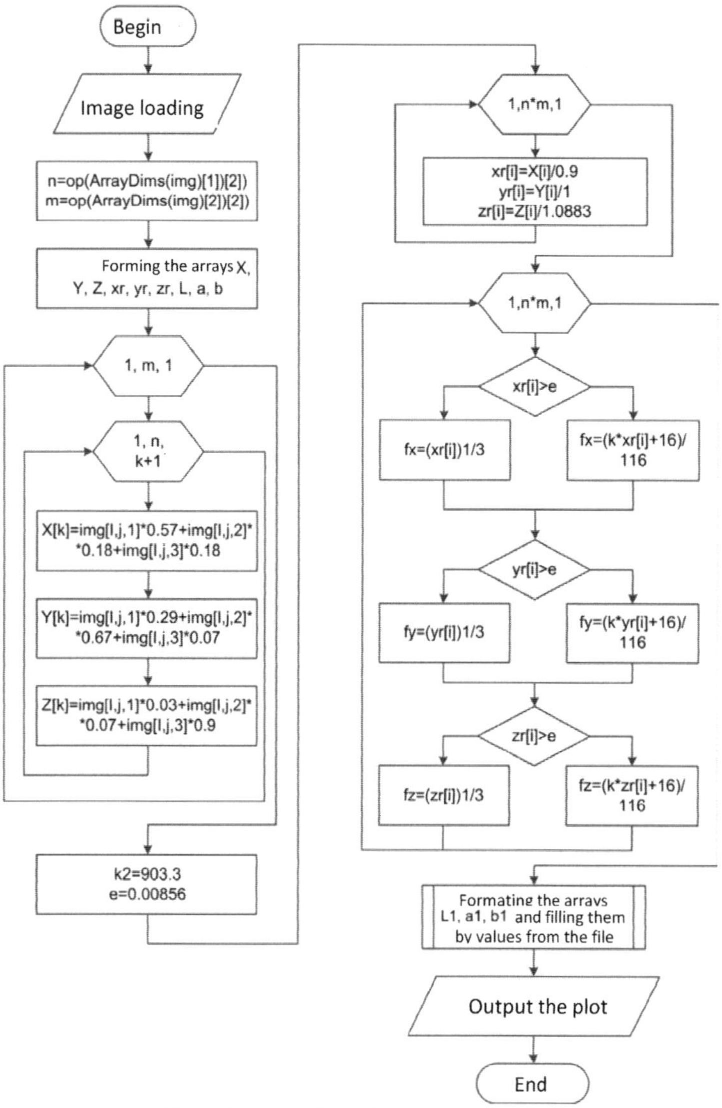

Figure 3. Simplified algorithm for estimating the color reproduction of a printing system

For the automation of the calculation process of the color gamut body volume, a software was developed. Computing the color gamut of the reproduction system is carried out using the Qhull (Quickhull) algorithm. The functions of plotting and calculating the body of color gamut are realized with the help of developed software [5]. The program fragment presented shows the construction of distributing the color samples in three projections on the L * a * b * graph using the scatter3 function.

```
figure ('Color' ,[0.5 0.5 0.5]); axes ('Color' , [0. 8 0.8 0.8]);
camproj('perspective'); view(-44,17); grid on; xlabel('a*');
ylabel('b*'); zlabel('L*'); hold on;
scatter3(a, b, L, 200,[ sR sG sB ],'filled'); hold off.
```

Practical realization of another part of printout is putted out in the construction of wire-frame model of a color gamut body surface.

```
[C,V] = convhulln([ a b L ]);
for i = l:size(C,l);
j = C(i,[1 2 3 1]); patch (X(j,1) ,X (j ,2) ,X (j , 3) , [0 . 8 0.8 0.8]);
end;
view(3), axis vis3d; axis([-100 100 -100 100 0 100]);
grid on; xlabel('a*'); ylabel('b*'); zlabel('L*'); hold on;
rotate3d on; camproj ('perspective') ; view (-44 ,17) ; figure (gcf).
```

The geometric representation of the data of color gamut body surfaces projections allows visualizing the color gamut of printing systems color reproduction by the brightness ranges and separate color sectors, and lets to compare systems qualitatively for the reproduction of certain colors in a particular range of brightness. As an example, the construction of the projections of color gamut body surfaces was carried out in highlights (brightness value – L = 80), midtones (brightness value – L = 50) and shadows (brightness value – L = 20). A cross-section of color gamut body surfaces was made. Section of L = 20 is a mixture of solid colors with white – muting by the color of printed material; proportional reduction of both colors. Section of L = 50 is the line of solid (most saturated) colors formed by the main ink sets and their binary mixtures in different quantitative ratios. Section of L = 80 is a mixture of solid colors with black – muting the binary ink mixture by black; proportional increase in the relative amount of black. The projections of color gamut body surfaces are shown in Fig. 4.

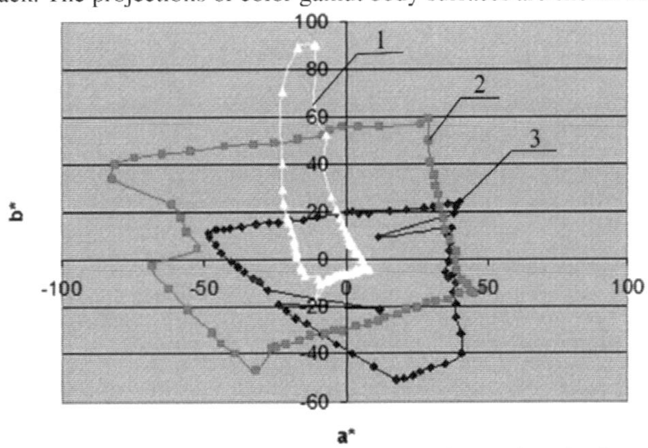

Figure 4. Projective representation of geometrical model of reproduced colors for the system of reproduction in highlights (L = 80) and shadows (L = 20) (1– L = 80, 2– L = 50, 3– L = 20)

This approach allows predicting the garbling (deviation) in color reproduction of separate colors in the number of reproduced tints of compared systems, for example, in highlights (graphs 1 of Fig. 3).

4. Resume

Thus, for the optimal selection of the components in ink-and-paper system ensuring the quality of multicolour image in accordance with the requirements of international standard ISO 12647-2: 2004, it is suggested to control the quality rating of paper, ink coating in accordance with the proposed methodology by stages, i. e. by carring out the burst of controlled systems on a number of independent subsystems (separate quality ratings). The process should be guided by the exercise control on the deviation principle. Thus, the distribution of ink layer on the printed substrate is proposed to characterize by so-called the deviation of the ink layer thickness on the sheet from the nominal surface.

Quality control on each stage in the "ink-and-paper" system is an integral part of a developed automated system for control and the prediction of a multicolor image quality. A distinctive feature of this approach is the possibility of mutual coordination of subsystems and optimization the management process as a whole, that provides the scientific novelty of the task solution.

References

[1] Korzhev V A and Shepitinovsky G V 1978 *Researches in Pulp and Paper Technology. Proceedings of CRIP* (Moscow) **15** p 92-99
[2] Koivula H, Preston J S, Heard P J and Toivakka M 2008 *Colloids and Surfaces A: Physicochemical and Engineering Aspects* **317 (1-3)** p 557-567
[3] Ozaki, Y and Kimura M 2000 *Appita J.* **(3)** 216-219
[4] Varepo L G, Nagornova I V and Trapeznikova O V 2015 *Proc. Eng.* **113** 357-36
[5] Varepo L G and Trapeznikova O V 2014 *A software for the estimating the color reproduction of printing system* Certif. of registr. the computer program No 2014619696 from 19.09.2014. (Moscow).FIPS

Applying an overstress principle in accelerated testing of absorbing mechanisms

V G Tsyss, M Yu Sergaeva, A A Sergaev

Omsk State Technical University, 11, Mira av., Omsk 644050, Russia

E-mail :mjsergaewa@mail.ru

Abstract. The relevance of using overstress test as a forced one to determine the pneumatic absorber lifespan was studied. The obtained results demonstrated that at low load overstress the relative error for the absorber lifespan evaluation is no more than 3%. This means that the test results spread has almost no effect on the lifespan evaluation, and this effect is several times less than that at high load overstress tests. Accelerated testing of absorbers with low load overstress is more acceptable since the relative error for the lifespan evaluation is negligible.

1. Introduction

Forced mode testing acceleration is achieved by intensifying the degradation processes by forcing the loading loads (stress, deformation, displacement amplitude, temperature, frequency, etc.). In this regard single loading mode change tests are of the greatest interest, when an overstressed (high) mode follows a low one or precedes it. Depending on the time of the "loaded" mode, such tests can be called high or low load accelerated overstress tests (figure 1). If an overstressed mode follows a low one, such forced testing is called high load overstress; if this mode precedes the low one, then it is low load overstress.

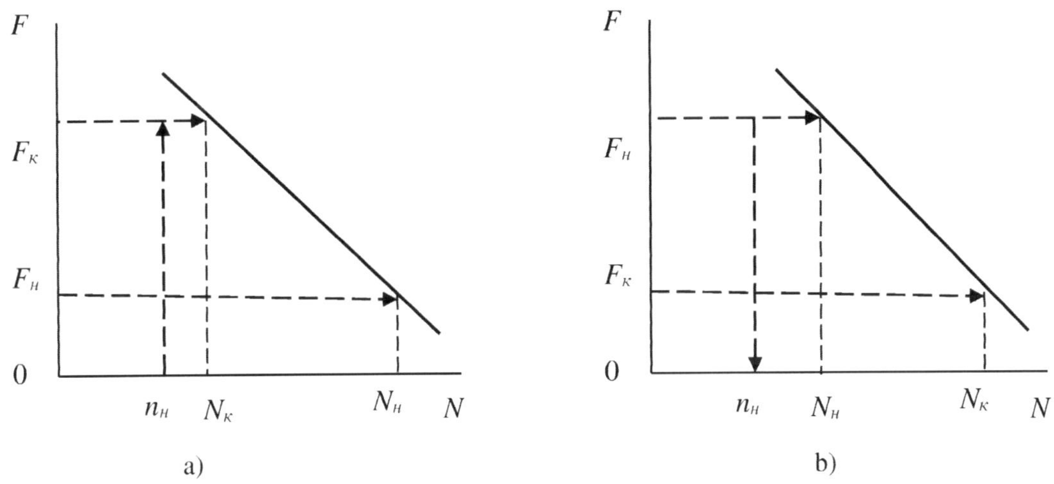

Figure 1. Forcing the overstress testing with high (a) and low (b) loads.

Overstress principle is as follows. No matter what conditions the product was in or was operated under, some fatigue changes accumulate in it. These changes do not disappear, and then they affect the fault-free operation time. Therefore, the testing should be constructed as a step loading, where for a number of cycles $n_н$ stress $\sigma_н$ is applied, and then for $n_к$ cycles the product is overstressed with even higher stress $\sigma_к$.

2. Problem statement

The objective is to determine the absorbing mechanisms lifespan $N_н$ at loading amplitude $F_н$, when firstly the sample is tested for $n_н$ cycles at amplitude $F_н$, and then it is overstressed at relatively high amplitude $F_к$.

The duration of the preloading $n_н$ is known. The residual lifespan $n_к$ at the overstress is observed during the experiment. The lifespan $N_к$ at the amplitude $F_к$ is easy to determine by testing several samples, as $F_к$ is profound and the testing time is short. If the functional relation between $n_к$, $n_н$, $N_к$ and unknown value $N_н$ is set, the latter can be calculated by the step testing results. Such functional relations based on various suppositions were proposed by different authors [1, 2, 3].

3. Theory

Functional relation for the dependence $n_к=n_к(n_к,N_н,N_к)$ can be identified by analysing physical nature of damages accumulation. Let the value of the damages accumulation be determined by the equation

$$D_p(n)=\int_0^n \dot{D}_p(x)dx, \tag{1}$$

where $\dot{D}_p(x)$ is the rate of the damages accumulation; $D_p(n)$ is the value of the damages accumulation.

At the sector $(D,\ n_н)$ stress $\sigma_н$ acts, and at the sector $(n_н,\ n_н+n_к)$ stress $\sigma_к$ does, and at the time moment $(n_н+n_к)$ the product failure occurs. Then, the value of the damages accumulated by the time moment $n_н$ is represented by the dependence

$$D_p(n_н)=\int_0^{n_н} \dot{D}_н(x)dx. \tag{2}$$

At the sector $(n_н,\ n_н+n_к)$ the damages accumulation rate depends on the stress $\sigma_к$. Let us denote it by $\dot{D}_к(n)$. Taking into account that the changes, accumulated in the product for the time $(0,\ n_н)$, affect the subsequent damages accumulation, the value of the damages accumulated at the second sector is represented by

$$D_p\left(n_н,n_н+n_к\right)=\int_{n_н^*}^{n_н^*} \dot{D}_к(x)dx, \tag{3}$$

where $n_н^*$ is some number of cycles representing the impact of the stress $\sigma_н$ on the lifespan at the stress $\sigma_к$.

Equations (2) and (3) enable us to write the failure occurrence condition for the time $(0,\ n_н+n_к)$ under the step load as follows

$$\frac{\int_0^{n_н} \dot{D}_н(x)dx}{\int_0^{N_н} \dot{D}_н(x)dx}+\frac{\int_{n_н^*}^{n_н^*+n_к} \dot{D}_к(x)dx}{\int_0^{N_к} \dot{D}_к(x)dx}=1 \tag{4}$$

or

$$\frac{\int_0^{n_н} \dot{D}_н(n)dn}{\int_0^{N_н} \dot{D}_н(n)dn} + \frac{\int_{n_н^*}^{n_н^*+n_к} \dot{D}_к(n)dn}{\int_0^{N_к} \dot{D}_к(n)dn} = 1, \tag{5}$$

where $N_н$ and $N_к$ are destroying cycles number under the stresses $\sigma_н$ and $\sigma_к$.

At constant damages accumulation rate, i.e. at the adherence of equation $\dot{D}(n) = \text{const}$, it follows from formula (5) that

$$D_н + D_к = 1, \tag{6}$$

where $D_н = \frac{n_н}{\overline{N}_н}$ is the damage measure at the stress $\sigma_н$;

$D_н = \frac{n_к}{\overline{N}_к}$ is the damage measure at the stress $\sigma_к$;

$n_к$ is the value of the residual lifespan at the stress $\sigma_к$;

$n_н$ is the number of loading cycles at the stress $\sigma_н$;

$\overline{N}_к$ is the mathematical expectation of the lifespan at the stress $\sigma_к$;

$\overline{N}_н$ is the mathematical expectation of the lifespan at the stress $\sigma_н$.

Ratio (6) connects values $n_н$, $N_к$, and $N_н$ and provides an opportunity to plan accelerated testing. From formula (6) one derives unknown value $\overline{N}_к$ (figure 1).

$$\overline{N}_к = \frac{n_к}{1 - D_н}. \tag{7}$$

Knowing the lifespan value at the high load $\overline{N}_к$, accelerated testing is planned as step loading. In this regard the loading is performed for $n_н$ cycles at the stress $\sigma_н$, and then the transition to the stress $\sigma_к$ is carried out (figure 1). Several tests are done at constant $n_н$ and $\sigma_к$, with the observed values $n_к^{(i)}$ noted and the mathematical expectation calculated for the value $\overline{n}_к = \frac{1}{m}\sum_1^m n_к^{(i)}$. The lifespan $\overline{N}_н$ estimation is found correspondingly as

$$\overline{N}_н = \frac{n_н}{1 - D_к}. \tag{8}$$

Accelerated testing with high load overstress is based on the fact that if the lifespan $\overline{N}_н$ is to be determined for the low mode (amplitude $S_н$) with the known lifespan $\overline{N}_к$ for the high mode (amplitude $S_к$), the product is preloaded for a few cycles $n_н$ at the low mode, and then it is overstressed at the high one. The reduced testing time is achieved, when the value $n_н$ is a small portion of the lifespan $\overline{N}_н$, and the time to determine the values $\overline{n}_к$ and $\overline{N}_к$ is short as the overstress mode is high enough.

During accelerated testing with low load overstress, the product under test is loaded for a large number of cycles $n_н$ at the high mode (amplitude $S_н$) when the lifespan $\overline{N}_н$ is known, and then it is overstressed at the low mode (amplitude $S_к$). The reduced testing time is achieved, when at large $n_н$ the lifespan $\overline{n}_к$ is a small portion of the lifespan $\overline{N}_к$, and the duration of the preloading and the lifespan $\overline{N}_н$ determination is negligible. Compared to ordinary tests, the time advantage is more if the value $n_н$ is more and the residual lifespan is less, consequently.

The biggest challenge is to find the means of fatigue damages accumulation, taking into consideration all real processes going on in the material at step loading with a single loading mode change. The following hypotheses can be applied for this purpose [5].

1. The Palmgren-Miner hypothesis

$$\bar{n} = \bar{N}\left(1 - \frac{n_н}{\bar{N}_н}\right). \tag{9}$$

2. The Manson-Nachtigall-Frisch hypothesis

$$\bar{n} = \bar{N}\left(1 - \frac{n_н}{\bar{N}_н}\right)^{\alpha}. \tag{10}$$

3. The hypothesis of Bolotina V. V.

$$\bar{n} = \bar{N}\left[1 - \left(\frac{n_н}{N_н}\right)^{\alpha}\right]. \tag{11}$$

4. The Henry hypothesis

$$\bar{n} = \frac{1 - \frac{n_н}{\bar{N}_н}}{1 + \frac{\alpha n_н}{\bar{N}_н}}. \tag{12}$$

5. The hypothesis of Kordonskyi Kh. B.

$$\bar{n} = \alpha n_н\left(\bar{N}^{1 - \frac{lg n_н}{lg \bar{N}_н}} - 1\right). \tag{13}$$

6. Parabolic relation

$$\bar{n} = \bar{N}\left(1 - \frac{n_н}{\bar{N}_н}\right)\left(1 + \alpha\frac{n_н}{\bar{N}_н}\right). \tag{14}$$

The comparison of predicted and experimental results showed the following:

- The Palmgren-Miner hypothesis has an error depending on material hardening or weakening degree, and it can be significant.
- The Manson-Nachtigall-Frisch hypothesis provides a sufficiently accurate agreement with practice at material hardening, but the error becomes considerable at weakening.
- At the absence of material ageing, the hypothesis of Bolotin V.V. provides an accuracy sufficient for practice, if $n_н$ > (20…30)% of $\bar{N}_н$.
- The Henry hypothesis can also be used at the absence of material ageing.
- The hypothesis of Kordonskiy Kh. B. provides a reliable estimation of the lifespan in most cases if $n_н$ > (15…20)% of $\bar{N}_н$.
- Parabolic relation also has a good agreement with the practice at the absence of material weakening.

All these models of recalculating experimental results, obtained when applying overstress test to normal mode, are based on the so-called physical reliability concept [4]. The concept is that product reliability relies on the value of the used lifespan and not on the way it was used. One of the physical reliability concept sequences in case of linear models is the ratio validity [6].

$$\sum_{j=1}^{k} \frac{M_{\tau_j}(\varepsilon_j)}{M_{\tau_l}(\varepsilon_j)} = 1, \tag{15}$$

where k is the number of loading mode steps;

$M_{\tau_j}(\varepsilon_j)$ is the mathematical expectation of fault-free operation time at single step testing in the mode ε_j;

$M_{\tau_j}(\varepsilon_j)$ is the mathematical expectation of operation time at j-th step in the mode ε_j.

Using ratio (15), it is easy to determine the dependence for mathematical expectation estimation of fault-free operation time when using overstress test.

4. Experimental results

To evaluate the relevance of applying overstress as a forced method, experimental tests on pneumatic absorber lifespan were conducted.

The results of calculations on the basis of formulae (7) and (8) are given in table 1 for low load overstress and in table 2 for high load overstress. These results are compared with experimental values for the absorber lifespans. Tables 1 and 2 show relative error of the absorber lifespan estimation for each test.

Table 1. Relative error of the absorber lifespan estimation at the low load overstress

Testing mode	$F_н$, mm	$F_к$, mm	$n_н \cdot 10^3$	$n_к \cdot 10^3$	$\overline{N}_н \cdot 10^3$	$\overline{N}_к^{оп} \cdot 10^3$	$\overline{N}_к^{р} \cdot 10^3$	$D_к$	$\Delta = \frac{\overline{N}_к^{оп} - \overline{N}_к^{р}}{\overline{N}_к^{р}} 100\%$
16→13	16	13	200	142	258	633.9	630	0.9995	+0.62
15→12	15	12	250	177.5	331	704.4	724.4	1.0072	-2.76

Table 2. Relative error of the absorber lifespan estimation at the high load overstress

Testing mode	$F_н$, mm	$F_к$, mm	$n_н \cdot 10^3$	$n_к \cdot 10^3$	$\overline{N}_н \cdot 10^3$	$\overline{N}_к^{оп} \cdot 10^3$	$\overline{N}_к^{р} \cdot 10^3$	$D_к$	$\Delta = \frac{\overline{N}_к^{оп} - \overline{N}_к^{р}}{\overline{N}_к^{р}} 100\%$
13→16	13	16	100	269.2	400.9	401.5	302	1.1224	-32.87

5. Results and discussion

The analysis of the obtained results shows that at the low load overstress the value of relative error Δ for the lifespan estimation is more than 30%. This can be explained by a profound impact of the fatigue strength values spread during testing and material hardening at changing deformation mode from low amplitudes to high ones.

The damage measure value in the testing equals $D_\Sigma = 1.1224$, signifying a non-linear nature of the fatigue damages summation at the high load overstress.

At the low load overstress, as seen from table 1, the relative error Δ for the lifespan estimation is not more than 3%. This demonstrates that the lifespan testing results spread has a minor effect on the jacket lifespan estimation and this effect is several times less than that at accelerated tests with the high load overstress. In this case, the damage measure value is close to one. The deviation from the fatigue damages linear summation in this case can be explained by the material weakening, when the deformation mode changes from higher amplitude to a lower one.

6. Conclusion

The obtained experimental results demonstrate that high (low) loading amplitudes overstress provides lifespan values close to those of accelerated tests.

Accelerated testing with low loading amplitude overstress is more preferable since the relative error value of lifespan estimation is inconsiderable, and this method allows preloading time $n_н$ to be set.

Relative total damage measure at high (low) loading amplitudes overstress lies in the ranges D_{Σ} = 0.999…1.122, and its mean value is close to one. This points out the possibility of applying fatigue damages linear summation hypothesis to evaluate real lifespan of absorbing mechanisms.

References

[1] Krinetskiy L N et al. 1989 *The Foundations of Flying Vehicles Testing*. (Moscow, Mashinostroenie). Russian.
[2] Sudakov R S 1988 *System Testing: Choosing Volumes and Duration.* (Moscow, Mashinostroenie). Russian.
[3] Kubarev A I et al. 1987 *Machinery and Equipment Reliability*. (Moscow, Izdatelstvo standartov). Russian.
[4] Zarenin Yu G 1970 *Reliability Compliance Tests*. (Moscow, Izdatelstvo standartov). Russian.
[5] Nelson W and Hahn G 1972 *Technometrics*. **14(2)** 235-244.
[6] Sudo H and Nakano Y 1985 *Microelectron. Reliab.* **25(3)** 525-540.

Methods of Measurement the Quality Metrics in a Printing System

L G Varepo[1], A Yu Brazhnikov[1], I V Nagornova[2] and O A Novoselskaya[3]

[1] Omsk State Technical University, 11, Mira av., Omsk, 644050, Russia
[2] Moscow Polytechnic University, 38, B. Semenovskaya str., Moscow, 107023, Russia
[3] Belarusian State Technological University, 13a, Sverdlova str., Minsk, 220006, Belarus

E-mail: larisavarepo@yandex.ru

Abstract. One of the main criteria for choosing ink as a component of printing system is scumming ability of the ink. The realization of algorithm for estimating the quality metrics in a printing system is shown. The histograms of ink rate of various printing systems are presented. A quantitative estimation of stability of offset inks emulsifiability is given.

1. Introduction
A distinctive feature of offset printing is that in different ink areas, each of which corresponds to a certain color gate, the thickness of the ink layer is contacting with a different amount of fountain [1-3]. Considering this, the degree of ink layer scumming in one inker is not the same and varies in different color zones [2-4]. The same ink with different degrees of emulsification can have unstable rheological properties that lead to destabilization of the color reproduction process. Accounting for divergence of quality metrics from the nominal value represents one of the priority problems.

2. Objects of Research
For quantitative estimation the degree of ink scumming and its effect on the formation of an ink layer on paper researches were carried out on a sheet-fed offset printing machine Man Roland 704. The main components of the printing system are the paper Titan gloss 115 gr/m2; printing ink of various manufacturers – LitiFlora FTX (FlintGroup) (1); Cristal (SunChemical) (2); Triumf (SunChemical) (3); ToyoInk TK Nex NV100 (4).

Test samples were obtained from test plates that included:

- bars, providing different inkload by each color and control 100 % solid plates for CMYK along the valve;
- 100 % solid plates and binary overprints with consecutive measurement of them, i. e. C+M, M+Y, C+Y;
- printing of 1000 sheets and measuring in two zones with different amount of inkload.

Measurements of the quality metrics in the printing system in accordance with the ISO were carried out using an X-Rite spectrophotometer, a Techkon 410 densitometer.
A quantitative estimation of ink absorption in the paper structure was evaluated on the basis of images obtained with a scanning electron microscope of the JSM7500F JEOL model at an accelerating voltage of 10 kV and a beam current of 2 • 10-9 A [5].

3. Results and Discussion

Experimental results are presented in Table 1 and in Figure 1.

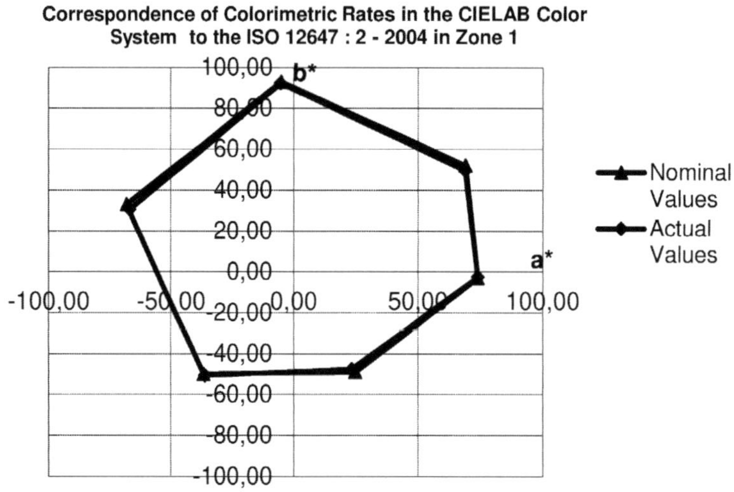

Figure 1. Correspondence of colorimetric rates in CIE LAB system.

Table 1. Optical Density and color difference dE

	black		**cyan**		**magenta**		**yelow**	
	D	dE^*_{ab}	*D*	dE^*_{ab}	*D*	dE^*_{ab}	*D*	dE^*_{ab}
LitoFloraFTX	1.7	1.82	1.45	1.06	1.38	0.69	1.24	1.21
Cristal	1.71	2.23	1.37	1.74	1.33	2.98	1.23	3.24
Triumf	1.64	3.23	1.46	2.56	1.42	1.58	1.24	1.65
ToyoInk	1.68	2.06	1.41	2.39	1.39	4.02	1.21	2.89

The analysis of the values of coloring gate opening by the color zones during printing the same sample on the same carrier presents the difference in ink rate comparatively to a number of inks in per cents. Graphical analysis of such researches is presented in Figure 2-4.

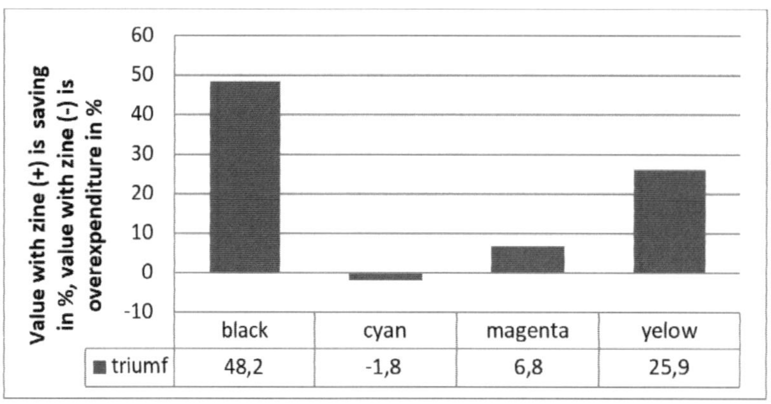

Figure 2. Histogram of Ink Rate of Sun Chemical Triumf to LitoFlora FTX.

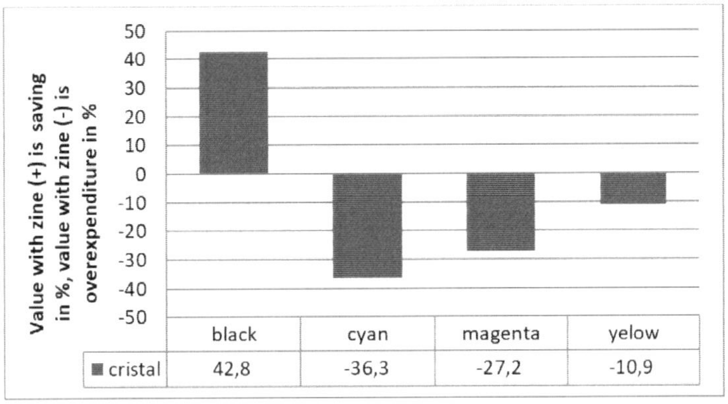

Figure 3. Histogram of Ink Rate of Sun Chemical Cristal to LitoFlora FTX.

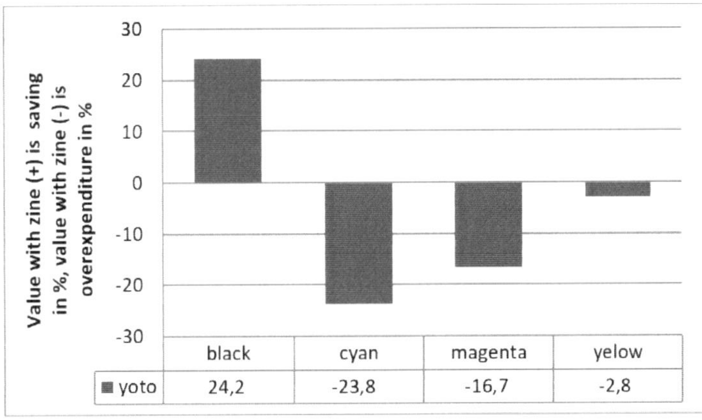

Figure 4. Histogram of Ink Rate ToyoInk to LitoFlora FTX.

The data of ink layer distribution in the surface and solid layers of the paper structure are well-formed with the results of estimating the ink rate. The visualization of the obtained data is shown in Figure 5. To estimate the emulsifiability of offset inks, an algorithm has been developed. It includes:

- Constructing the dependences of the degree of ink scumming on the under test parameter of the fountain solution in the form of a kinetic curve. A distinctive feature of this stage is that the mixing of ink with the fountain solution is realized in three cycles with different time periods; stirring rate is 400 rpm; the temperature of the fountain solution and the ink emulsion is kept at 25 degrees during mixing.
- Analysis of results. The stages boundaries on the kinetic curve of ink scumming were determined in the following order:

1. The initial and final points of ink scumming are determined. The origin is the point of intersection of the curve with the OX axis. The end point is the point on the curve equal to the final value of the ink scumming degree at the 300th second.
2. Two tangents to the two most clearly defined linear sections of the curve are drawn. As a result of the intersection of the tangents with the kinetic curve the segments that characterize the stage of active absorption (stage II) and the saturation stage (stage IV) are obtained. The segments on the kinetic curve characterizing the initial stage of ink scumming (stage I) and the deceleration stage (stage III) are determined.

IOP Conf. Series: Journal of Physics: Conf. Series **998**(2018) 012040 doi:10.1088/1742-6596/998/1/01204

Practical realization of the estimation of ink emulsifiability according to the developed procedure and at the concentration of isopropyl alcohol in the fountain solution from 0 to 5% showed that the SunChemical Cristal ink reduces the degree of saturation by 21.2% and maintains a tempered stability. The printing ink LitoFlora FTX reduces the degree of saturation by 2.4%. The percentage range of alcohol content in the fountain solution (in the referred above range) and in the ink emulsion will not destabilize the printing-technical properties of the LitoFlora FTX ink.

On the bases of analysis of the scanning electron microscopy results with SEM images it is found that the distribution of ink layer on the paper surface, ceteris paribus, is determining by the degree of ink scumming during printing that adequate mappingin Figure 5.

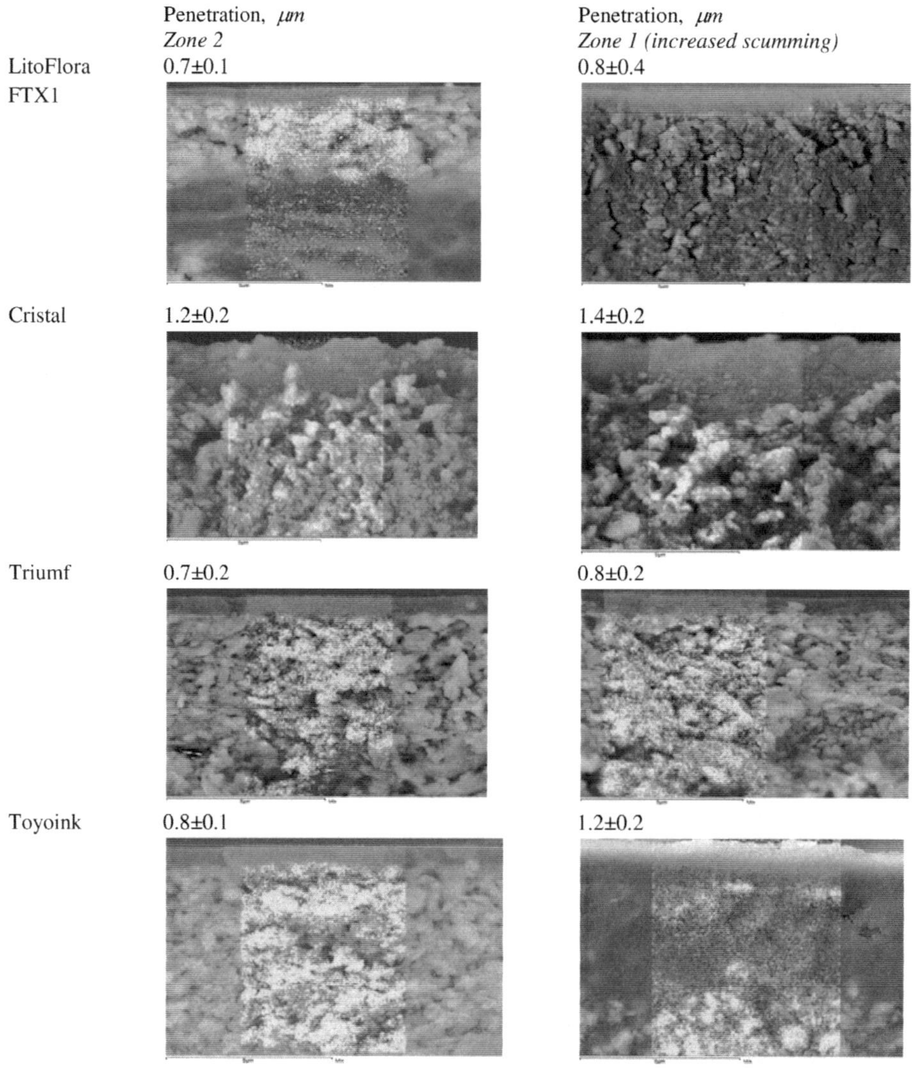

Figure 5. The influence of degree of ink scumming on the formation of an ink layer in binary overprints (C+M).

4. Conclusion

In conclusion, Optimum selection of the components of the "ink-substrate" system ensures the quality of printing in accordance with the requirements of ISO 12647-2: 2004, which is achieved through the diagnostics of ink. The algorithm of each stage of quality control in the "ink-substrate" system is an integral part of the development the automated system for predicting the quality of the ink coating.

References

[1] Verikas A, Lundström J, Bacauskiene M and Gelzinis A 2011 *J. Expert Systems with Applications* **38 (10)** 13441-13447
[2] Surland A 1983 *TAGA Proceedings* p 191-233
[3] Bertholdt U 2007 *European Coatings J.* **5** 184
[4] Kushlyk B 2013 *J. of Materials Science and Engineering* **3 (8)** 545-550
[5] Varepo L G, Nagornova I V and Trapeznikova O V 2015 *J. Procedia Engineering* **113** 357-361

Software for Quantitative Estimation of Coefficients of Ink Transfer on the Printed Substrate in Offset Printing

L G Varepo[1], **O V Trapeznikova**[1], **A V Panichkin**[2], **B A Roev**[3] **and G B Kulikov** [3]

[1] Omsk State Technical University, 11, Mira av., Omsk, 644050, Russia
[2] Sobolev Institute of mathematics, 13, Pevtsova st., Omsk, 644043, Russia
[3] Moscow Polytechnic University, 38, B. Semenovskaya str., Moscow, 107023, Russia

E-mail: larisavarepo@yandex.ru

Abstract. In the framework of standardizing the process of offset printing, one of the most important tasks is the correct selection of the printing system components, taking into account the features of their interaction and behavior in the printing process. The program allows to calculate the transfer of ink on the printed material between the contacting cylindrical surfaces of the sheet-fed offset printing apparatus with the boundaries deformation. A distinctive feature of this software product is the modeling of the liquid flow having free boundaries and causing deformation of solid boundaries when flowing between the walls of two cylinders.

1. Introduction

In a number of Russian and foreign papers, various approaches to the mathematical description of ink devices, the transfer and splitting of printing ink. They are mostly based on a system of algebraic equations that reflect the addition and division of ink layers in contact zones, the use of which allows solving the dynamic problem of transferring paint layers on the printout [1-4].

In connection with the computer hardware and software possibilities expansion, which greatly simplified the modeling of hydrodynamic processes, the works of H Rech, John MacPhee, B Patzelt and R Ruder, Y Ozaki., M. Kimura and others are of high scientific interest [5-7].

It should be noted that in most papers qualitative analysis of the process under investigation is given. Data on the paint splitting with a simplified half-cleavage with zero absorbency are given. It is conventionally accepted that the paint layer is divided in half along the separation angle. Another drawback of the works published on this issue is insufficient attention to the incorporation of ink absorption into the paper pores, along with its fixation on the surface and taking into account the characteristics of the surface microgeometry.

Along with the experimental methods, theoretical methods are intensively developed. Among them, the use of modern modeling theory, in particular computer modeling, is an important research tool. This approach provides the most accurate and maximum adequate real process to study the change that occur in practice when moving the ink layer between rotating cylinders of the printing apparatus and the interaction of the paper porous structure with the liquid medium.

2. Statement of the problem

When solving the problems defined, it is necessary to overcome significant mathematical difficulties due to both the nonlinearity and complexity of the equations themselves and the need to determine the free surface when solving the system of partial differential equations. The search for new solutions of this problem emphasizes its relevance.

3. The problem solution

To solve the problem, the authors developed an algorithm and its software implementation [8-10].
The program algorithm works with data sets describing the two-dimensional fluid motion and given in the calculated polar coordinate system Orθ in the form of a discrete set of points in which we apply the finite-difference equations in the transformed Oxy, coordinate system, which is polar and is presented for a more convenient consideration of the calculation results.

The program calculates the fluid flow between the contacting cylindrical surfaces with the boundaries deformation. A distinctive feature of this software product is the modeling liquid flow which has free boundaries and causes solid boundaries deformation when flowing between the walls of two cylinders. After the file opening and the introducing of data in the input lines of the file with their further preservation, the program allows to make calculations with the preservation of numerical data in a separate file of intermediate and output data and graphical visualization of the described process.

Figures 1-2 present the screen forms of files with input data and the calculation of speed arrays while modeling the transfer of ink from the cylinders of the offset printing machine to the substrate.

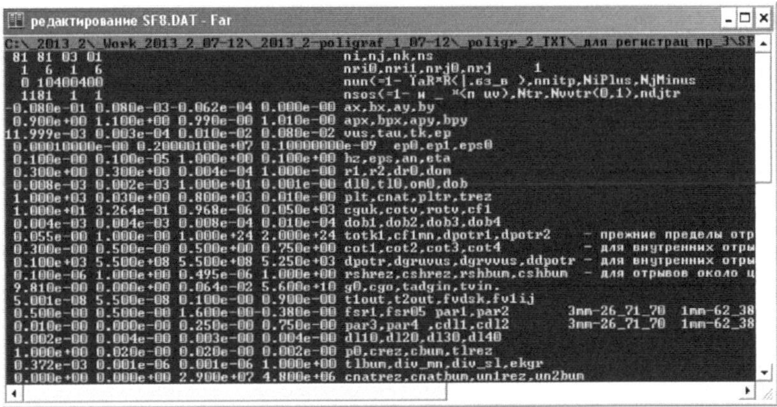

Figure 1. Data input interface.

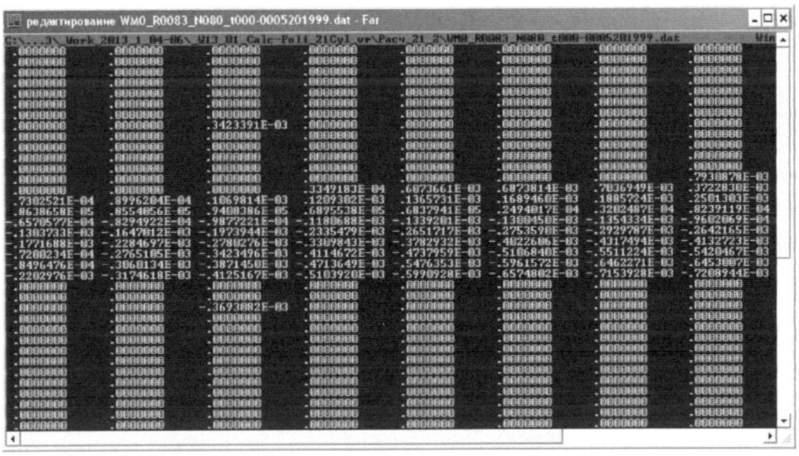

Figure 2. Interface for computing speed arrays when modeling the ink transfer to a substrate.

The algorithm for the numerical modeling of the flow (transfer) of a viscous incompressible fluid with free boundaries between rotating cylinders consists of the following blocks:

Block 1 is responsible for the description of all variables, including data arrays.

In block 2, a number of parameters and initial data, such as pressure values, velocities, cylinder radii, mean pore radii and others are set; the formation of all calculated arrays at time t = 0 is made.

The input of calculated data from the control point at a certain time point, which was final in the previous calculation or the transition to calculation with all initial data, is carried out in block 3.

In block 4, the values are determined at the time t and the values from the file are input for all necessary parameters at the modeling start or at the moment of the calculation continuation with the control point. Besides, the parameter for using the control point and the end time of this calculation with the creation of the next control point are set.

Additional arrays formation at the initial moment of time $t = 0$ or time instant t from the control point and transition to data with increment of time by the iteration step τ is performed in blocks 5 and 6, accordingly.

Block 7 calculates the movements of the free boundaries of the fluid and the boundaries of the cylinders as a function of the velocity field and the forms computed nodes that fall within the fluid boundaries.

The verification of fluid adhesion conditions on the cylinders is performed in block 8 .

In block 9, the iterations begin to calculate the pressure and determine the termination condition for the iterations.

The calculation of velocities, displacements of free fluid boundaries and markers for their geometric determination is performed in block 10.

The calculation of pressure, determination of the adhesion condition in the internal fluid computation nodes, including contiguity to the free boundary, verification of the condition for setting the iteration for the pressure and subsequent exit from the iteration cycle are performed by blocks 11 and 12, respectively.

In block 13, the amount of ink absorbed into the pores of the printed material is calculated.

In block 14, the curvature radii of the fluid free boundaries are calculated together with the determination of the surface tension forces for the boundary nodes.

In blocks 15 and 16, the condition for the termination of the iterative process with respect to time is checked (block 15) and the current lines are determined by the calculated velocity field (block 16).

Block 17 records all calculated data at the final point in time with the creation of a control point for possible subsequent continuation of the calculation.

Block 18 provides additional output of data to files and closing of files.

For numerical modeling, some programs were developed in the Fortran PowerStation v.4.0 software package [11-12]

The results of practical testing of the developed software algorithm are shown in Table 1. Angular rotational velocity of the cylinders is equal to 10 rad / s. P = Patm = 10^5 H / m^2 is the pressure of the environment, $\nu = 0.012$ m^2 / s is the kinematic viscosity. The initial dimensions of the liquid region were $\delta L = 4$ μm and $\delta S = 2$ μm, $\delta = 1$ μm represents the substrate thickness, the radii of the cylinders are $r_1 = r_2 = 0.13$ m.

Table 1. Numerical modeling results.

Paper sample number	Paper roughness, μm	Average pores radius, nm	Quantity of paint, transferred onto the paper, %		
			Penetrated into the paper structure	On the paper surface	Total paint quantity on the printout
1	0.250	60	7.49	47.15	54.65
2	0.450	110	5.28	47.13	52.31
3	0.350	1000	37.42	35.07	72.49

A uniform grid with the number of computed nodes with two coordinates and equal to 80 was used, which allowed to analyze the convergence and behavior of numerical solutions with an accuracy of 1% for Reynolds numbers about 1. The flow type is laminar.

A fragment of the program for calculating ink absorption into the pores of the printed substrate and determining the boundaries near the cylinders is shown in figure 3.

```
761 continue
    goto 755

    if(krasch1(i,j).eq.1)goto 762
    if((kdisk1(i,j+1).ne.1).and.(krasch1(i,j-1).ne.1))goto 762
    u1(i,j,2)=(u1(i,j+1,2)+u1(i,j-1,2))/2
    v1(i,j,2)=(v1(i,j+1,2)+v1(i,j-1,2))/2

    krasch1(i,j)=1
    ekgr1(1,i,j)=hx/2
    ekgr1(2,i,j)=-hx/2
    ekgr1(3,i,j)=hy/2
    ekgr1(4,i,j)=-hy/2
762 continue
750 continue
755 continue
    jn=jdisk1(i)
    do 758 j=jn,nj
    v1(i,j,2)= t_v1(i)
758 continue
759 continue
6753 continue

    ybum_d2=dabs(s1) !  maximum ink absorption
    do 729 i=2,nim1
    jdisk12(i)=jdisk11(i)
    do 730 j=1,njm1
    j1=nj-j

    if(j1.ge.jdisk11(i))goto 736
    if(krasch1(i,j1).ne.1)goto 734
    jdisk12(i)=j1
```

Figure 3. Fragment of the program for calculating ink absorption into the printed substrate pores.

When calculating fluid flows between the contacting surfaces, the deformation of the surface layers of rubber and paper resulting from the change in pressure in the fluid contacted with them was recorded. Taking into account the smallness of the relative motion velocities of the compression surfaces with the difference between the internal and external pressures (P_1-P_0) with the time-dependent deformation of the boundary Δr for the thickness of the paper or rubber layer H of area S having elasticity modulus E, the changes in the center x_c of the layer thicknesses were calculated according to the following equation:

$$\rho H S \ddot{x}_c/2 = -\frac{E}{H}(\Delta r/2 + x_c)\cdot S + (P_1 - P_0)\cdot S$$

Program implementation on calculation the deformation values was carried out according to the algorithm written in Fortran (Figure 4).

```
c  Calculating the boundaries deformation for the 1st and the 2nd disk
     do 442 i=1,ni
     r1i00(i)=r1
     r2i00(i)=r2
     do 442 j=1,nj
     dtxy1(i,j)=0
```

```
      dtxy2(i,j)=0
 442 continue
      do 127 i=2,nim1
      kdisk1(i,nj)=1
      kdisk1(i,1)=2
      do 127 j=2,njm1
      dtxy1(i,j)=0
      x3=r1i(i)
      x4=r2i(i)
      x1=dabs(p0+dobpw(j))
      x12=(  p1(i-1,j,2)+2*p1(i,j,2)+  p1(i+1,j,2)+
     * p1(i-1,j-1,2)+2*p1(i,j-1,2)+p1(i+1,j-1,2))/8
      if((kdisk1(i,j).ne.0).or.(kdisk1(i,j+1).ne.1))goto 311
      x5=cnatrez/plt*rom1(3,i,j)
      x71=tau*tau*(x12+x5-x1)*plt/(tlrez*pltr)
      x81=tau*tau*un1rez/(tlrez*tlrez*pltr)
      x83=x81*(r1-x3)/2
      x85=(x71 -x83)
      x87=dsqrt(x85*x85+tau*tau*dt_v1(i)*dt_v1(i)*coefs3)
      if(x85.lt.0)x87=-x87
      r1i00(i)=x3-2*(x85 + x87)/(1+x81)*crez
      if((r1-r1i00(i)).gt.tlrez/2)r1i00(i)=r1-tlrez/2
      if( r1i00(i) .gt.r1)r1i00(i)=r1
      r1i0(i)=x3
      dtxy1(i,j)=   (x3-r1i(i))
 311 continue
 if((kdisk1(i,j).ne.0).or.(kdisk1(i,j-1).ne.2))goto 312
.......
 312 continue
 127 continue
      do 443 i=2,nim1
      do 443 j=2,njm1
      if((kdisk1(i,j).eq.0).and.(kdisk1(i,j+1).eq.1))
     * dtxy1(i,j)=   (r1i0(i)-r1i(i))
      if((kdisk1(i,j).eq.0).and.(kdisk1(i,j+1).eq.1))
     *dt_v1(i)= (r1i0(i)-r1i(i))/tau
 443 continue
```

Figure 4. Fragment of the program for calculating the deformation boundaries.

For the considered modeling parameters for sample number of paper 1 in figure 5, the calculated visualization picture of the paint splitting cavity formation and growth at the initial stage with the appearance of strands and dusting.

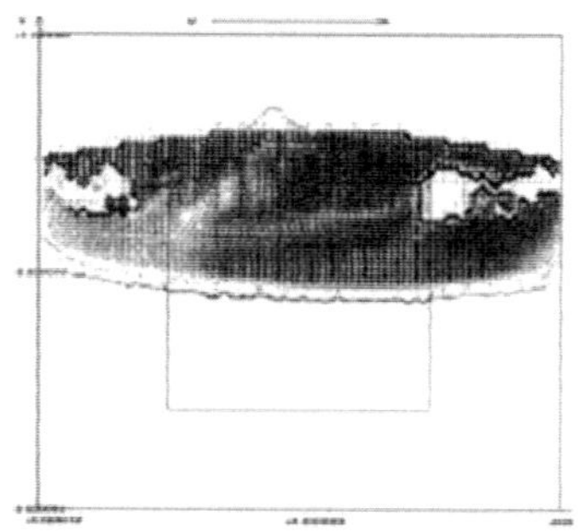

Figure 5. The ink splitting cavity appearing.

4. Conclusion

Software implementation of the developed algorithm allows:

- to calculate the transfer coefficients of ink on the substrate to be printed under offset printing, taking into account the deformation of the printing system components;
- to calculate the absorption of the paint in the porous surface of the printed material;
- to help automating the control of print quality indicators.

All the factors mentioned confirm the practical and scientific significance of this development and contributes to its improvement, taking into account the further evolution of computer technology.

References

[1] John MacPhee 1998 Fundamentals of Lihographic Printing p 365

[2] Vlachopoulos G, Claypole T and Bould D 2010 Ink mist formation in roller trains, in: Iarigai proceedings: Advances in printing and media technology. Montreal, Canada **37** p 227-234

[3] Claypole J, Williams P R and Deganello D 2012 Control of breakup of ink filaments in offset printing, in: Iarigai proceedings: Advances in printing and media technology. Ljubljana, Slovenia **39** p 207-211

[4] Ozaki Y and Kimura M 2000 *Appita J.* **3** 216-219

[5] Koivula H, Preston J S, Heard P J and Toivakka M 2008 *Colloids and Surfaces A: Physicochemical and Engineering Aspects* **317** (1-3) 557-567

[6] N C Reis Jr, Griffiths R F and Santos J M 2004 *J. Comput. Phys.*, **198** 747-770

[7] C Reis Jr, Griffiths R F and Santos J M 2008 *Applied Mathematical Modelling* **32** (3) 341–346

[8] Panichkin A V and Varepo L G 2014 The Numerical Calculation of a viscous Incompressible Fluid transfer onto Poros Surface between Rotating Cylinders. *Springer Proceedings in Physics* **154** 79-83

[9] Panichkin A V, Varepo L G and Trapeznikova O V 2016 *IOP: Materials Science and Engineering* **124** 1-6

[10] Varepo L G and Panichkin A V 2017 *IOP Conf. Series: Journal of Physics: Conf. Series* **858** p 1-6

[11] Varepo L G and Panichkin A V Certif. No 2016617529 of state registration of the program "Calculation of the forces and moments of the interaction forces of cylindrical surfaces during the transfer of a viscous incompressible fluid to a substrate. (Russian Federation); publ. 07.07.2016

[12] Varepo L G and Panichkin A V Certif. No 2016617873 of state registration of the program "Calculation of the transfer coefficients of a viscous incompressible fluid to a substrate between contacting cylindrical surfaces (Russian Federation); publ. 15.07.2016

IOP Conf. Series: Journal of Physics: Conf. Series **998** (2018) 012042 doi:10.1088/1742-6596/998/1/012042

The strength study of the rotating device driver indexing spatial mechanism

N V Zakharenkov[1] and I N Kvasov[1]

[1]Omsk State Technical University, 11, Mira av., Omsk, 644050, Russia

E-mail: znickbar@mail.ru

Abstract. The indexing spatial mechanisms are widely used in automatic machines. The mechanisms maximum load-bearing capacity measurement is possible based on both the physical and numerical models tests results. The paper deals with the driven disk indexing spatial cam mechanism numerical model at the constant angular cam velocity. The presented mechanism kinematics and geometry parameters and finite element model are analyzed in the SolidWorks design environment. The calculation initial data and missing parameters having been found from the structure analysis were identified. The structure and kinematics analysis revealed the mechanism failures possible reasons. The numerical calculations results showing the structure performance at the contact and bending stresses are represented.

Key words: rotary mechanism, cam mechanism, orthogonal mechanism

1. Introduction

The driven sprocket indexing spatial cam mechanism at the driving disk constant angular velocity is considered. Similar constructions are used, for example, in advertising installations [1]. The study subject (figure 1) consists of the driving disk (figure 1, reference 2) and the driven sprocket (figure 1, reference 1) and represents the driving mechanism providing the stop in three positions.

The disk-sprocket mechanism is the transmission comprising the complex spatial interaction between the driving and driven links with the point contact moving along the links surface [2]. Therefore, the numerical simulation in the Solidworks environment was chosen for the study. The similar models examining in the computer environment provides the study subject comprehensive information [3].

Most present day researches focuse on the spatial mechanisms kinematics studying [4, 5, 6] without taking into consideration the acting loads, that does not confirm their performance.

The presented in papers [7, 8, 9] structures analysis reveals that the main disadvantage is the parts working surfaces overwhelming sliding friction. This type of friction wears the parts surface thus reducing the thickness. Parts shape changing increases backlashes and gaps in the structure caused by wear that has a negative impact on the operation smoothness and develops the additional dynamic moments [4].

The research objective is to study the driving mechanism under the increased acting loads and to identify measures for reducing harmful factors emerging in operation.

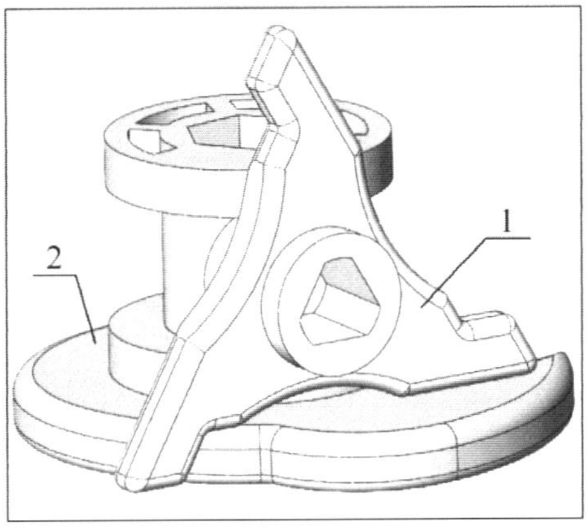

Figure 1. The computer model: 1 is the driven sprocket, 2 is the driving disk.

2. Problem statement

The aims of the study include the following:
- the analysis of the structures on the wear criteria, of the performance at high loads, of the mechanism seizure and jamming inability;
- the strength calculation according to the interacting surfaces contact and bending stresses.

The driving mechanism structure analysis was carried out. The driving mechanism structure analysis objective is to identify the factors influencing the parts operation.
Circumferential speed is defined by the following equation:

$$V = \frac{\pi \cdot d_{e1} \cdot n_1}{60000} = \frac{\pi \cdot 68 \cdot 12}{6000} = 0.042 \text{ m/s} \quad (1)$$

where d_{e1} is the sprocket contact point diameter. On the assumption of the speed the transmission is low-speed one, lubrication conditions are without lubrication or grease lubricant can be applied.
The backlashes occurring can be caused by the material compression at the environmental temperature changing. Furthermore, teeth jamming may occur at the temperature increase.

Table 1. The examined model material.

Properties	Armamid PA SV 30-2T
Density ρ, kg/m³	1380
Melting temperature, °C	220
Tensile strength, MPa	165
Bending stress at maximum load, MPa	233
Flexural modulus, MPa	8000
Heat deflection temperature is 1.8 MPa, °C	200
Water absorption in water (23 °C, 24 h), %	0.95
Transversal shrinkage in casting, %	0.7…0.9
Longitudinal shrinkage in casting, %	0.1…0.2
Shrinkage temperature, °C	260…270

IOP Conf. Series: Journal of Physics: Conf. Series **998** (2018) 012042 doi:10.1088/1742-6596/998/1/012042

3. Theory

For calculating the loads, the computer model making possible to obtain the systems behaviour approximate evaluation by performing computational experiments is used. Moreover, it is also necessary to define the missing initial data used in calculations and required for the model future research.

The driving disk shaft torque is set $M = 7.53$ N · m.

Taking into account the operating conditions possible changes the safety coefficient $C_s = 1.3$ was introduced. Then the transmitted moment by the mechanism is $M_1 \approx 10$ N · m.

The inertia moment is identified for the part mounted on the mechanism output shaft $J = 0.03248$ kg · m^2.

The driving disk and sprocket rotation frequency is ω: 72 ^{0}/s = 1.256 rad/s.

Since the angular velocity real change over time is unknown, then the assumption that the angular acceleration is $\varepsilon = \frac{d\omega}{dt}$, where ω varies from 0 to ω will be used.

In this case, taking into account the load application time 0.001 seconds (the non-linear analysis step in SolidWorks) the following equation is derived for the peak load: $\varepsilon = \frac{d\omega}{dt} = \frac{1.256 - 0}{2 \cdot 0.001} = 628$ rad/s^2

Then the transmitted dynamic torque is:

$$M_{dyn} = -J \cdot \varepsilon = 0.03248 \cdot 628 = 20.4 \text{ N} \cdot \text{m}.$$

While working for 5 seconds and having a standstill for 15 seconds, each tooth takes the stress once per minute. The service life in the stresses variation cycles for a period of one year is designated as N_{k1} and equals to: $N_{k1} = 525600$; of 3 years is: $N_{k3} = 1576800$.

The service life in the stresses variation cycles for a period of one year is: $N_{k1} = 525600$; of 3 years is: $N_{k3} = 1576800$.

4. Experimental results

The model studying is to define the stresses (figures 2, 3) and resource (figure 4) for the driving disk and driven sprocket.

In the case of pure shear, at von Mises stresses, the destruction occurs if $\tau_{max} = 0.577\sigma_t = 0.577 \cdot 139 = 80.2$ MPa.

The force limiting value is identified on the basis of the dynamic torque obtained value. The dynamic torque value is assumed to be transferred by the mechanism in operation, since the pair parts manufacturing poor accuracy results in the fact that the new mechanism initial dead range can already be compared with the permissible one [2]. The maximum force not causing the model shearing is: F = 600 N. The force obtained value corresponds to the transmitted torque of 20.4 N · m.

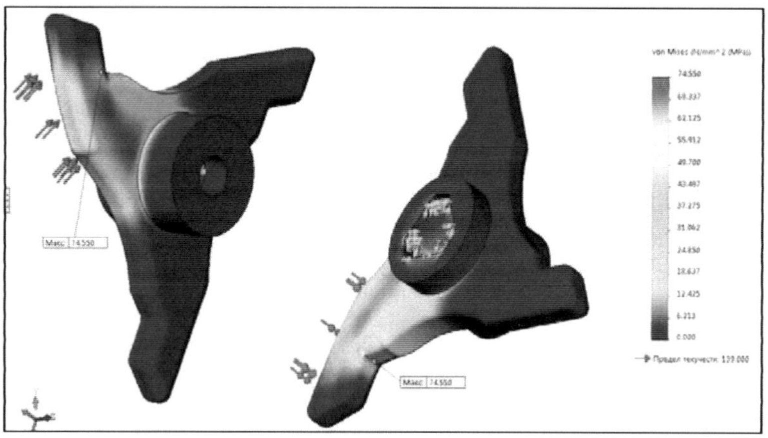

Figure 2. Sprocket stress diagrams on the basis of the maximum distortion energy theory (von Mises).

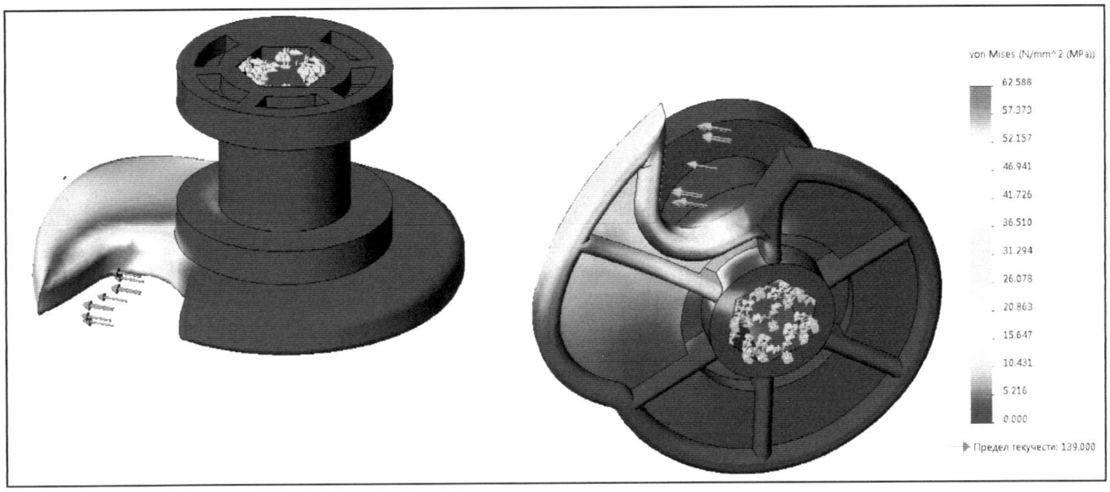

Figure 3. Driving disc stress diagrams on the basis of the maximum distortion energy theory.

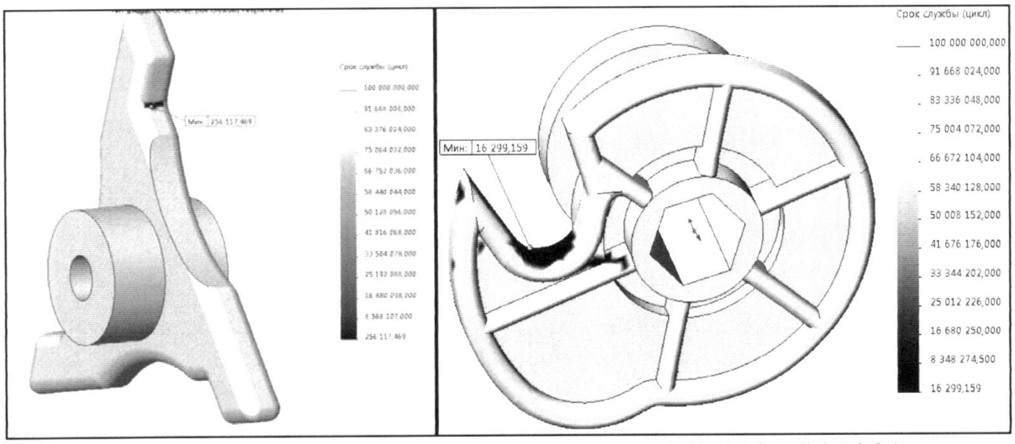

Figure 4. The model parts service life: the sprocket (left), driving disk (right).

5. Results discussion

The sprocket stresses value (figure 2) approximately equals to 75 MPa which is based on the assumption of the shearing absence. Hence, the limit contact force obtained as a result of modelling in accordance with the sprocket working surfaces to be not more than 600 N.

The constant amplitude material fatigue event is completely determined by the alternating stress and load cycles number. The sprocket load cycles number (figure 4) is 256117 cycles which corresponds to 6 months of the part operation.

The sprocket breakage point under the contact stresses influence (figure 2) coincides with the fatigue failure area (figure 4).

The engaged driving disk stresses value (figure 3) is somewhat less (≈63 MPa) than the sprocket one. However, the values (figure 4) obtained for the resource indicate disk surface failure approximately in half a month. These results suggest the surface failure and premature fatigue wear as physical modelling results do not imply the mechanism breakdown as a result of the disk failure [2].

6. Conclusions

As a consequence of the conducted research, the results of contact stresses were obtained, the parts strength and durability were checked.

The obtained modelling results show the model application for the given initial data to be unacceptable. Thus, it is manifested in the parts short service life for the most severe operating conditions. The working surfaces overwhelming sliding friction has the greatest influence. However, the model structure perfection is impossible to be improved, as the sprocket cranked beam forms cannot be avoided.

Although the simulation results for the specified life do not satisfy the required, they have shown good agreement with the experiments results obtained on the basis of the physical model [2]. The parts minimum model contacting surfaces wear life is half a month, but it should be specified by experimental methods. It should also be noted that the model service life based on the parts damage assumption for the most severe operating conditions, is rather short.

The rotary mechanism under consideration functionality enhancing is possible to be achieved by using the other forms of the indexing transmission and avoiding the sliding friction in favour of the rolling one.

References

[1] Advertising installation Patent Russia 18792, IPC G09F 11/02 No 2001108114/20, 30.01.2001

[2] Buchakov Ju V, Dobrovolskij V P and Kulakov A A 2004 Experimental study of the disc-sprocket rotation mechanism performance of the advertising installation Dynamics Proc. Int. Conf. vol I (Omsk: Omsk State Technical University) p 428-32

[3] Xu F, Nu N and He W 2009 Modeling and Simulation of Globoidal Indexing Cam Mechanism Based on CATIA and ADAMS Journal of Mechanical Transmission **33** 42-3

[4] Ji S, Zhao J and Zhang Y 2015 An application of geodesics to the calculation of the rib-thickness of globoidal cam mechanisms Mechanism and Machine Theory **87** 163-76

[5] Li L, Feng X, Zhang Z, Han X and Song Y 2010 3D Modeling and Simulation of a New Type of Globoidal Indexing Cam Mechanism *Applied Mechanics and Materials* **26-28** 931-35

[6] Zhaoa H, Caob Y and Guoc P 2011 Dynamics Simulation of Parallel Indexing Cam Mechanism Based on Rigid-flexible Coupled Modelling *Advanced Materials Research* **156-157** 172-76

[7] Hsieh J F 2014 Design and analysis of indexing cam mechanism with parallel axes *Mech. Mach. Theory* **81** 155–65

[8] Guilan T, Haibo F and Weiyi Z 1999 A new method of torque compensation for high speed indexing cam mechanisms *ASME Journal of Mechanical Design* **121** 319-23

[9] Gonzalez-Palacios M A and Angeles J 1994 Generation of Contact Surfaces of Indexing Cam Mechanisms a Unified Approach *ASME Journal of Mechanical Design* **116** 369-74

Hydrocarbons pipeline transportation risk assessment

A V Zanin[1], A A Milke[1] and I N Kvasov[1]

[1]Omsk State Technical University, 11, Mira av., Omsk, 644050, Russia

E-mail: zanin.andrew@gmail.com

Abstract. The pipeline transportation applying risks assessment issue in the arctic conditions is addressed in the paper. Pipeline quality characteristics in the given environment has been assessed. To achieve the stated objective, the pipelines mathematical model was designed and visualized by using the software product SOLIDWORKS. When developing the mathematical model the obtained results made possible to define the pipeline optimal characteristics for designing on the Arctic sea bottom. In the course of conducting the research the pipe avalanche collapse risks were examined, internal longitudinal and circular loads acting on the pipeline were analyzed, as well as the water impact hydrodynamic force was taken into consideration. The conducted calculation can contribute to the pipeline transport further development under the harsh climate conditions of the Russian Federation Arctic shelf territory.

Key-words: risks assessment, offshore pipeline, hydrodynamic load, mathematical model

1. Introduction

For the Russian Arctic shelf exploration the stationary offshore oil platforms with high ice loads resistance are mainly used. Risks assessment is one of the most important factors for the continental shelf deposits development. Offshore installations emergencies occurring in the process of extracted products drilling, production, treatment and processing, as well as when oil transporting may result in the vast human losses, apparatuses losses, serious environmental and economic consequences [16].
Offshore pipeline oil transportation from platforms is largely due to such characteristics as the distance from the coast, extreme weather events, seabed geological and hydro meteorological hazards, etc. High hazard when oil transporting from oil-rig platforms is confirmed by large-scale accidents that occurred during operation. Accidents with fires and explosions both on offshore oil platforms and during the pipelines operation can result in disastrous consequences, in which case human and material resources engagement at the level of one or several states may be required [4, 5].
The defined objectives are relevant and clear: firstly, the offshore pipeline laying risk is necessary to be understood and therefore calculations are required to be conducted, the optimal pipeline both with regard to materials and diameter to meet the harsh environmental conditions of the promising region where the work will be performed needs to be analized and defined [11].

2. Theory

With regard to the offshore pipelines, the conditions of the line construction from the oil production platform and to the major transport hub should be understood. Therefore, the Arctic shelf territory environmental conditions are worth studying and as the only developed field in this area is the Prirazlomnoye oil field, then all the initial data on the environmental conditions will be derived from the given territory [2, 6].

IOP Conf. Series: Journal of Physics: Conf. Series **998** (2018) 012043 doi:10.1088/1742-6596/998/1/012043

Transportation from this platform is inexpidient, as it is performed only by using the tankers. Pipeline transportation would be more profitable from the economic perspective as Arctic shelf reserves are very high and amount to 15 billion tons of oil, and pipeline transportation will allow to deliver oil to the nearest oil export terminal as fast as possible and to reduce both human labor in extreme conditions and quantity of the equipment used [13].
Before starting the given research, foreign and Russian scientific journals have been examined. The similar topic have been considered by Alireda Aljaroudi, Faisal Khan, Ayhan Akinturk, Mahmoud Haddara, Premkumar Thodi in the paper «Risk assessment of the offshore crude oil pipeline failure» of the scientific journal «Journal of Loss Prevention in the Process Industries» in which the authors provide methodologies when laying the pipeline on the seabed [12].
The authors of the mentioned papers did not take into account the projected conditions severe natural and climatic features. Based on findings and conclusions, the study taking into consideration the parameters required for achieving the objectives was possible to be carried out.
The offshore ice-resistant stationary platform is located on the south-eastern shallow marine shelf of the Barents Sea (Pechora Sea) in the exclusive economic zone, 55 km from the coastline within the plain with depths from 18.0 to 20.5 m. Sea depths are reduced to the lowest theoretical level. Onshore depths vary quite evenly, especially in 5 km from the coast in the locality of Varandey they reach 12-13 m [3, 7].
Hydro meteorological, ice, geological engineering conditions have been compiled from the geotechnic survey on the Barents Sea [8].
Based on the data obtained, it is possible to start the mathematical model calculations and development.

3. Results

For designing the offshore pipeline among other steels K2, K52, 09G2S, 12H18N10T, steel 17G1S was chosen on account of its increased resistance to hydrodynamic loads and in respect of the ultimate strength. Further, D_a pipelines five different diameters: 600 mm, 720 mm, 820 mm, 1020 mm, 1220 mm are considered. The pipeline class corresponds to G3. According to the project, the internal working pressure p_i is 5 MPa, however, it is necessary to obtain the pipeline calculated pressure p_0 by the formula [15, 17]:

$$p_0 = \left(p_i - p_{g\,\min}\right) + \Delta p \tag{1}$$

The minimum external hydrodynamic pressure on the pipeline p_{gmin}, MPa is calculated according to the formula:

$$p_{g\,\min} = (\rho_w g\left(d_{\min} - \frac{h_w}{2}\right) \cdot 10^{-6} \tag{2}$$

where ρ_w is the sea water density considered equal to 1080 kg/m^3;
$d_{\min}$ is the still water minimum level along the pipeline route which is assumed equal to 9.1 m taking into account tidal variations and surges with reliability of 10^{-2}*1/year;
h_w is the calculated wave height of the projected pipeline section which is considered equal to 10.1 m with reliability of 10^{-2}*1/year [1].
The additional imposed pressure value Δp, MPa considering the hydraulic impact phenomenon should be not less than the value determined by the formula:

$$\Delta p = V_{int}\sqrt{\frac{\rho_{int} \cdot E \cdot t_c \cdot K}{E \cdot t_c + D_{int} \cdot K}} \cdot 10^{-3}, \tag{3}$$

where V_{int} is the pipeline transported medium motion speed taken to be 1.0 m/s [18];
E is the pipes material elasticity modulus which is assumed to be 2.06 MPa;
K is the transported medium bulk modulus which is based on 1.1 MPa.
The steel pipeline wall thickness t_c, mm based on the local strength conditions is determined by the formula:

$$t_c = \frac{p_0 D_a}{2\sigma\varphi} + c_1 + c_2, \quad (4)$$

where σ is the pipe material permissible voltage MPa;
φ is the strength coefficient calculated depending on the pipes manufacturing method and is taken equal to $0.9 \cdot 10^{-3}$;
c_1 is the corrosion allowance assumed to be 3 mm;
c_2 is the allowance compensating the pipes manufacturing tolerance which is based on 2 mm.
The permissible voltage σ, MPa is assumed to be equal to the lowest value

$$\sigma = min\left(\frac{R_e}{n_e}, \frac{R_m}{n_m}\right), \quad (5)$$

where R_e is the yield tensile strength minimum value assumed to be equal to 510 MPa;
R_m is the pipes metal tensile strength minimum value which is based on 335 MPa;
n_e is the safety factor for the yield strength which is taken to be 1.25;
n_m is the safety factor for the tensile strength assumed to be 2.0.
The calculated pressure on the pipeline having been defined and t_c calculated for the loads acting on the pipe determining, the pipeline section (figure 1) of 10 m in size with the support spacing of 8 m on which the maximum hydrodynamic pressure p_{gmax}, vertical load per unit length $F_{w,v}$ as well as horizontal load $F_{w,h}$ are acting, is considered.
In the paper «An optimum design of on-bottom stability of offshore pipelines on soft clay» of the international journal «Architecture and Ocean Engineering», Su Young Yu, Han Suk Choi, Seung Keon Lee, Chang Ho Do, Do Kyun Kim constructed the projected pipeline on the clay. Based on their methods, the examined pipeline laying was projected, however, the conditions under which the calculations are performed in this study differ from ones described in the paper [10].

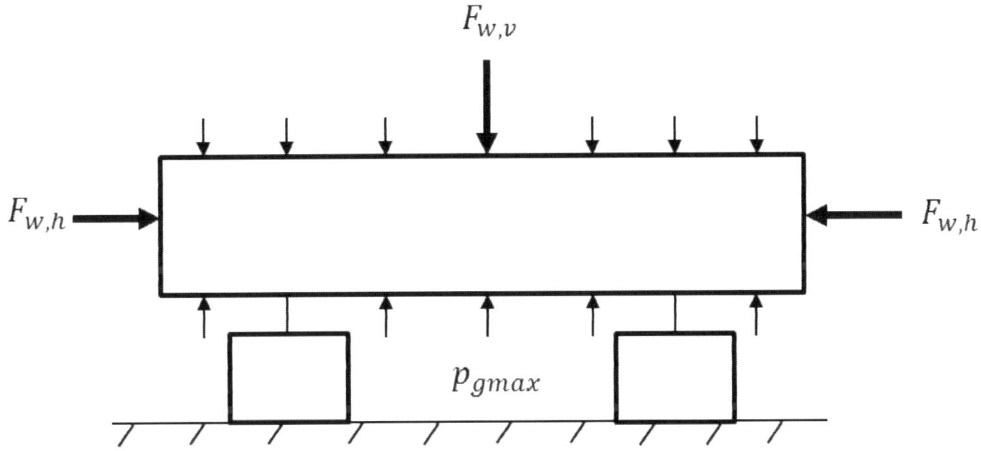

Figure 1. The calculation scheme of the loads acting on the pipeline.

The calculation of the maximum external water pressure on the pipeline p_{gmax}, MPa under the temperature coefficient impact γ is defined by the following formula:

$$p_{gmax} = \gamma \rho_w g\left(d_{max} + \frac{h_w}{2}\right) \cdot 10^{-6}, \quad (6)$$

The hydrodynamic pressure critical value p_p, MPa, at which the avalanche collapse extent is possible to occur is determined by the formula:

$$p_p = 24 \cdot R_e \cdot \left(\frac{t_c}{D_a} \right)^{2.4} \tag{7}$$

The avalanche collapse absence condition is the following inequality requirements fulfilment:

$$p_{gmax} < 1.2 \cdot p_p , \tag{8}$$

The hydrodynamic pressure maximum values p_{gmax} for the different pipeline diameters are shown in figure 2. Depending on the pipeline outer diameter D_a, p_{gmax} decreases, simultaneously the condition according to the formula (8) should be met. The hydrodynamic pressure critical values p_p for each diameter are illustrated in dotted lines. If the dotted line is above the maximum hydrodynamic pressure, then the pipe loses its stability with a further destruction.

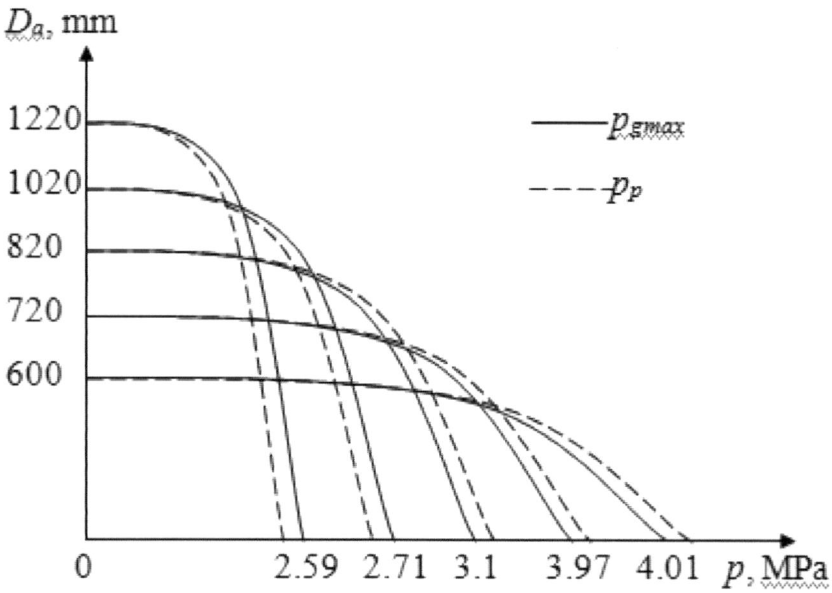

Figure 2. The action of hydrodynamic loads on pipelines.

The analysis of the dependencies presented in figure 2 showed the pipelines with the diameter of 600 mm, 720 mm. and 820 mm not to fulfill the condition according to formula (8). Consequently, further calculations will be carried out with the diameters of 1020 mm and 1220 mm.

Hence, the linear loads on the pipelines are calculated. The total horizontal wave load $F_{w,h}$ N/m is determined taking into account the wave action coefficient γ_1 by the formula:

$$F_{w,h} = \gamma\gamma_1 \sqrt{F_{w,s}^2 + F_{w,i}^2} \tag{9}$$

The vertical linear wave load $F_{w,v}$ N/m, is calculated considering the flow influence coefficient by using the formula:

$$F_{w,v} = \gamma\gamma_2 c_v \cdot \frac{\rho_w \cdot V_w^2}{2} \cdot D_a \tag{10}$$

Based on the calculations, the mathematical model development data taking into consideration the hydrodynamic loads having showed that not all diameters for the constructed offshore pipeline are suitable for the study further stages were obtained.

4. Discussion

The software product SOLIDWORKS application taking into account the calculations results made possible to develop the pipelines models having been applied hydrodynamic, linear, vertical and horizontal loads, which visualization is represented in figure 3. Hydrodynamic loads are wave loads, water column pressure is linear vertical load and seawater flow is horizontal loads. When developing the mathematical model the pipe weight, connections with fixed supports and liquid internal pressure variations along the pipeline length are taken into account.
Light shade indicates the places where the pipelines are under the maximum pressure, thus, based on the models it can be stated that pipelines with external diameters of 1020 mm and 1220 mm withstand all calculated loads.

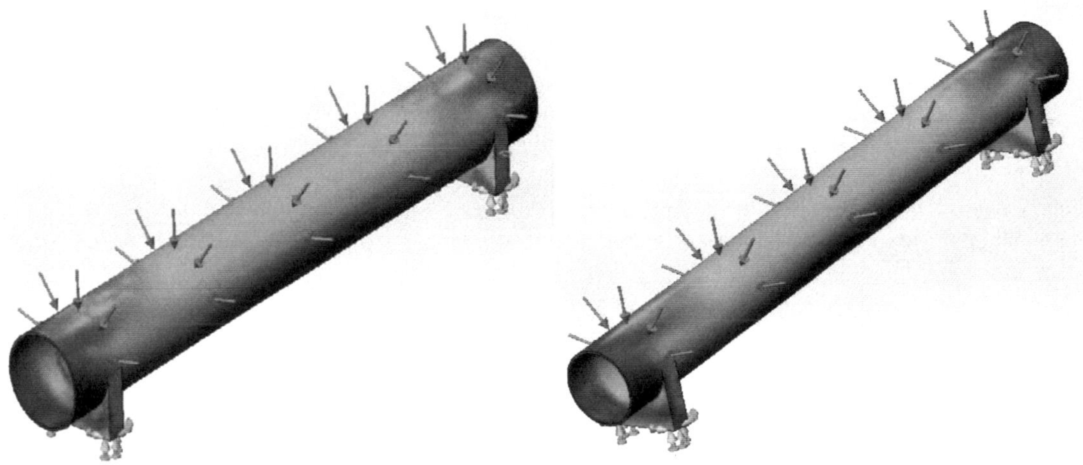

Figure 3. The pipelines models with diameters of 1220 mm and 1020 mm.

Based on the data obtained, it can be concluded that the pipelines with the given diameters are equally ready for operation, however, it is necessary to take into consideration both the risk assessment and economic component.
Despite the fact that the pipeline 1220 mm in the nominal diameter has greater performance than the pipeline with diameter of 1020 mm, the first pipe operation risk is higher due to the increased pressures in it. It is also necessary to consider a great amount of passable raw materials increasing the internal circular as well as longitudinal loads. Consequently, at the wave hydrodynamic and internal loads resonance, the pipeline can achieve its maximum acceptable condition, which will result in the pipe breakage [14].
Moreover, it is economically feasible to use the smaller amount of consumed materials during the given pipeline construction work. Even with a lower flow capacity of the pipeline system with diameter of 1020 mm, this cannot be compared with the materials costs taking into account the projected pipeline length of 50 km [9].

5. Conclusion

In the course of conducting the research the pipe avalanche collapse risks were examined, internal longitudinal and circular loads acting on the pipeline were analyzed, as well as the water impact hydrodynamic force was taken into consideration. Thus, based on the calculations the oil pipeline transportation risk assessment from the Arctic shelf was conducted. The sampling for the pipe material and diameter was carried out until the results whereby it is recommended to use the material for the pipeline - steel 17G1S, the outer diameter of the pipeline - 1020 mm, were obtained.

References

[1] ND N 2-020301-003 Rules for the Offshore Pipelines Classification and Construction
[2] Zanin A V 2016 The Arctic shelf development as the Russia's economic development factor *Proc. Conf. The current issues of economics and management* (Russia: Omsk) p 154-7
[3] Zanin A V 2016 Prospects for offshore oil and gas production in the Arctic shelf and transportation ways *Proc. Conf. Science and youth in the 21st century* (Omsk: OmSTU) 135-9
[4] Kvasov I N 2016 The main oil pipeline leak detection systems analysis *Proc. Int. Conf. Engineering: looking into the future* (Russia: Omsk) p 33-4
[5] Nikolaeva O 2017 *Security is the foundation Siberian oil* **141** 16-21
[6] Sviriz I 2013 *The main channel Siberian oil* **106** 42-7
[7] Tretjakov E 2011 *Everybody to the port Siberian oil* **83** 8-11
[8] Shavykin A A 2010 Integral vulnerability assessment of the Barents Sea from oil pollution (Murmansk: MMBI KNC RAN)
[9] Mark J Kaiser 2017 Offshore pipeline construction cost in the U.S. Gulf of Mexico *Marine Policy* **82** 147-166
[10] Su Young Yu, Han Suk Choi, Seung Keon Lee, Chang Ho Do and Do Kyun Kim, *International Journal of Naval Architecture and Ocean Engineering* **5** 598-613
[11] Øistein Harsem, Arne Eide and Knut Heen 2011 Arctic *Energy Policy* **39** 8037-45
[12] Alireda Aljaroudi, Faisal Khan, Ayhan Akinturk, Mahmoud Haddara and Premkumar Thodi *Journal of Loss Prevention in the Process Industries* **37** 101-9
[13] Mark J Kaiser 2007 *Energy Policy* **35** 3496-3525
[14] Yong Bai and Wei-Liang Jin 2016 *Marine Structural Design* (Oxford: Butterworth-Heinemann) p 735-63
[15] Alireza Bahadori 2017 *Oil and Gas Pipelines and Piping Systems, Gulf Professional Publishing* (Boston) p 1-27
[16] James G Speight 2015 *Subsea and Deepwater Oil and Gas Science and Technology*, (Boston: Gulf Professional Publishing) p 71-106
[17] Qiang Bai and Yong Bai 2014 *Subsea Pipeline Design, Analysis, and Installation* (Boston: Gulf Professional Publishing) p 405-33
[18] Bruschi R, Vitali L, Marchionni L, Parrella A and Mancini A *Ocean Engineering* **108** 369-92

Accuracy increase of the coordinate measurement based on the model production of geometrical parts specifications

O Yu Zlatkina

Omsk State Technical University, 11 Mira ave., Omsk, 644050, Russian Federation

E-mail: zlatkina.olga@gmail.com

Abstract. There is a relationship between the service properties of component parts and their geometry; therefore, to predict and control the operational characteristics of parts and machines, it is necessary to measure their geometrical specifications. In modern production, a coordinate measuring machine is the advanced measuring instrument of the products geometrical specifications. The analysis of publications has shown that during the coordinate measurements the problems of choosing locating chart of parts and coordination have not been sufficiently studied. A special role in the coordination of the part is played by the coordinate axes informational content. Informational content is the sum of the degrees of freedom limited by the elementary item of a part. The coordinate planes of a rectangular coordinate system have different informational content (three, two, and one). The coordinate axes have informational content of four, two and zero. The higher the informational content of the coordinate plane or axis, the higher its priority for reading angular and linear coordinates is. The geometrical model production of the coordinate measurements object taking into account the information content of coordinate planes and coordinate axes allows us to clearly reveal the interrelationship of the coordinates of the deviations in location, sizes and deviations of their surfaces shape. The geometrical model helps to select the optimal locating chart of parts for bringing the machine coordinate system to the part coordinate system. The article presents an algorithm the model production of geometrical specifications using the example of the piston rod of a compressor.

Key words: geometrical specifications, coordinate system, informational content of part component, coordinate measuring machine.

1. Introduction

The widespread use of CNC machines in the production of machine parts and devices increased the accuracy of the products, which increased the requirements for the control accuracy. There is a relationship between the service properties of component parts and their geometry; therefore, to predict and control the operational characteristics of parts and machines, it is necessary to measure their geometrical specifications. At the present time, a coordinate measuring machine (CMM) is the advanced measuring instrument of the products geometrical specifications. Most of control operations of complex shape modern products in science absorbing industry and high technologies can be carried out only on the CMM because of strict requirements for the measurement accuracy and efficiency of its implementation.

2. Statement of the problem

A large number of publications of leading scientists such as A.Yu. Kasparaitis, V.G. Lysenko, V.L. Solomakho [6], R.G. Wilhelml and K.D. Summerhays [5] are devoted to the development of the process models of coordinate measurements and studies of measurement errors in measuring geometrical specifications of component parts. The analysis of publications has shown that during coordinate measurements the problems of choosing locating chart of parts and coordination have not been sufficiently studied.

During measurements on a CMM measuring procedure can be divided into the following sub-procedures:

1) selection and production of the locating chart of parts for specifying the location of the part coordinate system in the CMM coordinate system;

2) selection of the necessary and sufficient number of measurement points;

3) identification of the coordinate system for performing auxiliary constructions, production of auxiliary systems which are necessary for the geometrical specifications of auxiliary bases sets and executive and free surfaces.

The selection of locating chart is the selection of the set of elements, mathematically representing physical surfaces which parameters are calculated based on the assumption of finite set of measured control points. The main elements are "straight line", "circle", "plane surface", "cylinder", "sphere" and "cone". Thus, the primary task is to select the most optimal locating chart that will provide the necessary positional accuracy and specify a coordinate system. However, the selection of the elements for locating is not always obvious; therefore, in order to increase the results reliability it is recommended to construct a geometrical model of the measurement object.

3. Theory

The construction of the analyzed part geometrical model allows us to visually reveal the relationship between the coordinates of the deviations in the location, size and shape of the parts surfaces and to develop an optimal measurements methodology. During the development of the coordinate measurements object geometrical specifications model the leading role is played by the generalized coordination which is materialized by a set of principal assembly bases. The accuracy of part positioning in the space is achieved by limiting the item by six degrees of freedom: three linear and three angular motions.

Depending on the surfaces type of the mating elementary items (plane, sphere, cylinder, cone, tore, multi-plane surface), sizes and number of elementary items and mating are classified into seven classes according to the number of degrees of freedom (from zero to six) which are composed of linear *(t)* and angular *(r)* constraints *(c)* [3]. According to [2], the sum of the degrees of freedom limited by the elementary item is called "informational content". Thus, during constructing the coordinate system of a part, it should be based on the element informational content. The informational content of each element determines the number of linear and angular coordinates that need to be normalized for its identical positioning in the part coordinate system [3]. The classification of elementary items by information content which allows determining the information content of an element more clearly depending on its mating pair is given in [3].

The development of a measurement object geometrical model is carried out using the following algorithm [1].

1. Identification of the measurement object coordinate system.

As a measurement object, consider the piston rod of the compressor. The piston rod transmits the force from the crankshaft to the piston and serves as the main link for converting the rotational motion of the crankshaft into reciprocating motion of the piston. The piston rod small end is connected to the piston via the piston pin; the piston rod big end serves to connect it to the crankshaft crank (figure 1).

IOP Conf. Series: Journal of Physics: Conf. Series 998(2018) 012044 doi:10.1088/1742-6596/998/1/0120

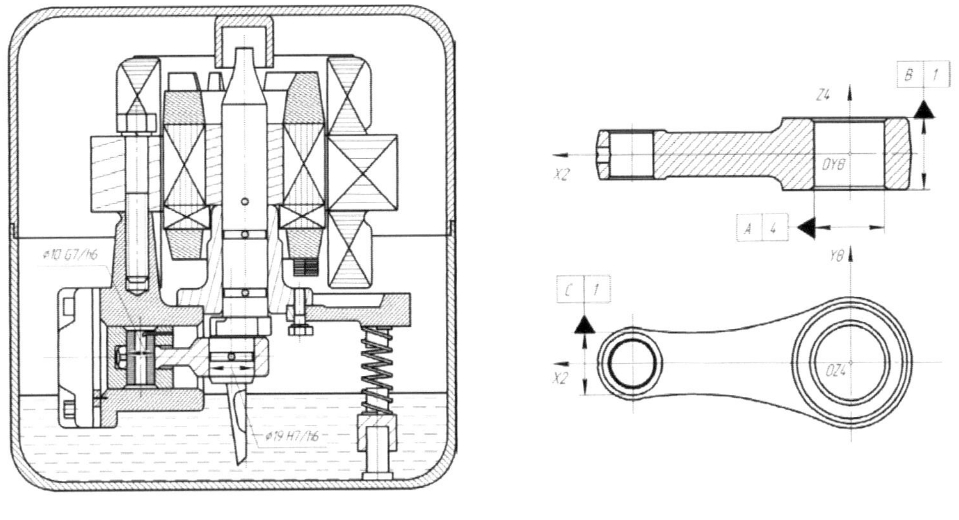

a) b)

Figure 1. Compressor: a) general assembly drawing; b) coordinate system of the piston rod

The set of principal assembly bases forms a rectangular coordinate system *0XYZ* and determines the position of the piston rod in the compressor:

A) the base symmetry axis of the inner tubular element *A4* in the piston rod end with informational content of 4 limits the piston rod in two linear movements, perpendicular directions to the axis and two angular rotations which are mutually perpendicular to the axis *A4* and forms the coordinate axis *Z4* with the informational content of 4 of the rectangular coordinate system;

B) the base plane of symmetry *B1* of the piston rod end prismatic element with informational content of 1 limits the piston rod in one linear motion along the axis *Z4*. The point of the base *B1* symmetry plane intersection with the coordinate axis *Z4* forms the origin 0 of the rectangular coordinate system;

C) the base axis of the inner tubular element *C1* in the piston rod end has informational content of 1, deprives the piston rod of rotation around the axis *Z4* and determines the direction of the axis *X2* with the informational content of 2 of the rectangular coordinate system;

D) the axis *Y0* has zero informational content, because its direction does not need an additional base which is the perpendicular reconstructed from the origin *0* to the coordinate plane *Z4X2*, in which the axes *Z4* and *X2* are located with a general informational content of 6.

2. Determination of the geometrical specifications of the measurement object

The rectangular coordinate system finally takes the form *0X2Y0Z4* taking into account the informational content of the axes. The geometric model of the piston rod is constructed in three projections on the contours of the nominal piston rod for their comparability. The constructed image of the generalized coordinate system *0X2Y0Z4* is similar to the construction on the nominal sketch of the piston rod. Let us define the geometrical specifications of the principal assembly base, starting with the base *A4*, which has the greatest informational content. The base *A4* does not have unused degrees of freedom because the maximum informational content of a tubular element is four *(2t + 2r = 4c)*. All four degrees of freedom are spent on materializing the coordinate axis *Z4*, providing informational content of 4 for the axes. The distance between the generators of the superimposed cylinder determines the first geometrical adjective of the element diameter D_I+ED_I, which will have a primary sizing error. Representing from the element body the real shape of the element surfaces with a wavy line, we can find the second geometrical adjective of the elementary item *A4* these are the primary form errors *EF1*. The form deviations of the base hole are included in the diameter deviations ED_I and increase the diameter due to the positive sign of the deviation.

Let us now continue with plain base *B1*. Since the base limits the piston rod of one translational motion *1t* along the axis *Z4*, two angular motions *(1t + 2r-1t = 2r)* are unused, having maximum informational content *(1t + 2r = 3c)*. Therefore, the base determines the position of the rectangular coordinate system *0*, origin remaining two angular degrees of freedom in the form of primary angular distortions around the axes *X2* and *Yθ* relative to the nominal angle of *90°* with the base axis *Z4*: $90°\pm AEX_2$ и $90°\pm AEY_2$. The prismatic element has its own width size w_2-Ew_2, and the primary form errors EF_2 of two planes that reduce the width size.

Then, continue with cylindrical base C1. The base has informational content *1r* and limits the piston rod from one angular rotation around the axis *Z4* due to losing of own linear degree of freedom *1t* on the axis *Yθ*. The maximum informational content of a cylindrical element is four *(2t+2r=4c)*, therefore, three degrees of freedom remain unused: 1 linear and 2 angular *(2t+2r-1t=1t+2r)*. This leads to the formation of the linear coordinate $X_3\pm EX_3$ of the center C_3 and two angular distortions $0°\pm AEX_3$ and $0°\pm AEY_3$ of axis C_1 relative to the coordinate axis *Z4* with the maximum informational content of 4. The diameter of the element will have a primary size error D_3+ED_3 and the primary form errors EF_3. The length of the element base axis of symmetry *C1* is limited by the width size w_4-Ew_4 of the piston rod end. The action of the base axis extends to the protruding length l_P, which is equal to the length of the piston pin. This length determines the maximum X_{3M} and minimum X_{3L} values of the coordinate $X_3\pm EX_3$.

Primary errors of free surfaces can not be found. On the piston rod model, these surfaces are shown in nominal position and of nominal shape relative to the appropriate coordinate system, in which module of elements they work.

The base hole *C* performs the functions of the main *C1* and the auxiliary *C4* assembly base, which limits the four degrees of freedom of the connected piston pin. They form an auxiliary coordinate system *0΄X΄2Y΄θZ΄4* together with the prismatic element *D1* of the piston rod end w_4-Ew_4.

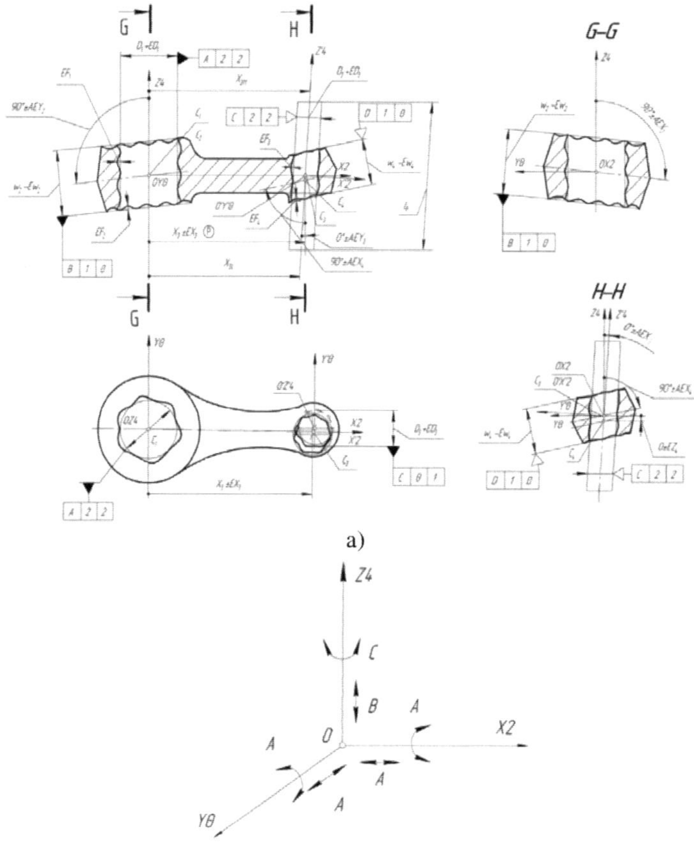

a)

b)

Figure 2. a) geometrical model of the compressor piston rod; b) indicator of the bases

3. Compiling the matrix of measurement object geometrical specifications.

Table 1. Matrix of piston rod geometrical specifications

Number and element designation, the center of the element, the reference system	The type, functionality, informational content and maximum informational content of element	Sizes and primary errors in the basic coordinate system	Bases, sizes, permissible error from the detail drawing
1 *A4* *0X2Y0Z4*	Cylindrical main base $N_{inf} = 2t+2r$ $N_{max} = 2t+2r$	$N_{crd}=2t+2r-2t-2r=0$ D_1+ED_1 EF_1 R_{Z1}	A4 ◀ $Ø19H7^{+0,021}$ ○ 0,006 = 0,006 RZ 3.2
2 *B1* *0X2Y0Z4*	Flat, prismatic main base $N_{inf} = 1t+0r$ $N_{max} = 1t+2r$	$N_{crd}=1t+2r-1t=2r$ $90°±AEY_2$ $90°±AEX_2$ EF_2 w_2-Ew_2 R_{Z2}	B1 ◀ 37±0.013 // 0,006 A4 $Ø10H6^{+0,009}$ ○ 0,006 = 0,006 RZ 1.6
3 *C1* *0X2Y0Z4*	Cylindrical main base $N_{inf} = 1r \Rightarrow 1t$ $N_{max} = 1t+2r$	$N_{crd}=2t+2r-1t=1t+2r$ $X_3±EX_3$ $0°±AEY_3$ $0°±AEX_3$ D_3+ED_3 EF_3 R_{Z3}	C1 ◀ ⊥ 0,19 Б1 9±0.011 RZ 3.2
4 *C4* *0X2Y0Z4* *0'X'2Y'0Z'4*	Flat, prismatic auxiliary base $N_{inf} = 0$ $N_{max} = 1t+2r$	$N_{crd}=1t+2r-0t-0r=1t+2r$ w_4-Ew_4 $0°±EZ_4$ $90°±AEX'_4$ $90°±AEY'_4$ R_{Z4}	C4 ▷ D0 ▷

4. Experimental results

The geometrical model of the compressor piston rod constructed in accordance with the recommendations is shown in figure 1. The hole with diameter D1 + ED1 deprives the greatest number of degrees of freedom – 4 (2 linear and 2 angular displacements) as it can be seen on the geometrical model. Thus, it is advisable to place the origin of the coordinate system in the center of this hole; therefore, to create a reference of a part to the CMM coordinate system, it is necessary to use the plane-cylinder-cylinder scheme. It is recommended to draw the table with a complete account of all bases, sizes and primary errors of the geometrical model (see table 2). This will allow developing the most effective measurement design. Thus, for example, it is more obvious to see that the difference between the values of X_{3L} and X_{3M} will be a deviation from parallelism.

Based on the constructed geometrical model at the KIM-1000 LLC "Lapik", the following studies were carried out and the following results were obtained:

- angle of axes misalignment is $0°±AEX_3 = 0°+0'47''$;
- angle of axes misalignment is $90°±AEX_4 = 90°+03'19''$;
- deviation of the hole C_4 real center relative to the nominal center C_3 $0±EZ_4 = 0{,}043$ mm;
- angle of axes misalignment is $90°±AEX_2 = 89°56'08'' = 90°-3'52''$;
- angle of axes misalignment is $0°±AEY_3 = 0°+0'48''$;
- angle of axes misalignment is $90°±AEY_2 = 89°52'00'' = 90°-8'$.

These primary errors can significantly affect the accuracy of the measurement result. However, without the model production of the geometrical specifications of the part, in most cases, the primary errors are not taken into account which leads to a decrease in the measurement accuracy. In support of this theory, studies were carried out to measure the coordinate X_3 of the axis of the piston rod base hole *B1* with a permissible error of ±0,013 mm (see table 2).

Table 2. Measurement results of coordinate X_3

The origin of the coordinate system at the center of the axis of the inner tubular element *B1* (detail drawing)	The results, mm	The origin of the coordinate system at the center of the axis of the inner tubular element *A4* (geometrical model)	The results, mm
X_{3M}	37.319	X_{3M}	37.299
X_{3L}	37.317	X_{3L}	37.294

Thus, the difference between the results was 0.020 mm for the maximum value X_{3M} and 0.023 mm for the minimum value X_{3L}. It is more than 75% of the permissible error for the coordinate size $TX_3=0.026$ mm.

5. Discussion of the results

The following statements are suggested for discussion:

- during coordinate measurement it is necessary to take into account the informational content of the elementary items and coordinate axes that allow us to identify coordinate systems without redundancy of the basing and helps identically specify the position of the connected part using three linear and three angular coordinates;
- the geometrical model construction of the coordinate measurements object taking into account the informational content of coordinate planes and coordinate axes increases the accuracy of the measurement results, since it allows to develop an optimal scheme of coordinate measurements.

6. Conclusions

The proposed approach for constructing geometrical models of parts based on the informational content of coordinate planes and coordinate axes for coordinate measurements makes it possible to scientifically and reasonably increase the reliability of measurement results and develop the most effective measurement scheme.

References

[1] Glukhov V I 2014 *Dynamics of Systems, Mechanisms and Machines (Dynamics), IEEE Conference Publications*, p 1-9. DOI: 10.1109 Dynamics 7005655.
[2] GOST 21495-76 1990 Basing and bases in mechanical engineering. 1977-01-01 (Moscow)
[3] Glukhov V I, Zlatkina O Yu, Ivleva I A 2016 *Dynamics of systems, mechanisms and machine* **1** **1** p 285-290
[4] Ramaswami H, Acharya S B, Kovvur Y and Anand S 2006 *Journal of Manufacturing Systems* **25(2)** p 77-94
[5] Summerhays K D, Henke R P, Baldwin J M , Cassou R M and Brown C W 2002 *Precision Engineering* vol. **26(1)** p 105-121
[6] Solomakho V L, Tsitovich B V , Sokolovsky S S and Solomakho D V 2012 The theory, computation and designing of optimized methods of coordinate control in mechanical (Minsk)
[7] Arambula K, Siller H R, De Chiffre LRodriguez, C A and Cantatore A 2012 *Int. J. Metrol. Qual. Eng.* **3** p 55–62

Institute of Physics
Dirac House, Temple Back
Bristol BS1 6BE UK

ISSN: 1742-6588
ISBN 978-1-5108-6167-1

CURRAN ASSOCIATES INC.
proceedings.com